Lecture Notes in Physics

Lecture Notes in Physics

Edited by J. Ehlers, München, K. Hepp, Zürich
R. Kippenhahn, München, H. A. Weidenmüller, Heidelberg
and J. Zittartz, Köln
Managing Editor: W. Beiglböck, Heidelberg

134

γγ Collisions

Proceedings of the International Workshop
(Journées d'Etudes Internationales)
Held at Amiens, France, April 8–12, 1980

Edited by G. Cochard and P. Kessler

Springer-Verlag
Berlin Heidelberg GmbH 1980

Editors

Gérard Cochard
Laboratoire de Physique Théorique des Particules
Université de Picardie
33, rue Saint-Leu
F-80039 Amiens Cedex

Paul Kessler
Laboratoire de Physique Corpusculaire
Collège de France
11, place Marcelin-Berthelot
F-75231 Paris Cedex 05

ISBN 978-3-540-10262-5   ISBN 978-3-540-38383-3 (eBook)
DOI 10.1007/978-3-540-38383-3

© by Springer-Verlag Berlin Heidelberg 1980

Originally published by Springer-Verlag Berlin Heidelberg New York in 1980

2153/3140-543210

AVANT-PROPOS

Le présent volume contient les Comptes-Rendus des "Journées d'Etudes Internationales sur les Collisions Photon-Photon" ("International Workshop on γγ Collisions") qui se sont tenues à l'Université de Picardie (Amiens) du 8 au 12 avril 1980.

Les collisions photon-photon (collisions entre spectres de photons virtuels, réalisées essentiellement auprès des anneaux de stockage d'électrons) constituent un domaine relativement récent de la physique des hautes énergies. Sur le plan théorique, les premiers travaux dans ce domaine, dûs à F. LOW et à F. CALOGERO et Ch. ZEMACH, datent de 1960. C'est à partir de 1969-70 qu'une impulsion considérable fut donnée à la physique des collisions γγ grâce aux études théoriques détaillées effectuées en particulier par trois groupes (Collège de France, Novosibirsk, Cornell-SLAC). A l'heure actuelle, c'est surtout dans le contexte des vérifications de la chromodynamique quantique que ce type de réactions rencontre un intérêt croissant auprès des théoriciens.

Sur le plan de l'etude expérimentale des collisions photon-photon, les premières expériences, effectuées au cours de la période 1970-1973 à Novosibirsk et à Frascati, constituaient essentiellement des tests de l'électrodynamique quantique. Par contre, les expériences les plus récentes, réalisées auprès des anneaux SPEAR (Stanford), PETRA (Hambourg) et D.C.I. (Orsay), ont apporté des résultats importants voire spectaculaires du point de vue de la physique hadronique.

Si les collisions γγ sont appelées, sans aucun doute, à tenir une place non négligeable en physique des particules élémentaires, les expériences correspondantes sont toutefois difficiles à exécuter et à analyser. Cela justifie l'organisation des Journées d'Etudes d'Amiens, dont les promoteurs ont voulu réunir un certain nombre de physiciens compétents et intéressés, en vue de discuter l'ensemble des problèmes théoriques et expérimentaux qui se posent actuellement en ce domaine. Il s'agissait de la troisième réunion internationale consacrée aux collisions photon-photon, après celle du Collège de France (sept. 1973) et celle de Lake Tahoe (août 1979). Ces Journées d'Etudes ont attiré environ 90 participants, venus de seize pays d'Europe Occidentale et d'Amérique.

Au nom du Comité d'Organisation, nous désirons exprimer toute notre gratitude aux organismes qui ont accepté de financer ces Journées d'Etudes, à savoir: l'Université de Picardie, le Conseil Municipal d'Amiens, le Conseil Régional de Picardie, et la Division des Recherches et Etudes Techniques (D.R.E.T.). Nous sommes également reconnaissants à la Société Française de Physique de nous avoir accordé son patronage.

Nous remercions d'autre part très vivement les personnalités qui ont accepté de faire partie du Comité International de Parrainage, les orateurs invités, les présidents des séances plénières, les responsables des discussions qui se sont tenues dans le cadre des séances parallèles, et enfin, l'ensemble des participants qui, par leur intérêt et leur ardeur, ont assuré la réussite de cette réunion internationale.

IV

Notre gratitude particulière va au Professeur M. FROISSART pour ses encouragements, ses conseils et son aide.

Il convient de mentionner spécialement les étudiants de l'Université de Picardie qui ont assuré les tâches de secrétariat avec dévouement et efficacité.

G. COCHARD - P. KESSLER

FOREWORD

This volume contains the Proceedings of the "International Workshop on $\gamma\gamma$ Collisions" ("Journées d'Etudes Internationales sur les Collisions Photon-Photon") held at the Université de Picardie (Amiens, France) April 8-12, 1980.

Photon-photon collisions (i.e., collisions between virtual photon spectra, basically performed with electron storage rings) are a relatively new field in high-energy physics. The first theoretical papers in this field, by F. LOW and by F. CALOGERO and Ch. ZEMACH, were published in 1960. From 1969-70 on, the physics of $\gamma\gamma$ collisions was considerably stimulated by the extensive theoretical studies performed in particular by three groups (Collège de France, Novosibirsk, Cornell-SLAC). At present, many theorists are interested in this type of reaction, in the context of checking quantum chromodynamics.

As regards the experimental investigation of $\gamma\gamma$ collisions, the first experiments, performed at Novosibirsk and Frascati in the period 1970-73, were mainly tests of QED. On the other hand, the most recent experiments, performed with the storage rings SPEAR (Stanford), PETRA (Hamburg) and D.C.I. (Orsay), have provided some significant and even spectacular results in hadronic physics.

Although it may be taken for granted that $\gamma\gamma$ collisions will play an important role in future particle physics, the corresponding experiments are difficult to carry out and to analyze. This was the motivation of the Amiens Workshop; it appeared desirable, indeed, to bring together a number of competent and interested people in order to discuss all the theoretical and experimental problems that are presently arising in this field. Our Workshop was the third international meeting on photon-photon collisions, after the Collège de France Colloquium (September 1973) and the Lake Tahoe Conference (August 1979). That Workshop was attended by about 90 participants from 16 different countries of Western Europe and America.

On behalf of the Organizing Committee, we wish to express our deep gratitude to the various French bodies that agreed to support our Workshop financially, the Université de Picardie, the Town Council of Amiens, the Regional Council of Picardie, and the Division des Recherches et Etudes Techniques (D.R.E.T.). We are also very obliged to the Société Française de Physique for having decided to sponsor our meeting.

In addition, we wish to warmly thank those physicists of various countries who accepted our invitations to be members of our International Committee, as well as the invited speakers of the Workshop, the chairmen of plenary sessions, the discussion leaders of parallel sessions, the authors of short contributions presented in the parallel sessions, and finally, all participants who had a share , shown by their interest and their active contributions, in the success of that international meeting.

We are particularly grateful to Prof. M. FROISSART for his encouragement, advice, and help.

We would specially like to mention the students of the Université de Picardie who performed all the tasks of the secretariat with great devotion and efficiency.

G. COCHARD   -   P. KESSLER

LIST OF PARTICIPANTS

ALONSO J.L.	University of Saragossa (Spain)
ALTARELLI G.	University of Roma (Italy)
ANASTAZE G.	Division des Recherches et Etudes Techniques, Paris (France)
ARTEAGA-ROMERO N.	Collège de France, Paris (France)
BALAND J.F.	University of Mons (Belgium)
BARBIELLINI G.	CERN, Geneva (Switzerland) & INFN, Frascati (Italy)
BERGER Ch.	I. Phys. Institut der Techn. Hochschule Aachen (Germany)
BERGSTRÖM L.	Royal Institute of Technology, Stockholm (Sweden)
BOESTEN L.	University of Hamburg (Germany)
BORDES G.	Collège de France, Paris (France)
BURGER J.D.	Massachusetts Institute of Technology, Cambridge, MA (U.S.A.)
BURGER J.	Deutsches Elektronen Synchrotron, Hamburg (Germany)
BURKE D.L.	Stanford Linear Accelerator Center, Stanford, CA (U.S.A.)
BUSSEY P.J.	University of Glasgow (Great Britain)
CALVA-TELLEZ E.	IPN, Mexico (Mexico)
CARIMALO C.	Collège de France, Paris (France)
CHAVANON A.	Collège de France, Paris (France)
COCHARD G.	University of Picardie, Amiens & Collège de France, Paris (France)
COIGNET G.	Laboratoire d'Annecy de Physique des Particules, Annecy (France)
COURAU A.	Laboratoire de l'Accélérateur Linéaire, Orsay (France)
CROZON M.	Collège de France, Paris (France)
DAINTON J.	University of Glasgow (Great Britain)
DAVIER M.	Laboratoire de l'Accélérateur Linéaire, Orsay (France)
DEFRISE M.	Vrije Universiteit Brussel, Brussels (Belgium)
DERIKUM K.	University of Siegen (Germany)
DUNBAR I.H.	University of Sussex, Brighton (Great Britain)
EISNER A.	University of California, Santa Barbara, CA (U.S.A.)
ERNE F.C.	NIKHEF, Amsterdam (Netherlands)
FALVARD A.	University of Clermont-Ferrand (France)
FIELD J.H.	Deutsches Elektronen Synchrotron, Hamburg (Germany)
FRIDMAN A.	Centre d'Etudes Nucléaires de Saclay, Gif-sur-Yvette (France)
FROISSART M.	Collège de France, Paris (France)
FRONTEAU M.	University of Orléans (France)
FULDA F.	Laboratoire de l'Accélérateur Linéaire, Orsay (France)
GOLDBERG H.	Northeastern University, Boston, CA (U.S.A.)
GOLDBERG M.	University of Paris-VI (France)
GOLDZAHL L.	Centre d'Etudes Nucléaires de Saclay, Gif-sur-Yvette (France)
GRARD F.	University of Mons (Belgium)

International Patronizing Committee

N. CABIBBO (Rome)

F. CALOGERO (Rome)

M. FROISSART (Collège de France)

E. GABATHULER (CERN)

H. HARARI (Weizmann Institute)

L. LEDERMAN (Fermilab)

E. LOHRMANN (DESY)

F.E. LOW (MIT)

H. PIETSCHMANN (Vienna)

J.J. SAKURAI (Los Angeles)

H. SCHOPPER (DESY)

S. TING (MIT)

M. VIVARGENT (LAPP Annecy)

Ch. ZEMACH (Los Alamos)

A. ZICHICHI (CERN)

Organizing Committee

G. BARBIELLINI (CERN & INFN Frascati)

G. COCHARD (Amiens & Collège de France)

A. COURAU (Orsay)

P. KESSLER (Collège de France)

J.C. MONTRET (Clermont-Ferrand)

F. VANNUCCI (LAPP Annecy)

P. WALOSCHEK (DESY)

GRECO M.   INFN, Frascati (Italy)

GUNION J.   Stanford Linear Accelerator Center, Stanford, CA & University of California, Davis, CA (U.S.A.)

HAISSINSKI J.   Laboratoire de l'Accélérateur Linéaire, Orsay (France)

HEPP V.   University of Hamburg (Germany)

HILGER E.   University of Bonn (Germany)

HOFFMANN H.   CERN, Geneva (Switzerland)

ICHOLA A.   University of Picardie, Amiens & Collège de France, Paris (France)

JENNI P.   CERN, Geneva (Switzerland)

JÖNSSON L.   University of Lund (Sweden)

JOUSSET J.   University of Clermont-Ferrand (France)

KAJANTIE K.   University of Helsinki (Finland)

KAWABATA S.   Deutsches Elektronen Synchrotron, Hamburg (Germany)

KESSLER P.   Collège de France, Paris (France)

KIRKBRIDE G.   Stanford University, Stanford, CA (U.S.A.)

LILLESTOL E.   University of Bergen (Norway)

LLEWELLYN SMITH C.H.   University of Oxford (Great Britain)

MARIN P.   Laboratoire de l'Accélérateur Linéaire, Orsay (France)

MISSONNIER G.   University of Clermont-Ferrand & Collège de France, Paris (France)

MONTAROU M.   University of Clermont-Ferrand (France)

MONTRET J.C.   University of Clermont-Ferrand (France)

MURTAS G.P.   INFN, Frascati (Italy)

NEUFELD H.   University of Vienna (Austria)

NICOLAIDIS A.   Collège de France, Paris (France)

ONG S.   University of Picardie, Amiens & Collège de France, Paris (France)

PALOMBO F.   University of Milano (Italy)

PARISI J.   Collège de France, Paris (France)

PECK C.W.   California Institute of Technology, Pasadena, CA (U.S.A.)

PHAM T.N.   Ecole Polytechnique, Palaiseau (France)

PIETSCHMANN H.   University of Vienna (Austria)

PIRE B.   Ecole Polytechnique, Palaiseau (France)

QUERROU M.   University of Clermont-Ferrand (France)

ROST M.   University of Siegen (Germany)

SANDER H.G.   I. Phys. Institut der Techn. Hochschule Aachen (Germany)

SCHILDKNECHT D.   University of Bielefeld (Germany)

SCHREMPP B.   University of Hamburg (Germany)

SCHREMPP F.   University of Hamburg (Germany)

SKARD J.A.   University of Bergen (Norway)

SMITH J.   State University of New York, Stony Brook, NY (U.S.A.)

SMITH J.R.	University of California, Davis, CA (U.S.A.)
SNELLMAN H.	Royal Institute of Technology, Stockholm (Sweden)
SPINETTI M.	INFN, Frascati (Italy)
SRIVASTAVA Y.	Northeastern University, Boston (U.S.A.)
STEUER M.	CERN, Geneva (Switzerland)
VAN HUELE J.F.	Vrije Universiteit Brussel, Brussels (Belgium)
VANNUCCI F.	Laboratoire d'Annecy de Physique des Particules, Annecy (France)
VERMASEREN J.	CERN, Geneva (Suisse)
WALOSCHEK P.	Deutsches Elektronen Synchrotron, Hamburg (Germany)
WALSH T.	Deutsches Elektronen Synchrotron, Hamburg (Germany)
WEDEMEYER R.	University of Bonn (Germany)
WINTER K.	CERN, Geneva (Switzerland)
WRIEDT H.	University of Lancaster (Great Britain)
ZEMACH Ch.	Los Alamos Scientific Lab., Los Alamos, NM (U.S.A.)
ZORN G.	University of Maryland, College Park, MD (U.S.A.)

TABLE OF CONTENTS

GENERAL INTRODUCTION

P. Kessler

Laboratoire de Physique Corpusculaire, Collège de France, Paris, France

Summary

A brief account is given on the history of photon-photon collisions for the last 20 years, and on problems arising in that field of high-energy physics.

History

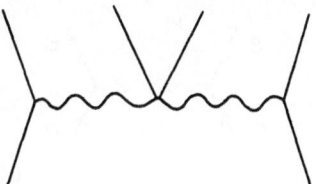

Fig. 1. Feynman diagram for photon-photon collision processes

Processes of the type shown in fig. 1 were studied by some of the pioneers of quantum electrodynamics already in the early thirties [1], but their application to electron colliding-beam physics started with two papers published in the "Physical Review" in 1960, namely the papers of LOW [2] and CALOGERO & ZEMACH [3].

LOW suggested measuring the lifetime of the π° through the reaction $e\,e \longrightarrow e\,e\,\pi^{\circ}$ (fig. 2).

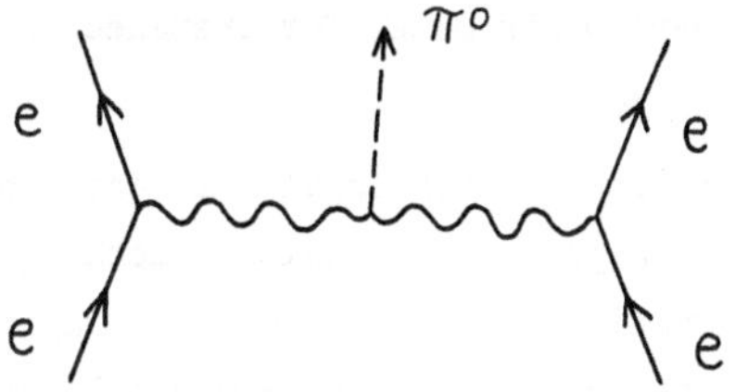

Fig. 2. Feynman diagram for $e\,e \longrightarrow e\,e\,\pi^{\circ}$

For the calculation of this process, LOW used a formula of extreme simplicity and elegance, where both electron beams were replaced by equivalent photon spectra:

$$\sigma_{ee \to ee\pi^\circ} = \int N(k_1) \, N(k_2) \; \sigma_{\gamma\gamma\to\pi^\circ}(k_1,k_2) \; dk_1 \, dk_2$$

with

$$N(k) = \frac{\alpha}{\pi} \frac{1}{k} \frac{E^2 + (E-k)^2}{E^2} \ln \frac{E}{m_e} \qquad (E = \text{beam energy, } k = \text{photon energy})$$

$$\sigma_{\gamma\gamma\to\pi^\circ}(k_1,k_2) = \frac{8\pi^2}{m_\pi \tau} \delta(4\,k_1\,k_2 - m_\pi^2) \qquad (\tau = \text{lifetime of the } \pi^\circ)$$

As for CALOGERO & ZEMACH, they suggested studying pair production in e e colliding beams, in particular e e $\longrightarrow$ e e $\pi^+\pi^-$.

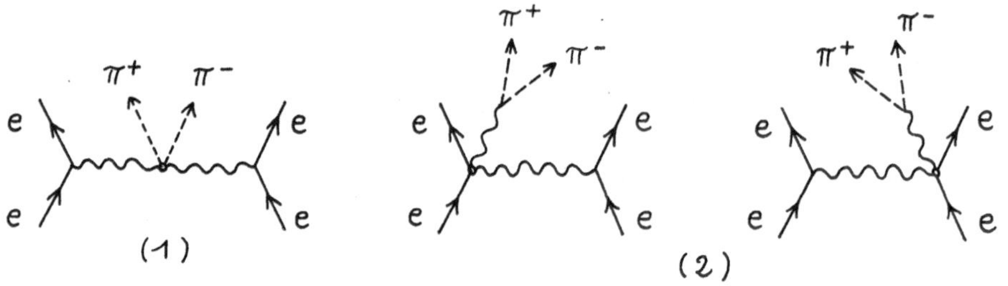

Fig. 3. Feynman diagrams for e e $\longrightarrow$ e e $\pi^+\pi^-$

That paper also contained a number of quite prophetic ideas:

(i) That process provides the possibility of studying the pion-pion interaction in the absence of any spectator hadrons.

(ii) Electrons scattered in the forward direction may be "tagged" as soon as they have lost a fraction of their energy, since their curvature in the magnetic field will be modified.

(iii) Cross sections are not too small, since log E/m_e factors partly make up for the additional α factors (when a comparison is made, for instance, with large-angle Møller scattering).

(iv) In fig. 3, diagram (1) gives rise to even angular momentum states of the pion pair, whereas diagrams (2) produce a pure J = 1 state. Something should be done in order to eliminate the less interesting "odd" diagrams. One way to do that is to stick to the configuration where the pion pair has zero total momentum (i. e. where the lab frame and the $\gamma\gamma$ c.m. frame coincide).

After those two fundamental papers appeared, they were to a large extent forgotten for almost ten years. The reason was, obviously, that electron storage rings of the first generation (the first machine at Stanford, VEPP-2 at Novosibirsk, ACO at Orsay) would not have allowed, both for lack of energy and of luminosity, to investigate such processes seriously.

In 1969, DE CELLES & GOEHL published a study on the production of σ (S-wave pion-pion resonance) in electron-electron collisions [4].

In the meantime, our group at the Collège de France had started the study of 2-photon exchange diagrams, as an application of a generalized (four-dimensional) helicity formalism that I had worked out a few years before for Feynman diagram computation in QED [5].

It soon became obvious to us that [6]

(i) when the electrons are tagged at 0° (i. e. in a small cone with an opening angle of a few milliradians or at most a few degrees), the theoretical background (i. e. the other diagrams in α^4) becomes totally insignificant;

(ii) under such conditions, the double equivalent-photon approximation can be applied with high accuracy; thus the external vertices in the diagram of fig. 1 may be "factorized out" and "forgotten", and <u>one is really studying photon-photon collisions</u>;

(iii) with the new generation of storage rings then planned or under construction (ADONE at Frascati; COPPELIA at Orsay, later replaced by D.C.I.; DORIS at DESY; SPEAR at SLAC) of beam energy $\approx$ 1.5-3 GeV and expected luminosity $\approx 10^{32}$ cm^{-2} s^{-1} (actually, the latter expectation later appeared over-optimistic), counting rates would be high enough (even taking account of realistic cuts in acceptance) to allow the study of such reactions as $\gamma\gamma \rightarrow e^+ e^-, \mu^+\mu^-, \pi^+\pi^-, \pi^o, \eta$.

Photon-photon collisions became popular after the Kiev Conference in 1970, with the reports of BRODSKY et al. [7] and BALAKIN et al. [8]. (At Novosibirsk, the "discovery" of $\gamma\gamma$ collisions followed from an experiment where a large number of coplanar but non-collinear particle pairs were observed).

In the years 1970-1973, $\gamma\gamma$ collisions became a very fashionable field of theoretical activity, while experiments in that area were performed at Frascati (they will be described in some detail by the next speaker, Prof. BARBIELLINI). Several hundreds of papers, almost all of them theoretical ones, were published in this period (see reference lists in the extensive reports by TERAZAWA [9] and by BUDNEV et al. [10], and also in the Proceedings of the International Colloquium that took place at the Collège de France in 1973 [11]).

The theoretical ideas discussed in those papers included:

(i) Applications of PCAC and low-energy theorems of current algebra to soft-pion production.

(ii) Study of meson-pair production, applying crossing symmetry, dispersion relations, Regge-pole theory, the quark model.

(iii) Study of multihadron production, using vector dominance, Regge-pole theory, the quark model.

(iv) Study of resonance production, applying quark and other models, and connecting it with multihadron production via duality.

(v) Study of deep-inelastic configurations (Q^2 or/and Q'^2 large), using the quark-parton model or light-cone algebra.

From 1974 on, however - and in particular after the discovery of the J/ψ - the interests of high-energy physicists were focussed on other processes and problems, and little attention was spent to $\gamma\gamma$ physics (only a handful of papers was published in that field between 1974 and 1978).

In 1978, a new stimulus was provided when BRODSKY et al. [12], in particular, showed that $\gamma\gamma$ processes at high energy may be considered an ideal place for checking QCD which, in the meantime, had become <u>the</u> theory of strong-interaction physics. Indeed, in these processes, due to the absence of any spectator hadrons, and to the extent that hadronic jets in the final state may be identified with quarks or gluons, we have to do only with <u>electrons, photons, quarks and gluons</u>. Therefore, as far as photons are assumed to have a pointlike behaviour, <u>everything is calculable</u> from QCD combined with QED. Since there is no doubt about QED being a good theory, decisive checks of QCD may thus be performed.

All that is true as well for $e^+ e^-$ annihilation processes; however, it appears that QCD checks to be performed in $\gamma\gamma$ reactions might be of even greater variety and interest; e. g. (see fig. 4) testing the quark propagator in (a) $\gamma\gamma \rightarrow q \bar{q}$ (two high-p_T jets), or testing the photon's or electron's structure function in (b) $e\, e \rightarrow e\, e\, q\, \bar{q}$ with one electron scattered at small angle and the other one at large angle (in that case, one should have one beam-pipe jet and one high-p_T jet).

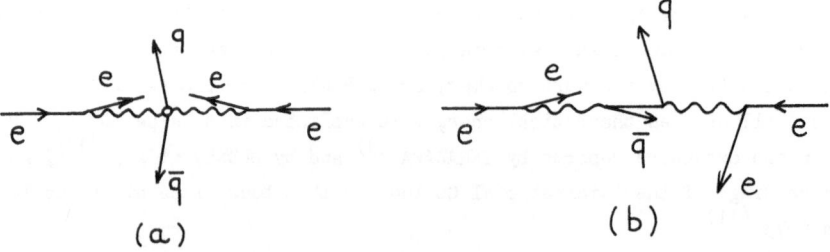

Fig. 4. Various configurations for QCD checks through $e\, e \rightarrow e\, e\, q\, \bar{q}$

Approximately at the time when these theoretical suggestions were made, new photon-photon experiments were started at D.C.I. (Orsay) [13], SPEAR (SLAC) with the Mark-II detector [14], and PETRA (where the first group to obtain $\gamma\gamma$ data was the PLUTO Col-

laboration [15]). Results of those experiments were reported and discussed at the Lake Tahoe Conference last year, and we shall hear more about them at this meeting.

However, at present, there is obviously still a wide gap left between theoretical
speculations and experimental possibilities. At least partly, this is due to the fact
that problems of analysis are of particular complexity in $\gamma\gamma$ processes.

Present problems

Problems of analysis in photon-photon collisions were studied in some detail in a
recent report written by our Collège de France group with the collaboration of A.
COURAU [16].

As is well known nowadays, the saturation of tagging counters by beam-beam and beam-
gas bremsstrahlung makes it very difficult, for high-energy electron storage rings,
to have tagging systems at 0°. Therefore such tagging systems were not foreseen for
storage rings of the third generation (PETRA, PEP, and ulteriorly LEP).

Non-tagging experiments may be performed in some cases (i. e. for simple $\gamma\gamma$ processes such as production of some resonances, or pair production at low energy); in general, however, problems of background rejection, and even more of event reconstitution, will make the analysis quite hazardous in such experiments.

On the other hand, $\gamma\gamma$ experiments with finite-angle tagging counters ($\theta_{min} \simeq 20$ mrad
at PETRA or PEP) allow for several options, but all of them have their shortcomings.

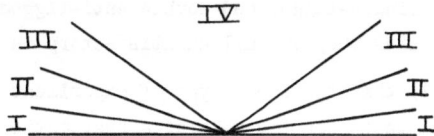

Fig. 5. Division of phase-space of the scattered electrons

We may divide the phase-space of scattered primary electrons as shown in fig. 5. In
region I, there is no tagging. Region II corresponds to "small-angle tagging", in the
sense that the condition Q (or Q') $\ll$ M/2 (where Q, Q' are the absolute values of the
virtual-photon masses, and M the invariant mass produced in the $\gamma\gamma$ collision) is satisfied in the average. Region III corresponds to "large-angle tagging", in the sense
that the condition Q (or Q') $\ll$ M/2 is not satisfied in the average. Finally, in region IV (in principle, the range covered by the central detector), there are practically no scattered electrons left. The condition specified for separating region II
from III, namely Q (or Q') $\ll$ M/2, must be satisfied, as has been shown [17], if one
wants to be able to "back-factorize", i. e. to extract the $\gamma\gamma$ cross section from the

measurement by simply using the double equivalent-photon approximation.

The options allowed for are the following:

(i) <u>Double-tagging</u>. Counting rates will in general be weak since the region I, where the main contribution lies, is lost on both sides. In addition, contributions from region III on both sides should not be used for the analysis if one wants to be able to apply back-factorization.

(ii) <u>Single-tagging</u>. Regions II and III on one side are combined with region I on the other side (where now the tagging counters are used as veto counters); the contribution of region III should however be ignored if one wishes to back-factorize. Nevertheless, counting rates are much larger than in case (i) above. On the other hand, since - by tagging only one of both outgoing electrons - one misses part of the information, background rejection and (even more) event reconstitution become more uncertain.

(iii) <u>Double anti-tagging</u>. Using the tagging counters as veto counters on both sides, one keeps only the region I on either side. One here gets high counting rates. Actually, double anti-tagging is an improved version of non-tagging, in the sense that - at the price of loosing a moderate fraction of the cross section - the "good" photons (i. e. the quasi-real ones) are selected, so that the double equivalent-photon approximation may here be used with high accuracy. On the other hand, the difficulties of background rejection and of event reconstitution remain the same as in the non-tagging case.

One may assume that, in a measurement performed with tagging counters, all types of events found (double-tagged, single-tagged and double anti-tagged ones) will be used for the analysis; but, as I said, they are all unsatisfactory in some sense.

One may conclude (see Table 1) that the ideal type of experiment remains double-tagging at 0°. Therefore I think that, for the sake of a systematic and quantitative study of photon-photon collisions, perhaps one should try to make available, some day in the future, a "$\gamma\gamma$ machine", i. e. an electron colliding-beam facility with the specific technology allowing for tagging at 0°.

Table 1. Options of measurement, and criteria of appreciation, for $\gamma\gamma$ collision processes

Options \ Criteria	Counting rates	Back-factorization	Background rejection	Event reconstitution
Non-tagging	maximal	1 term (double E.P.A.) (rough)	possible (with kinematic cuts), but perhaps not in all $\gamma\gamma$ channels	uncertain, except for specific channels
Double anti-tagging	almost maximal	1 term (double E.P.A.) (highly accurate)		
Single-tagging	still large	6 or 4 terms (when Q^2 not $\ll M^2$)	reasonably good	still uncertain, except for specific channels
Double-tagging	small	36 or 20 terms (when Q^2, Q'^2 not $\ll M^2$)	optimal	good
Double-tagging at 0° (assuming that a solution is found for bremsstrahlung saturation)	still large	1 term (double E.P.A.) (highly accurate)		

Finite angle tagging

<u>References</u>

(1) See e. g.: F. Perrin, Comptes Rendus Acad. Sc. <u>197</u>, 1302 (1933). L. Landau and E. Lifshitz, Phys. Zs. Soviet-Union <u>6</u>, 244 (1934). E. J. Williams, Nature <u>135</u>, 66 (1935). H. J. Bhabha, Proc. Roy. Soc. <u>152</u>, 559 (1935).

(2) F. E. Low, Phys. Rev. <u>120</u>, 582 (1960).

(3) F. Calogero and Ch. Zemach, Phys. Rev. <u>120</u>, 1860 (1960).

(4) P. C. De Celles and J. F. Goehl, Jr., Phys. Rev. <u>184</u>, 1617 (1969).

(5) P. Kessler, Cahiers de Physique <u>20</u>, 55 (1966); Report PAM 68-05 (1968); Nucl. Phys. <u>B15</u>, 253 (1970).

(6) N. Arteaga-Romero, A. Jaccarini and P. Kessler, Comptes-Rendus Acad. Sc. <u>269 B</u>, 153 and 1129 (1969); Report PAM 70-02 (1970). The same, plus J. Parisi, Lett. Nuovo Cim. <u>4</u>, 933 (1970).

(7) S. J. Brodsky, T. Kinoshita and H. Terazawa, Phys. Rev. Lett. <u>25</u>, 972 (1970).

(8) V. E. Balakin, V. M. Budnev and I. F. Ginzburg, Zh. Exp. Teor. Fiz. Pis'ma <u>11</u>, 559 (1970) (JETP Lett. <u>11</u>, 388 (1970)).

(9) H. Terazawa, Rev. Mod. Phys. <u>45</u>, 615 (1973).

(10) V. M. Budnev, I. F. Ginzburg, G. V. Medelin and V. G. Serbo, Phys. Rep. <u>15 C</u>, 181 (1975).

(11) Proceedings of the International Colloquium on Photon-Photon Collisions in Electron-Positron Storage Rings, Journal de Physique <u>35</u>, Coll. C-2, Suppl. to nr. 3 (1974).

(12) S. J. Brodsky, T. De Grand, J. Gunion and J. Weis, Phys. Rev. Lett. <u>41</u>, 672 (1978); Phys. Rev. D <u>19</u>, 1418 (1979). See also: R. N. Cahn and J. F. Gunion, Phys. Rev. D <u>20</u>, 2253 (1979). K. Kajantie and R. Raitio, Nucl. Phys. <u>B159</u>, 528 (1979). K. Kajantie, Acta Physica Austriaca, Suppl. XXI, 663 (1979). M. Abud, R. Gatto and C. A. Savoy, Phys. Rev. D <u>20</u>, 2224 (1979).

(13) A. Courau et al., Phys. Lett. <u>84 B</u>, 145 (1979).

(14) G. S. Abrams et al., Phys. Rev. Lett. <u>43</u>, 477 (1979); Preprint SLAC-PUB 2421 (1979).

(15) Ch. Berger et al., Phys. Lett. <u>89 B</u>, 120 (1979).

(16) N. Arteaga-Romero et al., Report LPC 80-06 (1980).

(17) See: C. Carimalo, P. Kessler and J. Parisi, Phys. Rev. D <u>20</u>, 1057 (1979).

THE EARLY EXPERIMENTS ON PHOTON-PHOTON COLLISIONS

G. Barbiellini
Laboratori Nazionali INFN
Frascati, Italy

and

CERN, Geneva, Switzerland

In this note I will try to trace the early development of the experimental investigation of the physics of photon-photon collisions at electron-positron colliding rings. Since in this survey I will consider the results obtained with the first e^+e^- colliders, I will not discuss some beautiful experiments, such as the Delbruck scattering performed at DESY or the trident production from lepton (e,μ) beams performed at the Fermi Laboratory and at the CERN SPS. Nonetheless, the results of these reactions are important experimental contributions to the progress achieved in the last ten years in the understanding of γ-γ physics.

The existence of this meeting at Amiens, where we can study and understand in detail the physics of colliding photons at the storage rings, is a result of many factors; a very important one is the long-term work of our theoretical colleagues. They have computed all the graphs of fourth order in QED which contribute to the γ-γ processes bringing, in this last decade, the pioneering work of Landau, Fermi, Weizacker and Williams to a level of accurate knowledge of the relative importance of each diagram in the different conditions of the final states (kinematic and charge conjugation).

During the experimental sessions of this workshop you will notice that the quality of the most recent experimental results on photon-photon physics is impressive (e.g. the PLUTO group has collected evidence for jet production in γ-γ collisions). The progress made since the pioneering work of the early experiments is a result of:

1) the increase in the energy available to the photon-photon system: $E_{beam}^{VEPP} \simeq 0.7$ GeV; $E_{beam}^{PETRA} \simeq 16$ GeV;

2) the higher luminosity of the new machines: $L_{early} \simeq 10^{28-29}$ cm^{-2} sec^{-1}; $L_{now} \simeq 10^{30-31}$ cm^{-2} sec^{-1};

3) the construction of sophisticated central detectors with good momentum analysis.

Hopefully this trend will continue in the future.

THE EARLY EXPERIMENTS

Table I summarizes the situation of photon-photon experiments done with the first e^+e^- colliders which have contributed to γ-γ physics.

Table I

Machine	E_{beam} GeV	L $cm^{-2}sec^{-1}$	Reaction	Year
VEPP 2	~ 0.7	10^{28}	$e^+e^- \rightarrow e^+e^-e^+e^-$	1970-71
ADONE	1.2	$2-4 \cdot 10^{29}$	$e^+e^- \rightarrow e^+e^-e^+e^-$ $\mu^+\mu^-$ $\pi^+\pi^-$ nh	1971-74
SPEAR	2.2	10^{30-31}	$e^+e^- \rightarrow e^+e^-$ $\mu^+\mu^-$ $\pi^+\pi^-$	1974

The first results on γ-γ physics from DCI and DORIS did not come much later than those quoted in Table I but these results will be presented in a special experimental session.

We will now review the main features of e^+e^- colliders and of the experimental set-ups involved in the early γ-γ experiments.

γ-γ PHYSICS AT VEPP II[1]

The main parameters of the Novosibirsk e^+e^- collider VEPP II are:

E	$\simeq$ 0.7 GeV
I^+	$\simeq$ 40 ma
I^-	$\simeq$ 70 ma
L	$\simeq 10^{28}$ cm^{-2} sec^{-1}
L_{int} (for this experiment)	16 ± 1 nb^{-1}
	13 ± 1 nb^{-1}

The experimental apparatus used at VEPP II is shown in Figure 1. The candidate events for the reaction

$$e^+e^- \rightarrow e^+e^-e^+e^-$$

are presented in Table II. Figures 3a, b and c show the acoplanarity distribution for the two sets of data on the $e^+e^- \rightarrow e^+e^-e^+e^-$ reaction (Figs 3a and b) and the same distribution for multi-hadron production in e^+e^- annihilation in multihadrons $e^+e^- \rightarrow$ nh.

Table II

Energy 2E, MeV	1020		1180-1340
Luminosity integral, 10^{33}cm^2	8.5 ± 0.4		13.2 ± 0.6
Interval of the ranges, g/cm^2	6.4 - 16		10.5 - 20
Region of angles $|\Delta\theta|$	0°-40°	40°-90°	0°-90°
Effect	150	71	20
Background (Normalized)	53	70	5
Admixture of the other processes investigated	13 ± 5	0.2	< 0.1
"Pure" effect	84 ± 19	1 ± 18	15 ± 6
Calculation according to Baier and Fadin	65 ± 13	22 ± 5	15.4 ± 3.2

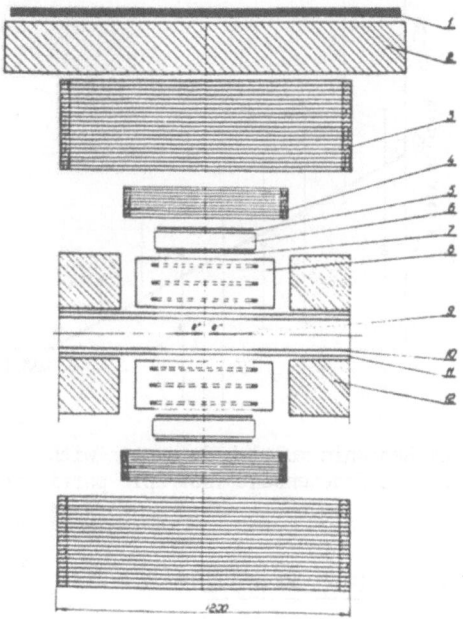

Fig. 1   The geometrical lay-out of the experiment (1970): 1) anticoincidence scintillation counter; 2) lead; 3) optical range spark chamber; 4) optical shower spark chamber; 5),7) scintillation counters; 6) water Cerenkov counter; 8) coordinate wire spark chambers; 9) interaction region; 10),11) internal and external vacuum chambers; 12) storage ring magnet.

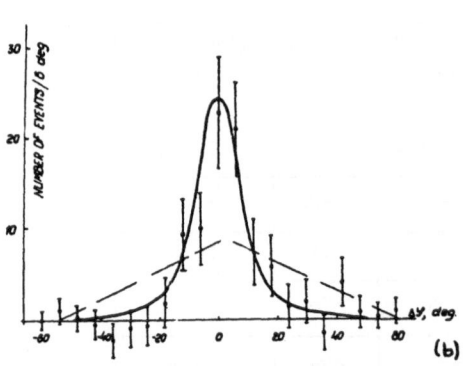
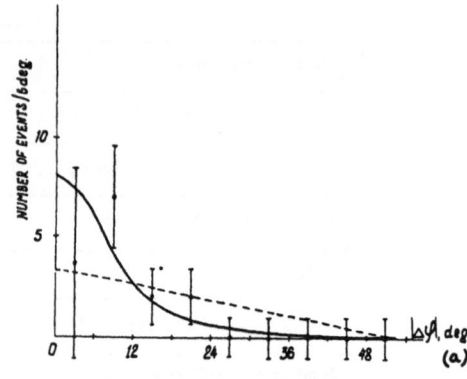

Fig. 2a   Distribution of events of the first experiment (1969) with respect to $\Delta\phi$. The solid curve was obtained with the Baier and Fadin formulae, the dashed one was calculated for the process with independent isotropic particle distribution.

Fig. 2b   Distribution of events of the second experiment (1970), shown as in Fig. 2a.

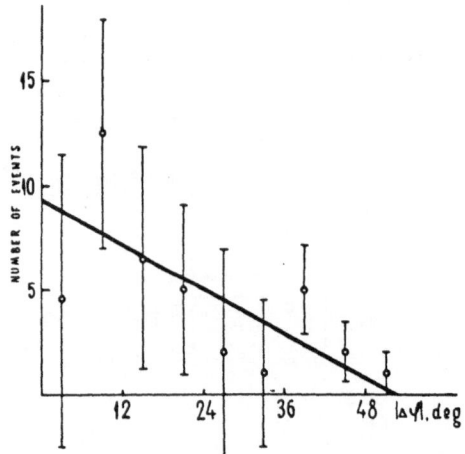

Fig. 2c   Distribution of hadronic multi-body events with respect to $\Delta\phi$.   The solid line corresponds to independent isotropic particles.

THE EARLY PHOTON-PHOTON COLLISIONS AT ADONE

ADONE is the Frascati e^+e^- storage ring that began physics operations in 1963. The main parameters of ADONE are:

E_{beam} $\simeq$ 1 - 1.5 GeV

L $\simeq$ $4 \cdot 10^{29}$ cm^{-2} sec^{-1}

$I^+ \simeq I^-$ $\simeq$ 30 ma

The first evidence at ADONE on the reaction

$$e^+ e^- \rightarrow e^+ e^- e^+ e^-$$

has been obtained[2] using a zero degree tagging system where the machine bending magnet was used as a momentum analyzer (Fig. 3). The photon energy acceptance of this tagging is limited to $K \simeq (0.1 - 0.3)E_b$. The energy acceptance of this system has been greatly extended in successive experiments[3],[4],[5].

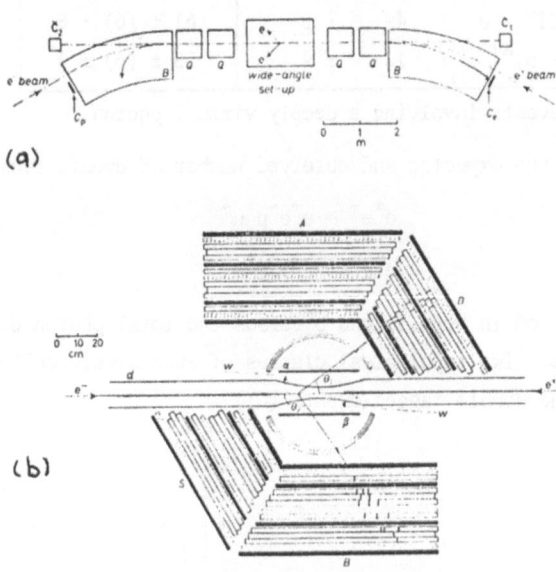

(a)

(b)

Fig. 3 a) Schematic view of the first ADONE tagging system. The machine elements are: bending magnets and quadrupoles (B,Q); the wide angle set-up is shown in more detail in b).

Figure 4 shows the large angle apparatus and the tagging system used for the first detection of muon pairs from light-light collisions[3].

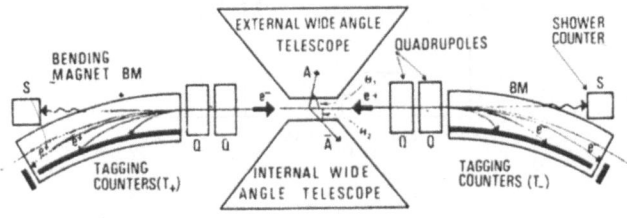

Fig. 4 Schematic view of the general set-up. The "shower counters", S, were used to events involving photons from real bremsstrahlung. A and Ā are WA particles.

Table III

Type of event	Expected number	Observed number
DT - μ	10.9 ± 1	14 ± 4
ST - μ	27.9 ± 2.6	20 ± 5
(ST + DT) - μ	38.3 ± 3	34 ± 6
DT - e	8.0 ± 1	12 ± 4
ST - e	41 ± 5	49 ± (6) ± 7
(ST + DT) - e	49 ± 5	61 ± (6) ± 8
(ST - e)[a]	18 ± 9	15 ± (6) ± 4

[a] e events involving a deeply virtual photon.

Table III shows the expected and observed number of events from the reactions

$$e^+e^- \to e^+e^-\mu^+\mu^-$$
$$e^+e^- \to e^+e^-e^+e^-$$

These events are plotted in Figs 5a and b versus the total photon energy, K, and the acoplanarity angle $\Delta\phi$. Two topological classes of events were collected: single tagged events (ST) and double tagged events (DT).

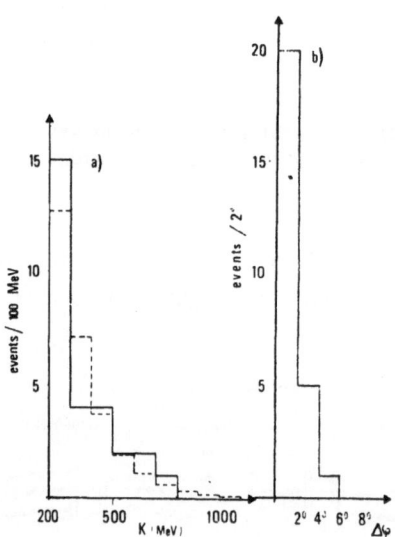

Fig. 5a Photon momentum distribution relative to the DT-μ events. The dashed line is derived from a Monte Carlo calculation based on the equivalent photon approximation.

5b Acoplanarity angle distribution for all recorded DT events.

A similar set of data has been presented by the other ADONE group[4] operating with a similar tagging system but different wide angle apparatus. The double tagging system at zero degrees measures the electron and positron energy with a resolution $\Delta p/p \simeq 8\%$ so that the mass of the hadron system produced in the reaction

$$\gamma\gamma \rightarrow nh$$

can be measured with similar accuracy.

The two ADONE experiments, equipped with zero degree tagging, have collected a few events with three charged hadrons detected in the wide angle apparatus with

$$0.8 \leq M\gamma\gamma \leq 1.5 \text{ GeV}$$

From these candidates, the measured integrated luminosity and the calculated acceptance it is possible to derive an upper limit on the η'-$\gamma\gamma$ coupling

$$\Gamma_{\eta'\rightarrow\gamma\gamma} \leq 16 \text{ KeV} \quad (95\% \text{ C.L.})$$

THE ANALYSIS OF $\gamma\gamma$ INTERACTIONS AND THE EARLY SPEAR DATA

The recent SPEAR results from the MARK I apparatus will be presented in a later session by P. Jenni. In this session I will recall the early SPEAR $\gamma\gamma$ analysis from MARK I. These data are contained in an extensive work by J.E. Zipse, prepared for his Ph.D thesis (Berkeley publication, LBL 4281).

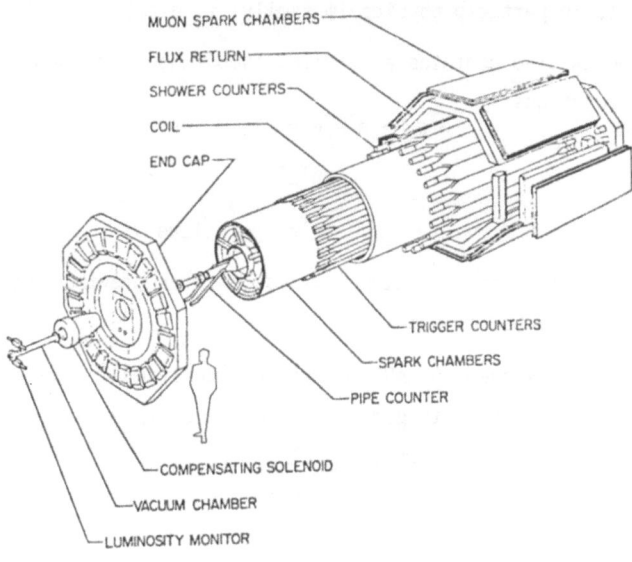

Fig. 6a Artistic view of Mark I apparatus

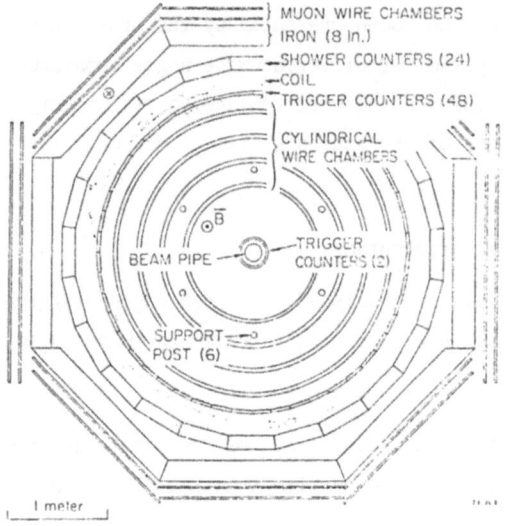

Fig. 6b Cross-section of Mark I apparatus normal to the beam axis

The main parameters of SPEAR are:

$$E_b \simeq 2.4 \text{ GeV}$$

$$L \simeq 10^{30-31} \text{ cm}^{-2} \text{ sec}^{-1}$$

Figures 6a and b show two views of the celebrated MARK I detector on which this $\gamma\gamma$ data was collected, together with the annihilation events that have brought so many fundamental results in particle physics (ψ family, τ, etc.)

The Zipse thesis also contains a detailed comparison between the exact QED calculation for the reactions

$$e^+e^- \rightarrow e^+e^-e^+e^-$$

$$\mu^+\mu^-$$

$$\pi^+\pi^- \text{ pointlike}$$

and the equivalent photon approximation that was used in previous experiments to evaluate the expected number of events.

The two-prong events seen in MARK I have been analyzed in two ways: without tagging and with 25 mrad single tagging. The result of this analysis on the reactions

$$e^+e^- \rightarrow e^+e^-e^+e^-$$

$$\mu^+\mu^-$$

$$\pi^+\pi^-$$

are summarized in Table IV.

Table IV

Cross-sections in nanobars	No "tagged" electron	One "tagged" electron
Calculated cross-sections:		
$e^+e^- \to e^+e^-e^+e^-$ (C = +)	0.314 ± 0.047	0.0145 ± 0.0022
$e^+e^- \to e^+e^-u^+u^-$ (C = +)	0.245 ± 0.036	0.0132 ± 0.0020
$e^+e^- \quad e^+e^-\pi^+\pi^-$ (C = +, point)	0.049 ± 0.007	0.0031 ± 0.0005
Sum of calculated cross-sections:	0.608 ± 0.075	0.0308 ± 0.0038
Measured cross-sections:		
measured two prongs with background subtracted	1.120 ± 0.066	0.0310 ± 0.0090
minus hadronic cross-section	-0.435 ± 0.021	---
minus accidental "tags"	---	-0.0009 ± 0.0003
Measured two photon cross-section:	0.685 ± 0.069	0.0301 ± 0.0090

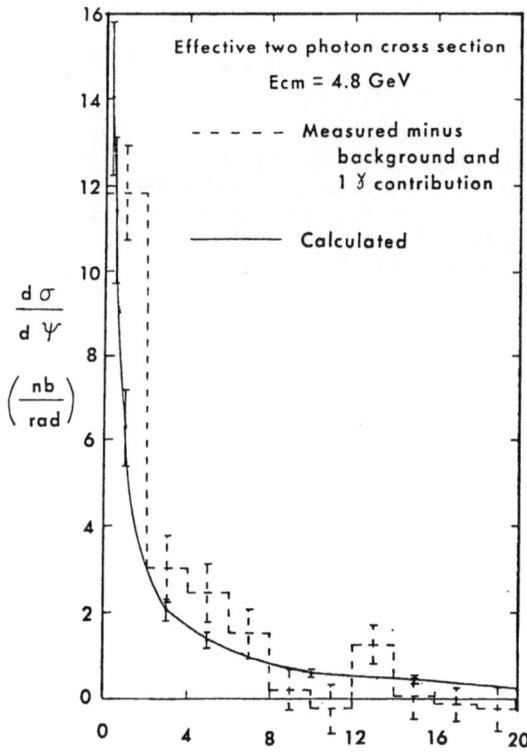

Fig. 7 Coplanarity angle

Figure 7 shows the coplanarity angle of the two prong events. The conclusions of this analysis are:

1) the fourth order QED terms are checked experimentally and agree with the theoretical expectation;

2) the $\gamma\gamma \to \pi\pi$ cross-section in the explored energy region in not very different from the pointlike cross-section.

CONCLUSIONS FROM THE EARLY $\gamma\gamma$ EXPERIMENTS

1) The fourth order QED calculation agrees with the experiment and the exact calculation differs from the equivalent photon approximation by less than 30%.

2) Two methods of separating $\gamma\gamma$ events from annihilation events have been experimentally checked: the tagging of the forward e^+e^- and no tagging. Both methods have been successful.

3) The first indication of hadron production from $\gamma\gamma$ shows that the cross-section for $\gamma\gamma \to \pi\pi$ is not very different from the pointlike cross-section.

4) From the reaction $\gamma\gamma \to nh$ an upper limit on η' radiative decay has been derived: $\Gamma_{\eta' \to \gamma\gamma} \leq 16$ KeV, 95% C.L.L.

* * *

REFERENCES

1) V.E. Balakin et al., Phys. Lett. 34B (1971).
2) C. Bacci et al., Lett. Nuovo Cimento 3 (1972) 709.
3) G. Barbiellini et al., Phys. Rev. Lett. 32 (1974) 385.
4) L. Paoluzi et al., Lett. Nuovo Cimento 10 (1976) 435.
5) R. Baldini et al., Phys. Lett. 86B (1979) 239.

TAGGING AND CONSTRAINTS IN γγ EXPERIMENTS

A. Courau

Laboratoire de l'Accélérateur Linéaire, 91405 ORSAY CEDEX (France)

INTRODUCTION

Experimentally there are different ways of studying γγ processes, with or without tagging of the photons by detection of the scattered incident electrons. Tagging can be performed on one side or on both sides (single or double), in a cone near $0°$ ($0 \leqslant \theta^e_{scatt} < \theta_o$) or at finite angle ($\theta_{min} < \theta^e_{scatt} < \theta_{max}$). Without tagging there is not always a clean signature of the γγ processes, except, may be, for some specific channels and provided sharp experimental cuts are introduced.

Zero degree tagging needs a machine with a good duty-cycle and a specific magnetic structure ; for instance, such a system is possible at D.C.I and DORIS. Those two ring machines should also allow one to perform e^+e^+ or e^-e^- collisions.

For higher energy machines, with only one very long ring, tagging at zero degree is hardly feasible and anyway unusable for the designed luminosities : Bremsstrahlung events would indeed saturate such a system at each bunch crossing. Tagging at finite angle then remains the only way to tag, taking account of the fact that the tail of scattered electrons from γγ events ($\sim 1/\theta$) is less strongly peaked than that from bremsstrahlung events ($\sim 1/\theta^3$).

I here limit myself to studying the influence of various experimental requirements, leading to conclusions that are sometimes different from - or even opposite to - those provided by intuitive theoretical considerations. I would like in particular to contribute to clarifying in a visually transparent way questions about the influence of cuts, the types of γγ processes that can be studied with various machines and set-ups, and experimental improvements that should be tried.

NOTATIONS

We use the following definitions (energies and angles being defined in the Lab. frame, i.e. the collision C. of M. frame).

- E is the incident beam energy,
- P_i, E_i, Θ_i, Φ_i are respectively the four-momentum vector, the energy, the polar and azimuthal angle of either scattered electron (i = 1.2),
- q_i, ω_i, θ_i, φ_i are respectively the same quantities for either virtual photon associated with a scattered electron (i = 1,2),
- W, $E_{\gamma\gamma}$, β are respectively the invariant mass, the energy and the absolute value of the velocity of the γγ system in the Lab. frame,
- ΔΘ is the acolinearity angle between the two scattered electrons (ΔΘ = 0 when the two electrons are back to back).

In addition we define the following dimensionless quantities :

$$X_i = \omega_i/E \quad ; \quad Z = W/2E \quad ; \quad V = E_{\gamma\gamma}/2E \quad ; \quad Q_i = \sqrt{-q_i^2}/E$$

$$\Delta = 2 \sin(\Delta\Theta/2) \quad ; \quad t_i = 2 \sin(\Theta_i/2) \quad ; \quad u_i = \frac{1-X_i}{X_i}\frac{E}{m} t_i$$

Let me just notice that in practice, as long as Θ remains small ($\lesssim .5$), we have :

$$\Delta \simeq \Delta\Theta \quad ; \quad t_i \simeq \Theta_i$$

This is obviously true for tagged $\gamma\gamma$ events and for the dominant part of the $\gamma\gamma$ events, where for $Q \ll X$ we have also :

$$u_i = \frac{E}{m} \sin \Theta_i$$

KINEMATIC RELATIONS

In the following, we assume relativistic electrons :

$$\frac{m}{E} \ll 1, \quad 1 - X \qquad \text{(m is the electron mass)}$$

Then we obtain the relations :

$$V = (X_1 + X_2)/2$$

$$Z^2 = X_1 X_2 - \frac{1}{4}(1 - X_1)(1 - X_2)\Delta^2$$

$$\beta^2 = \frac{(X_1 - X_2)^2 + (1 - X_1)(1 - X_2)\Delta^2}{(X_1 + X_2)^2}$$

For $\Theta_i \gg \frac{m}{E}$ we have $Q_i^2 = (1 - X_1)t_i^2$.

For $\Theta_i \simeq 0$ we have $Q_{i\,min}^2 = \frac{X_i^2}{1 - X_i}\left(\frac{m}{E}\right)^2$

Then since m/E is always very small, we write for the whole Θ_i range :

$$Q_i^2 = \frac{X_i^2}{1 - X_i}\left[\left(\frac{1 - X_i}{X_i} t_i\right)^2 + \left(\frac{m}{E}\right)^2\right]$$

or :

$$Q_i^2 = \left(\frac{m}{E}\right)^2 \frac{X_i^2}{1 - X_i}(1 + u_i^2)$$

Let us notice that, due to the condition $0 \leqslant Z, \beta \leqslant 1$, one gets the following kinematic limitation :

$$\Delta^2 < \frac{4 X_1 X_2}{(1 - X_1)(1 - X_2)}$$

THE CASE OF QUASI-REAL PHOTONS

In the $\gamma\gamma$ processes, because of the occurrence of the propagators $1/q_i^2$, small q_i^2 values play a dominant role : those small q_i^2 values are obtained, whatever ω_i may be, when the corresponding electrons are scattered at very small angles. If that is the case of both electrons, the cross-section to be computed may be easily factorized, i.e. equivalent-photon spectra may be used (according to the Williams-Weizsäker approximation) for both virtual photons, which are then called "quasi-real".

We shall stick to that dominant case, and only say a few words about the deep inelastic configuration (highly virtual photons).

The condition of validity for the W-W approximation (condition of quasi-reality) may be written as follows :

$$- q_i^2/W^2 = \frac{1}{4} Q_i^2/Z^2 \leqslant \varepsilon \ll 1,$$

which leads to :

$$t_i \leqslant Z\sqrt{\frac{4\,\varepsilon}{1 - X_i}}$$

When the maximum experimental tagging angle is small, $\Theta_0 \stackrel{<}{\sim} .5$, the quasi-reality condition imposes the restriction :

$$\Theta_i \leqslant \mathrm{Min}\left(\Theta_0,\ Z\sqrt{\frac{4\,\varepsilon}{1 - X_i}}\right)$$

If Θ_0 is small enough such a limitation arises only for small values of X_i; then, if we arbitrarily choose a value of $\varepsilon \simeq .25$, this restriction becomes :

$$\Theta_i \leqslant \mathrm{Min}\ (\Theta_0,\ Z)\ .$$

The quasi-reality condition applied to both photons leads, for the acolinearity angle, to the inequality :

$$\Delta^2 < \Delta_0^2 < \frac{4\,Z^2}{(1 - X_1)(1 - X_2)}\ \varepsilon\ (\sqrt{1 - X_1} + \sqrt{1 - X_2})$$

Since we choose $\varepsilon \ll 1$, the kinematic limitation on Δ is always satisfied and, on the other hand, we can derive the following simplified kinematic relations :

$$Z^2 \simeq X_1 X_2 \qquad ; \qquad \beta \simeq \left|\frac{X_1 - X_2}{X_1 + X_2}\right|$$

Then, if we submit the electron scattering angles to the quasi-reality restriction, we may integrate separately both photon spectra over the Θ_i acceptance, so that each spectrum will be a function only of the corresponding X_i and possibly on Z (whenever the quasi-reality condition is effective , i.e. $Z \stackrel{<}{\sim} \Theta_0$).

We assume for simplicity, in order to provide analytic integrations, that

i) All detectors have a cylindrically symmetric acceptance, and we do not look for any Φ-dependance of the cross-section.

ii) The central detector is symmetrical with respect to the beam crossing point. The polar acceptance in the Lab. frame is given by $-\psi_o \leqslant \psi \leqslant \psi_o$.

iii) The $\gamma\gamma$ cross-section depends only on W. We neglect any q^2 dependance, as for instance a V.D.M. form factor.

Then, in the expression of the counting rate

$$\tau = \mathcal{L}(ee) \int_{X_1} \int_{X_2} \int_{\Omega_1} \int_{\Omega_2} \int_{\Omega_L} dn_1 \cdot dn_2 \cdot d\sigma_{\gamma\gamma} \quad ,$$

we may integrate separately and analytically

(1) dn_1 over the tagging acceptance on one side (Ω_1),

(2) dn_2 over the tagging acceptance on the other side (Ω_2),

(3) $d\sigma_{\gamma\gamma}$ over the central detector's acceptance (Ω_L).

Thus there remain only two independent variables : X_1, X_2 or Z, β, or V, β ...

Let us consider the integral (1) or (2) using the equivalent-photon spectrum, after integration over $\Delta\Phi = 2\pi$. We set (dropping the subscript i = 1,2)

$$\int_\Omega dn = \frac{\alpha}{\pi} \frac{dX}{X} \int_\Theta dS$$

with dS = $[(1 - X + X^{2/2}) - (1 - X) Q^2_{Min}/Q^2] dQ^2/Q^2$.

Integrating over Θ from 0 to Θ_o, equivalently over u from 0 to u_o, one gets :

$$S = \left\{ \left[(1 - X + X^2/2) \, Ln \, (1 + u_o^2)\right] - (1 - X)\left(1 - \frac{1}{1 + u_o^2}\right) \right\}$$

This expression is valid (quasi-reality condition) for

$$u_o = \frac{E}{m} \frac{1 - X}{X} \, Min \left(\Theta_o, Z\sqrt{\frac{4\,\epsilon}{1 - X}}\right)$$

When $u_o \gg 1$ (i.e. $\Theta_o \gg m/E$), S is reduced to :

$$S = 2 (1 - X + X^2/2) \, Ln \, u_o - (1 - X)$$

S(X,Z) will be called in the following the "tagging acceptance function", whereas we shall call dX/X (which is independent of the acceptance Ω) the "flux factor".

As for the integral (3), we write :

$$\int_{\Omega_L} d\sigma_{\gamma\gamma} = \epsilon_{cd} \, \sigma_{\gamma\gamma} \quad ,$$

where ϵ_{cd} is the "acceptance factor of the central detector", defined as :

$$\epsilon_{cd} = \int_{\Omega_L} \frac{1}{\sigma_{\gamma\gamma}} \frac{d\sigma_{\gamma\gamma}}{d\Omega_L} d\Omega_L = \int_{\Omega^*} \frac{1}{\sigma_{\gamma\gamma}} \frac{d\sigma_{\gamma\gamma}}{d\Omega^*} d\Omega^*$$

where Ω^* is the acceptance in the $\gamma\gamma$ C. of M. frame ; it is a function only of Ω_L, W and β. In other words, at fixed E and Ω_L, Ω^* depends only on X_1, X_2 (or Z,β or V,β). We then obtain, for a given ee luminosity $\mathcal{L}(ee)$, the following expression of the counting rate :

$$\tau = \mathcal{L}(ee) \left(\frac{\alpha}{\pi}\right)^2 \iint \frac{1}{X_1 X_2} S_1(X_1,X_2) \cdot S_2(X_2,X_1) \cdot \varepsilon_{cd}(X_1 X_2) \sigma_{\gamma\gamma}(X_1 X_2) \, dX_1 dX_2$$

that we may also write

$$\tau = \int d\ell_{\gamma\gamma}(X_1 X_2) \cdot \varepsilon_{cd}(X_1 X_2) \, \sigma_{\gamma\gamma}(X_1 X_2) \, dX_1 dX_2$$

where we define a double differential $\gamma\gamma$ luminosity. Since this expression can be also written as a function of Z,β or V,β, we have :

$$d\ell_\gamma(X_1 X_2) = \left(\frac{\alpha}{\pi}\right)^2 \mathcal{L}(ee) \cdot \frac{1}{X_1 X_2} S_1(X_1 X_2) S_2(X_2 X_1)$$

$$d\ell_{\gamma\gamma}(Z,\beta) = \left(\frac{\alpha}{\pi}\right)^2 \mathcal{L}(ee) \cdot \frac{1}{Z} \frac{2}{1-\beta^2} S_1'(Z,\beta) S_2'(Z,\beta)$$

$$d\ell_{\gamma\gamma}(V,\beta) = \left(\frac{\alpha}{\pi}\right)^2 \mathcal{L}(ee) \cdot \frac{1}{V} \frac{2}{1-\beta^2} S_1''(V,\beta) S_2''(V,\beta)$$

STUDY OF THE INFLUENCE OF THE VARIOUS FACTORS AND CUTS

We shall study the behaviour of the factors involved in the counting rates and luminosity expressions, taking account of experimental requirements and cuts.

Flux factor

In first approximation the $\gamma\gamma$ luminosity is given by the flux factor $\frac{dX_1}{X_1} \frac{dX_2}{X_2}$. In order to visualize its influence and to assess more or less directly the effects of the other factors and of cuts, we are showing a diagram on a double logarithmic sale in X_1 and X_2. Then any area defined in a given $X_1 X_2$ range is directly proportional to the corresponding weight of the flux factor, since $dX/X = d(\text{Log } X)$. On the other hand, curves corresponding to constant values of β and Z simply become straight lines respectively parallel to the first and to the second bissectrix (see fig. 1). We notice that one can directly verify in this diagram the well-known theoretical fact that the $\gamma\gamma$ luminosity $\left(\sim \frac{d\beta}{1-\beta^2} \frac{dZ}{Z}\right)$ tends to favour large β and small Z values.

Let me give (fig. 2) two examples where we can see directly in this diagram the influence of various parameters.

We have plotted curves corresponding to the same R value where

$$R = (\Delta W/W)/(\Delta E'/E') = \sqrt{(1-X_1)^2/X_1^2 + (1-X_2)^2/X_2^2}$$ is the ratio between the resolution on W, obtained with a double tagging measurement, and the resolution of the tagging system.

We have also plotted the limits of quasi-reality for $\varepsilon = .25$, $\Theta = 100$ mr and $\Theta = 10$ mr. We can directly see for a finite angle tagging ($10 \leqslant \Theta \leqslant 100$ mr) the area where the experimental acceptance should be integrated on an angle smaller than the experimental one in order to satisfy "quasi-reality", and also the area where there remain no quasi-real photons any more. Let me just notice that, independently of the

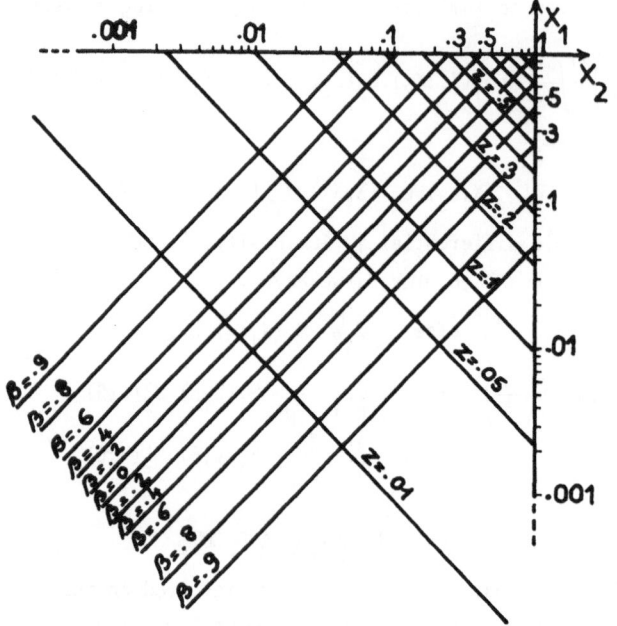

Fig. 1 : X_1X_2 Diagram

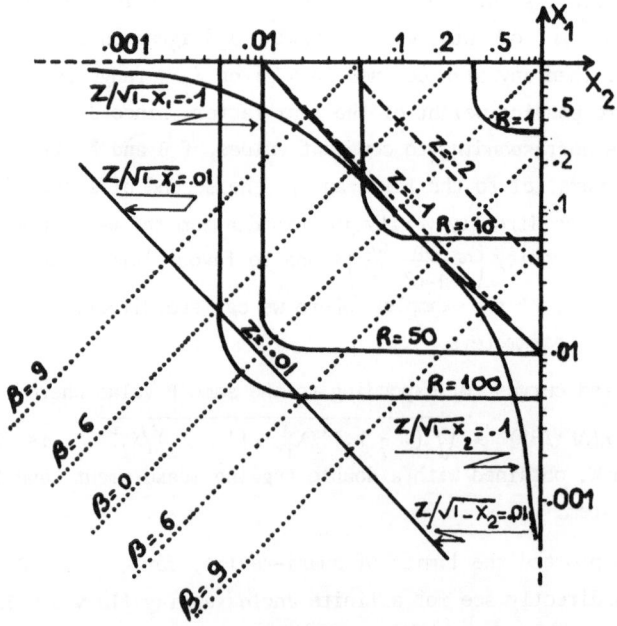

Fig. 2

vague   quasi-reality notion, the same curves limiting those areas   also correspond
to the absolute kinematic limitation of single tagged events with Θ = 200 mr and
20 mr, when the untagged electron is assumed to be emitted at $0°$.

Central Detector Acceptance

The acceptance of the central detector is the first thing to consider if one
wants to study the experimental importance of $\gamma\gamma$ processes as background of annihilation or to analyse those processes for themselves. In this last case, it is
obvious, from an experimental point of view, that the central detector plays a primary role, and that the tagging efficiency is meaning ful only for those events which
are detected in the central apparatus.

Presently most of the central detectors built or planned have a solenoidal magnet.
This involves a transverse momentum limitation for triggering ($p_T > p_o$) and a less
than 4π polar acceptance ($|\psi| > \psi_o$). These two constraints respectively lead to two
cuts $W > W_o$ and $\beta < \beta_o$, with an acceptance function decreasing when β increases.

As an example, we have plotted in the diagram of the fig. 3 the absolute

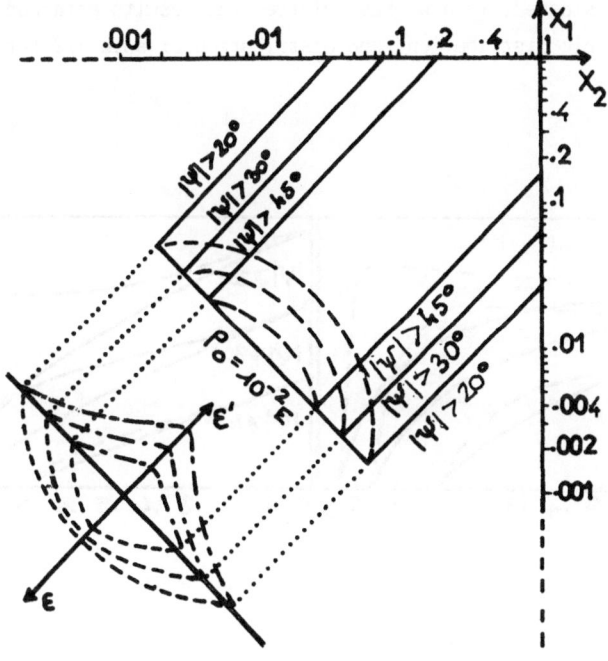

Fig. 3 : Central Detector's Acceptance

limitations for the production of two relativistic particles : $W > p_0/2$ and $\beta < \cos \psi_0$ obtained for different values of p_0 and ψ_0. The dashed line shows the area where, for small Z and β, the trigger limitation decreases the acceptance more than the geometrical acceptance. Our diagram directly shows the strong influence of the central detector involving a dramatic loss of counting rate due to not reaching the very small Z or W values. It also shows the importance of going to very small values of ψ_0.

In fact, the acceptance is not only defined by those cuts ; it is a function of β. In order to have an optimal visualization of the influence of the acceptance, we have plotted on the same figure the shape of the acceptance (in the area where it is only defined by the geometrical acceptance), for the production of two relativistic bodies :

$$\varepsilon_{cd} = \int_{-\psi}^{\psi} \frac{d\sigma}{d\Omega} d\Omega = 2 \int_0^{\cos\theta^*} \frac{d\sigma}{d\Omega^*} d\Omega^*$$

with $\cos\theta^* = \dfrac{\cos\psi_0 - \beta}{1 - \beta\cos\psi_0}$.

Of course, the acceptance function depends on the C. of M. angular distribution of the reaction considered. Then we have plotted the results obtained for two different cases : Isotropic distribution, and distribution of spin 1/2 fermion pair (leptons or quarks).

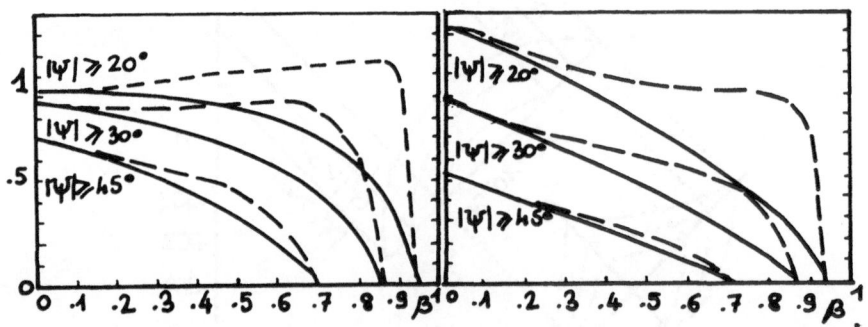

Fig. 4

Fig. 4a

Two Body   Isotropic Decay

——— $\varepsilon = 1/\sigma_{\gamma\gamma} \int_\Omega d\sigma_{\gamma\gamma}$

— — — $1/(1-\beta^2)\varepsilon$

Fig. 4b

Fermion Pair (Leptons or Quarks)

——— $\varepsilon' = W^2/(4\pi\alpha^2) \int_\Omega d\sigma_{\gamma\gamma}$

— — — $1/(1-\beta^2)\varepsilon'$

The figure 4 is also showing a convolution of the acceptance with the factor $d\beta/(1 - \beta^2)$ stemming from the $\gamma\gamma$ luminosity. Let us remark, incidentally, that experimentally the β distribution is dominated by the acceptance effect as long as ψ_o is not extremely small. Small values of β are more favoured than, may be, is intuitively expected.

We have studied the case of 2 relativistic particles. It is obvious that, for more particles produced, the central detector's acceptance should give even stronger constraints, assuming as a general principle (except perhaps in the double tagging case) that the experimentalist wishes to see all particles.

Tagging acceptance function

In all tagging configuration here considered, we shall study the behaviour of $S(X)$ (or $S(Z)$) using a logarithmic scale in our plots in order to take account of the $1/X$ (or $1/Z$) flux factor. One should notice , in addition, that very small values of Z (or indirectly X) are eliminated anyway by the central detector's acceptance.

"Acceptance function without any tagging system" : the usual expression corresponding to the E.P.A. is obtained by assuming the condition $Q \ll X$ which is not very different from the quasi-reality condition $Q \ll Z$ when β is not too close to 1 .

Since in this case $u \simeq \dfrac{E}{m} \sin \theta$, integration over θ up to $\sin \theta_o = 1$ ($u_o = \dfrac{E}{m}$) leads to :

$$S(X) = 2 (1 - X + X^2/2) \, \text{Ln} \, \frac{E}{m} - (1 - X)$$

We shall leave aside the question of validity of that formula which has been checked, through comparison with exact calculations, by many authors. Let us only make the following remark.

The upper limit of integration over θ was defined rather vaguely. This is not very relevant, because the very dominant part of the contribution is due to very small angles. Nevertheless the integration up to $u_o = \dfrac{E}{m} \sin \theta_o = \dfrac{E}{m}$ is derived from the relation $X \sin \theta = (1 - X) \sin \Theta$ when $Q \ll X$. Therefore a slighly more correct expression would be :

$$S(X) = 2(1 - X + X^2/2) \, \text{Ln}\left(\frac{E}{m} \, \eta\right) - (1 - X)$$

with $\eta = \text{Min} \left(1, \dfrac{X}{1 - X}\right)$.

The shape of S(X) is plotted in figure 5 with respect to log(X). We can see the slight difference near X $\simeq$ 1 introduced by our last remark. This difference is of course not very significant since the phase space allowed in this range is very small.

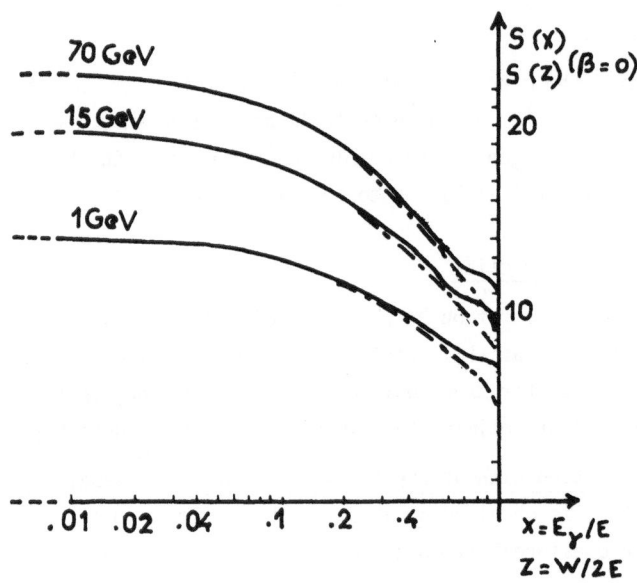

$$\underline{\text{Fig. 5}} : S(X) = 2(1 - X + X^2/2)Ln\left(\frac{E}{m}\eta\right) - (1 - X)$$

$$\underline{\hspace{2cm}} \quad \eta = 1 \quad ----\cdot- \quad \eta = \frac{1-X}{X}$$

To conclude we notice that :

S(X) increases with X decreasing.

S(X) increases with the energy like $Ln\left(\frac{E}{m}\right)$.

"Acceptance function for $0°$ tagging". When the experimental tagging angle is defined by $0 \leqslant \theta \leqslant \theta_0$ one has :

$$S_{\theta_0} = 2(1 - X + X^2/2) \, Ln \, u_0 - (1 - X)$$

That we can write :

$$S_{\theta_0} = 2(1 - X + X^2/2) \, Ln\left[\frac{E}{m}\frac{1-X}{X} t_0\right] - (1 - X)$$

where the quasi-reality condition is expressed as :

$$t_0 = Min\left(\theta_0, Z\sqrt{\frac{4\varepsilon}{1-X}}\right)$$

In practice θ_0 is always very small, and on the other hand very low values of Z are excluded because the detection of electrons (emitted at $\theta \sim 0°$) outside the closed orbit involves a lower limit on X. Then the quasi-reality condition is always satisfied, and we have :

$$S_{\Theta_o} = 2(1 - X + X^2/2) \ Ln \ \frac{E}{m} \frac{1-X}{X} \Theta_o \ - (1 - X) \ .$$

The corresponding shape of $S_{\Theta_o}(X)$ is plotted in figure 6. We here notice that :

$S_{\Theta_o}(X)$ increases with X decreasing

$S_{\Theta_o}(X)$ increases with the energy like $Ln \ \frac{E}{m}$

$S_{\Theta_o}(X)$ has necessarily a cut on X (or Z) which is experimentally involved in order to tag the electron outside the closed orbit.

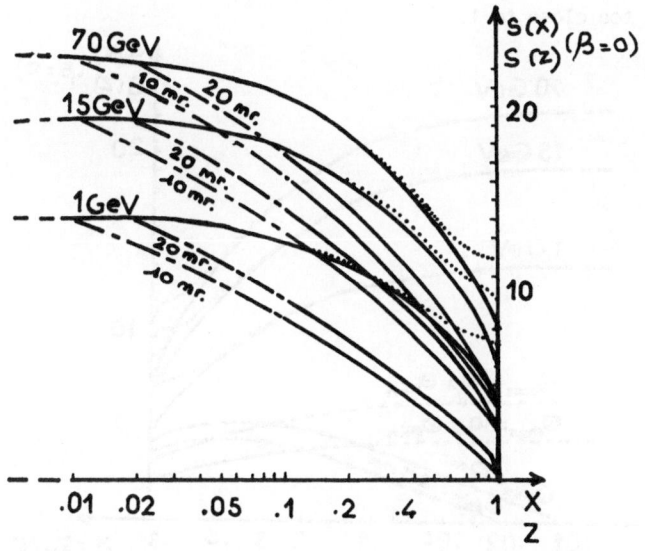

Fig. 6 : S(X) and $S_{\Theta_o} = 2(1 - X + X^2/2) Ln\left(\frac{E}{m} \ \frac{1-X}{X} \Theta_o\right) - (1 - X)$

In the X range left over, the tagging efficiency $S_{\Theta_o}(X)/S(X)$ increases

1) with rising Θ_o and E,

2) with X decreasing.

"Acceptance function at finite angle tagging". If the experimental tagging angle is defined by $\Theta_m < \Theta < \Theta_M$ we may write :

$$S_{\Omega} = S_{\Theta_M} - S_{\Theta_m}$$

and we obtain

$$S_{\Omega} = 2(1 - X + X^2/2) \ Ln \ (u_M/u_m) = 2(1 - X + X^2/2) \ Ln \ (t_M/t_m)$$

Then the quasi-reality condition becomes :

$$t_{m,M} = Min\left(\Theta_{m,M}, Z\sqrt{\frac{4\epsilon}{1-X}}\right)$$

The behaviour of S_Ω depends on the choice of ϵ on one hand and on the other hand on both X and Z. In figure 7 we have chosen ϵ = .25 and plotted $S_\Omega(Z)$ on a logarithmic scale assuming X $\sim$ Z. This last assumption is not very critical since anyway the X dependance (i.e. $(1 - X + X^2/2)$) is not violent. The behaviour of $S_\Omega(Z)$ remains about the same whatever Xshould be. On the other hand, for a comparison with the acceptance functions for others configurations, one may prefer consider the X (rather than the Z) dependance of S_Ω. It would be practically given by the same curve, as long as β is not too close to 1.

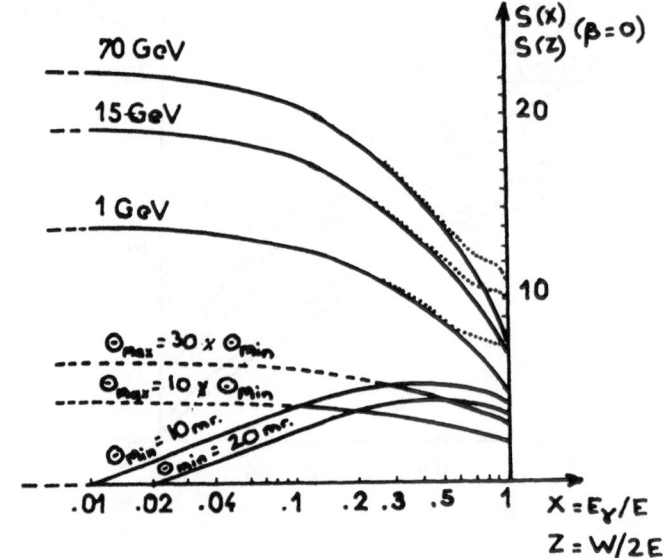

Fig. 7 : S(X) and $S_\Omega = (1 - X + X^2/2)Ln(u_M/u_m)$

----- $u_M/u_m = \Theta_M/\Theta_m$ —— $u_{M,m} = Min(\Theta_{M,m},Z)$ with Z$\sim$X

We here conclude that : S_Ω is independent of the beam energy.

The quasi-reality condition plays a very important role :

    it leads to a decrease of S_Ω when Z (and X) decreases,

    it introduces a cut on Z (when $\Theta_{min} = Z\sqrt{\frac{4\epsilon}{1-X}}$, we have t_2/t_1=1, S_Ω= 0).

As for the tagging efficiency S_Ω, in contradistinction to the 0° tagging efficiency, it

    decreases when X is increasing,

    decreases when E is increasing.

S_Ω/S increases more strongly when Θ_m decreases than when Θ_M increases. In other words, keeping the ratio Θ_M/Θ_m constant, such a tagging efficiency increases when Θ_m is decreased.

At this point, we would like to make a remark on the quasi-reality limit. One might object that this is an arbitrary limit, since certainly there still remain $\gamma\gamma$ events once one goes beyond. Nevertheless, that limit is important because it defines the range where we are allowed to write down explicitly the tagging acceptance function. In particular such a limit should be experimentally imposed if one wants to extract a $\gamma\gamma$ cross-section from the data. On the other hand, even if it were legal to analytically extrapolate the E.P.A. beyond its limits of validity, such an extrapolation would anyway lead us into a range where the angular phase space is sharply s rinking (see the kinematic limit on $\Delta\Theta$). More precisely, when an electron is emitted at a rather large angle, the angular phase space allowed for the second one becomes very small around the same large polar angle value. The two electrons must be pratically back to back at large angle, so that the probability to reach such events becomes exceedingly small.

In the case of single tagging where the non-tagged electron is considered as emitted at zero degree, the kinematic limit imposed to the tagged electron is pratically defined just like the quasi-reality limit, simply by taking $\varepsilon = 1$ instead of $\varepsilon = .25$ and changing Z to $\sqrt{X_1 X_2}$, (here Z is not necessary equal to $\sqrt{X_1 X_2}$) in the expression of S_Ω. More precisely $Z\sqrt{\frac{4\varepsilon}{1-X}}$ must be remplaced by

$$2\sqrt{\frac{X_1 X_2}{(1-X_1)(1-X_2)}} \; .$$

ENERGY BEHAVIOUR

It is a wide-spread belief that $\gamma\gamma$ processes increase with the beam-energy. For instance a well-known expression for the production of a pair of spin 1/2 particles of mass M (leptons or quarks) is :

$$\sigma \sim \frac{1}{4M^2} \; Ln^2 \; \frac{E}{m} \; Ln \; \frac{E}{M}$$

This expression is obtained from the integration of

$$Ln^2 \; \frac{E}{m} \; \frac{1}{W^2} \; \frac{d\beta}{1-\beta^2} \; \frac{dW}{W} \quad ,$$

over the total $X_1 X_2$ phase-space (i.e. $0 \leqslant \beta < \frac{1-Z}{1+Z}$ and $2M < W < 2E$) without taking in account the Z and β dependance due to the acceptance functions.

In fact, the angular acceptance of the central detector involves an other factor $\varepsilon_{cd} = Ln \left(\frac{1-\beta}{1+\beta} \; ctg^2 \; \psi/2 \right)$, and the integration over β is limited to $\beta_M = (1-Z)/(1+Z)$ only for $Z > tg^2 \; \psi/2$ and to $\beta_M = \cos \psi$ (independent of Z) for $Z < tg^2 \; \psi/2$ (see fig.3). The energy dependance of the counting rate will be then decreased with respect to the previous expression, when E increases beyond $M/tg^2\ \psi/2$.

There is also always a threshold W_t due to the triggering of the central detector. Then, even without taking in account the angular acceptance, the expression should be

$$\sigma \sim \frac{1}{W_o^2}\ Ln^2\ \frac{E}{m}\ Ln\ \frac{2E}{W_o}$$

with $W_o = MAX(2M,\ W_t)$.

If W_o is larger than 2M and to some extent energy dependent (as may be expected), the energy behaviour will be changed. In particular if $W_o (> 2M)$ is proportional to the beam energy, one gets :

$$\sigma \sim \frac{1}{E^2}\ Ln^2\ \frac{E}{m_e}$$

I should be noticed that, even without considering any restrictions due to triggering the central detector, the energy behaviour is affected anyway by the tagging system whenever there is one.

In the case of double $0°$ tagging, a minimal value is obviously imposed on Z, thus $W_{MIN} \sim E$, due to the cut on X_1, X_2 that is necessary in order to tag the electrons outside the closed orbits. In the case of single $0°$ tagging a Z_{MIN} is imposed in a different way namely through the tagged X_{MIN} and β_{MAX} (the latter restriction being due to the central detectors limited acceptance). In addition it is well known that $0°$ tagging is hardly usable when the machine energy becomes very large.

With finite angle tagging, the $Ln\ \frac{E}{m}$ dependance is lost. On an other hand a lower cut on Z (or at least a suppression of low Z values pratically equivalent to a cut) is resulting again from different considerations, namely from the quasi-reality condition and, in any case, from the kinematic limit. In double finite angle tagging the $\gamma\gamma$ luminosity function becomes :

$$F(Z) \sim Ln^2\left[\frac{MIN(2Z\sqrt{\varepsilon}\ ,\ \theta_{MAX})}{MIN(2Z\sqrt{\varepsilon}\ ,\ \theta_{MIN})}\right]$$

that function sharply decreases in the range

$$\theta_{MIN}/2\sqrt{\varepsilon} < Z < \theta_{MAX}/2\sqrt{\varepsilon}$$

For single finite angle tagging events, one has :

$$F(Z) \sim Ln\left[\frac{MIN(2Z\sqrt{\varepsilon}\ ,\ \theta_{MAX})}{MIN(2Z\sqrt{\varepsilon}\ ,\ \theta_{MIN})}\right]\ Ln\ \frac{E}{m}$$

Now that function is less decreasing than before; however it involves the same cut $Z_{MIN} \sim \theta_{MIN}/2\sqrt{\varepsilon}$ as before.

Thus it may be concluded that, in any tagging case, one has an effective threshold $W_{MIN} \propto E$. It results that the E dependance of the counting rate should be the same (apart from possible log E/m factors) as the dependance of $\sigma_{\gamma\gamma}$ on W, when the energy increases with respect to the threshold of the given reaction beyond a certain value.

We shall say a word apart on resonance production. If one compares such a process at constant Z and different energy E (for instance producing an η at E = 5.5 GeV and an η_c at E = 30 GeV), it is obvious, since $d\mathcal{L}/dW = 1/2E \, d\mathcal{L}/dZ$, that the counting rate $n \propto \Gamma \left(\dfrac{d\mathcal{L}}{dW} \right) W_R$ varies (apart from possible log E/m factors) like E^{-1}.

On the other hand, let us consider the production of a given resonance, thus keeping W constant at various energies. In double and single finite angle tagging the counting rate should also decrease with increasing energy, since we have $d\mathcal{L}/dW = 1/W \, F(Z)$, where $F(Z) = F(W/2E)$ goes down with Z decreasing.

CONCLUSIONS

In this paper, we are using simplified formulas for discussion, but we also show some more precise formulas to be used for computing counting rates or for γγ-event-simulating. In practice, either when integrating step-by-step over X, or when choosing events according to a probability law, one may take account of variables other than X_1 and X_2 - i.e. of angular variables - involved in the γγ interaction. Let us mention as examples : (i) polarization terms, involving azimuthal angles in the γγ c.m. frame ; (ii) form factors, involving $Q^2(X,\theta)$; (iii) the precise calculation of W and β in the case of finite angle tagging.

We would like now to draw the experimental conclusions of our study for the case of higher and higher beam energies, assuming that the e^+e^- luminosity remains constant.

The increase of the energy of electron-positron storage rings has the obvious advantage of allowing the study of γγ processes involving higher and higher invariant masses.

Nevertheless the study of resonances becomes more and more difficult for higher masses, even increasing the beam energy. On the other hand, when the energy becomes large with respect to the threshold of a given reaction, the counting rate does not necessary increase with the energy and even may decrease just as in annihilation processes (E^{-2}).

The best energy behaviour is obtained, practically, in the non tagging case. That case involves some short-comings as far as event identification and reconstruction is concerned. It should be noticed that at least the γγ signal-to-noise is improved at higher energies : the overlap with the annihilation background becomes less and less important ; on the other hand, the overlap of beam-gas background is also reduced, since in the latter processes the invariant mass produced scales only like $\sqrt{E}$ whereas it scales like E in the $\gamma\gamma$ collisions. In addition, at increasing energies, the β of beam-gas goes to 1. Therefore a large part of those events can be rejected. At the LEP energy, measurements without any tagging should possibly be the most efficient in most cases.

The finite angle tagging has a small efficiency decreasing with the energy. Nevertheless single finite angle tagging remains the only way to study the structure of the non-tagged quasi-real photon thanks to the large θ scattered electron (deep inelastic or $e\gamma$ study). Notice that it should be impossible to study the structure function which contains the cos 2ϕ dependance, since ϕ, which is the angle between the two planes of scattered electrons, cannot be experimentally measured.

To conclude, we shall state briefly what should be improved to make $\gamma\gamma$ physics more promising when increasing the energy of the machines:

1) The acceptance of the central detector should be increased. In particular the minimum angle of detection should go down to the smallest values of ψ as far as possible.

2) The minimum angle of tagging should be also decreased, as far as background problems allow such a decrease.

2γ PHYSICS VERSUS 1γ PHYSICS AND WHATEVER LIES IN BETWEEN

J.A.M. Vermaseren

CERN, Geneva, Switzerland

1. INTRODUCTION

In this talk I would like to present the results of some calculations of electromagnetic 2γ reactions in various configurations and compare them with corresponding
1γ signals or signals from processes that in their turn can be a rather bad background
to 2γ physics.

To this end I will start with outlining the method by which these calculations
were performed. The next section of this talk will be devoted to the study of various
final state configurations of the reaction $e^+e^- \rightarrow e^+e^-\mu^+\mu^-$ and, where possible, comparison with 1γ final states.

In the fourth section I will discuss problems that can arise in double-tag experiments owing to the presence of Bhabha scattering and its radiative corrections.
The contribution to the reaction $e^+e^- \rightarrow e^+e^-\mu^+\mu^-$ due to diagrams of the non-peripheral
type is investigated in the fifth and final section.

2. A COMPUTER PROGRAM FOR 2γ PHYSICS

The general behaviour of 2γ processes is rather different from that of 1γ physics.
This means that experimental acceptance effects are very different in both cases.
Meaningful comparisons can therefore only be made if one can calculate both types of
processes within a full set of experimental cuts. To appreciate this point one can
think of Bhabha scattering of which the total cross-section is infinite. This is
luckily not all observable. Also $e^+e^- \rightarrow e^+e^-\mu^+\mu^-$ has a very large total cross-section
of $\simeq$ 120 nb at $\sqrt{s}$ = 30 GeV, which is 1200 units of R at this energy. Again, the
fraction of this cross-section that is observable in a detector that covers 90% of 4π
is only minute.

To do calculations with experimental cuts one is forced to use Monte Carlo type
integration programs as these cuts can usually not be expressed in very simple formulae
of the integration variables. The use of the computer will allow many advantages over
approximative calculations like the various Weizsäcker-Williams schemes[1]. First of
all the computer deals with complicated formulae just as easily as with simple formulae if one only gives it more time. The biggest advantage is, however, that once
the program is set up properly many 2γ calculations can be done in very little time.
We will see examples of this in the later sections. The main disadvantage is that a
Monte Carlo calculation always has a statistical error associated with it which can be
rather big for an exotic set of cuts.

Computer programs of the type described here consist usually of three parts:
i) The matrix element. ii) The kinematics, by which I mean that part of the program
in which the phase-space integral is transformed into an integral over a unit cube of
sufficient dimension. The choice of variables is also important here. iii) The integration program and the routines for extraction of information. These parts will be
described in order.

i) The matrix element

Throughout this talk I will use the reaction $e^+e^- \rightarrow e^+e^-\mu^+\mu^-$ as a typical example
of a 2γ reaction. Many other reactions can be studied by taking the program for this
reaction and changing some parameters, e.g. $m_\mu \rightarrow m_\tau$ for $e^+e^- \rightarrow e^+e^-\tau^+\tau^-$, as J. Smith
will explain in a later talk[2]. In principle, 12 diagrams contribute to the reaction
$e^+e^- \rightarrow e^+e^-\mu^+\mu^-$ in lowest order. They can be divided into three groups (see Fig. 1).

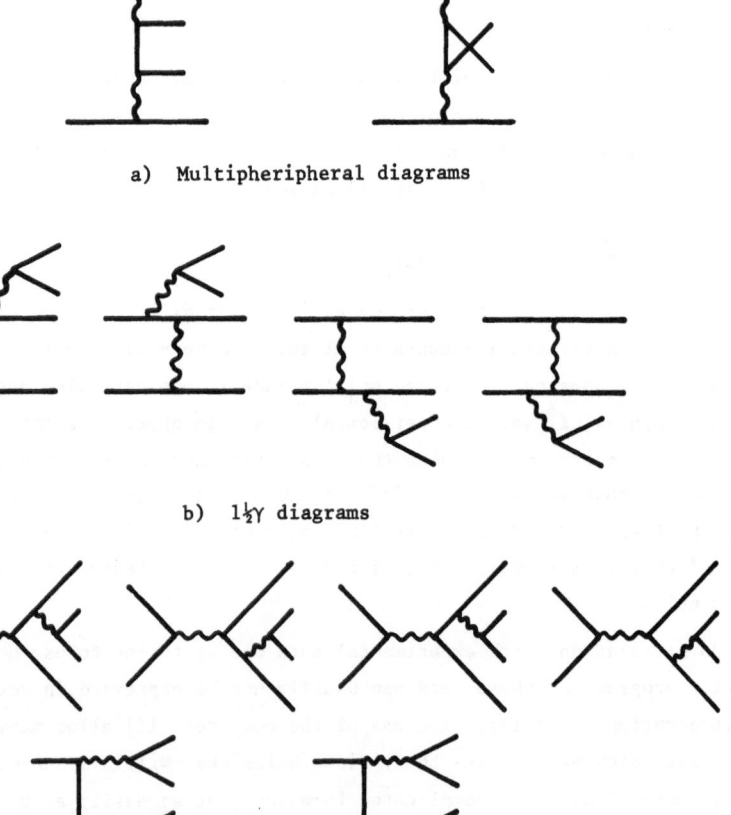

a) Multipheripheral diagrams

b) $1\frac{1}{2}\gamma$ diagrams

c) Annihilation diagrams

Fig. 1

The annihilation diagrams have a character so different from the others that we will not consider them here. In this section we will only consider the two multi-peripheral diagrams, while the contributions from the $1\tfrac{1}{2}\gamma$ diagrams will be studied in Section 5. The notation is given in Fig. 2.

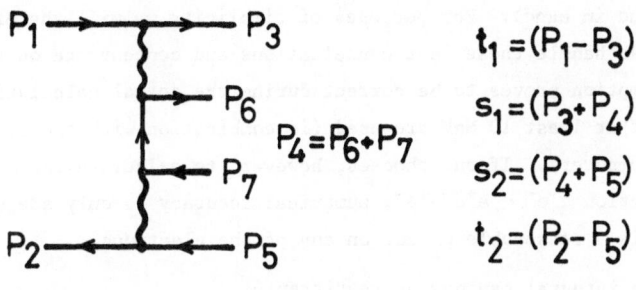

$$t_1 = (P_1 - P_3)$$
$$s_1 = (P_3 + P_4)$$
$$P_4 = P_6 + P_7$$
$$s_2 = (P_4 + P_5)$$
$$t_2 = (P_2 - P_5)$$

Fig. 2

First the full trace for the two diagrams is evaluated using an algebra program like SCHOONSCHIP[3]. After this the troubles begin, as the now obtained formula is numerically rather unstable (up to 14 digits loss of accuracy at $\sqrt{s} = 30$ GeV). To remove these cancellations we have to trace their origin. We know that QED is renormalizable owing to its gauge invariance. On the other hand, renormalizability dictates that σ_{tot} can at most increase with powers of ln s. A consequence of this is that the leading terms in the matrix element with respect to the variables t_1 and t_2 must be proportional to $1/t_1t_2$. After integration over t_1 and t_2 they give terms in $(\ln s)^2$.

An inspection of the matrix element reveals, however, terms in $1/t_1^2 t_2$, $1/t_1 t_2^2$ and even $1/t_1^2 t_2^2$ which can give powers of s in the cross-section if integrated individually. The gauge invariance and the resulting renormalizability demand therefore that all terms of this type cancel to some high degree. Ignoring electron masses for the moment the coefficient of $1/t_1^2 t_2^2$ can be written as $(sp_4^2 - s_1 s_2)^2$. Only if $\Delta \equiv sp_4^2 - s_1 s_2$ is small can the $1/t_1 t_2$ term be the leading one, as also the $1/t_1^2 t_2$ and $1/t_1 t_2^2$ are proportional to Δ. During integration one finds indeed that $\Delta/s_1 s_2$ is typically of the order 10^{-7}-10^{-8}, when t_1 and t_2 are small (the boundaries of Δ depend on t_1 and t_2). It is the cancellations in Δ that are responsible for the numerical instability of the matrix element. To remove this instability we will use Δ as an integration variable, instead of s_1 as described in the kinematics part. The value of s_1 can then be obtained via the relation $s_1 = (p_4^2 s - \Delta)/s_2$. If there is any physics hidden in this "trick" Δ should also fulfil some role in the kinematics. This is indeed the case as we will see.

ii) Kinematics

The kinematics of a process in which electrons or positrons (also called electrons from now on) can emerge at very small angles and very large energies has to be treated rather carefully. The potential for numerical cancellations is very large as will be shown. The philosophy used here is that the worst cancellations come from the $e e \gamma$ vertices, while the peaking of the fermion propagator in the subreaction $\gamma^* \gamma^* \rightarrow \mu^+ \mu^-$ is much less owing to the muon mass being so much larger than the electron mass (as shown in the previous subsection peaking in a variable and the presence of numerical cancellations go hand in hand). For purposes of simplicity we will therefore trust that the computer can handle these last cancellations and concentrate on the t-channel photons. This assumption proves to be correct during the actual calculations as long as fermion masses of at least 10 MeV are used (in combination with the single precision accuracy of a CDC computer). If one chooses, however, to calculate the multiperipheral diagrams of the reaction $e^+ e^- \rightarrow e^+ e^- e^+ e^-$, numerical accuracy is only adequate in a region where there is a reasonable p_T cut on any of the electrons.

The phase-space integral can now be rewritten:

$$\frac{1}{(2\pi)^8} \int \frac{d^3 p_3}{2E_3} \frac{d^3 p_5}{2E_5} \frac{d^3 p_6}{2E_6} \frac{d^3 p_7}{2E_7} \delta^4 (p_1 + p_2 - p_3 - p_5 - p_6 - p_7)$$

$$= \int \frac{dp_4^2}{2\pi} \frac{1}{(2\pi)^5} \int \frac{d^3 p_3}{2E_3} \frac{d^3 p_4}{2E_4} \frac{d^3 p_5}{2E_5} \delta^4 (p_1 + p_2 - p_3 - p_4 - p_5)$$

$$\times \frac{1}{(2\pi)^2} \int \frac{d^3 p_6}{2E_6} \frac{d^3 p_7}{2E_7} \delta^4 (p_4 - p_6 - p_7)$$

$$= \int \frac{dp_4^2}{2\pi} \frac{1}{(2\pi)^5} \frac{\pi}{16 \lambda^{1/2} (s, m_e^2, m_e^2)} \int \frac{dt_1 dt_2 ds_1 ds_2}{\sqrt{-\Delta_4 (p_1, p_2, p_3, p_4)}}$$

$$\times \int \frac{d\Omega^*}{32\pi^2} \sqrt{1 - \frac{4m_\mu^2}{p_4^2}} \quad ,$$

where Ω^* is Ω in the p_6, p_7 c.m. frame and Δ_4 is the 4×4 Gram determinant. The ϕ angle around the beam axis has already been integrated over. This can be done as long as there are no beam polarizations. All numerical problems are now related to the $dt_1 ds_2 dt_2 ds_1$ integrals of which the boundaries can be determined by the requirement that the Gram determinant should be negative. Severe peakings in the variables t_1 and t_2 owing to the $1/t_1 t_2$ behaviour of the matrix element can now be removed easily by switching variables again

$$\int dt_1 ds_2 dt_2 ds_1 = \int t_1 d(\ln t_1) ds_2 t_2 d(\ln t_2) \frac{d\Delta}{s_2}$$

$$= \int t_1 t_2 d(\ln t_1) d(\ln s_2) d(\ln t_2) d\Delta \quad .$$

The matrix element shows surprisingly little structure in terms of the variables $\ln t_1$, $\ln s_2$, $\ln t_2$, and Δ, and the Monte Carlo integral converges very rapidly. In the presence of strong cuts the asymptotic behaviour can change, as is shown in the next section, and then it might be advisable to keep t_1 and/or t_2 as variables instead of their logarithms. This depends, however, entirely on the cuts.

The really difficult problems arise when one wants to calculate from the above variables what the corresponding laboratory energies and angles are for all particles. This is best illustrated by an example: Suppose we have already calculated E_3 and want to know θ_3. For this we use the relations:

$$p_1 \cdot p_3 = m_e^2 - \frac{1}{2} t_1$$
$$= E_1 E_3 - p_1 p_3 \cos \theta_3 \ .$$

These reduce to

$$\cos \theta_3 = \frac{E_1 E_3 - p_1 \cdot p_3}{p_1 p_3} = \frac{E_1 E_3 - m_e^2 + \frac{1}{2} t_1}{p_1 p_3} \ .$$

The result of this calculation can be extremely close to 1 and to determine $\sin \theta_3$ or θ_3 from this needs very many digits accuracy. A better method is therefore to calculate:

$$1 - \cos \theta_3 = \frac{p_1 p_3 + m_e^2 - E_1 E_3 - \frac{1}{2} t_1}{p_1 p_3}$$

$$= \left\{ -\frac{1}{2} t_1 - \frac{m_e^2 (s_2 - m_e^2)^2}{4s(E_1 E_3 + p_1 p_3 - m_e^2)} \right\} (p_1 p_3)^{-1} \ .$$

This formula is very stable and shows that typical values for $1 - \cos \theta_3$ are of the order of m_e^2/s. During the actual integration of the total cross-section, values of 10^{-15} for $1 - \cos \theta_3$ were not very rare!

Similar methods can be used to rewrite most formulae for the calculation of laboratory variables. When calculating $\cos \theta_4$ one finds that the quantity Δ is needed to find a stable formula. This means that the cancellation in the $1/t_1^2 t_2^2$ terms is closely related to the direction of the dimuon system.

The most difficult formulae to derive are those for the ϕ angles of the e^+ and the e^- (p_4 defines the xz plane). For the sake of simplicity they will not be shown here as they are very messy. In the end, the last ϕ angle is overdetermined, so by calculating $\cos \phi_5$ and $\sin \phi_5$ by independent methods one can get a good idea of the numerical accuracy of the whole phase-space calculation.

The laboratory variables of the muons can now be obtained by a simple Lorentz boost and some rotations.

The application of cuts is now very simple: when the laboratory variables of a point in phase space do not meet the cut requirements one sets the corresponding phase-space Jacobian equal to zero, which means that this point carries zero weight in the integration. The rest is a matter of choosing the phase-space points in an adequate way.

iii) Integration and information extraction

The actual integration is done by a learning Monte Carlo routine. It samples a unit cube and will eventually put more points with smaller weights in regions where the integrand is large or varies very rapidly. The one used here is VEGAS written by G.P. Lepage[4], because it is rather good at dealing with the discontinuities introduced by the cuts.

One could, of course, improve the convergence by doing some of the integrals analytically (specifically $d\Omega^*$). In such a case the integration program could do a better job because it has to consider fewer variables. The drawback of this method is, however, that one loses information about where the muons are, so cuts cannot be applied any more.

Another advantage of a Monte Carlo integration is that one can obtain any number of differential cross-sections at the same time that one calculates the total cross-section within a given acceptance. This is done by calculating the variable for which one needs a differential cross-section for each Monte Carlo point and histogramming it, taking weights due to the matrix element, Jacobians, and the learning Monte Carlo into account.

3. SOME APPLICATIONS

As a first application I would like to discuss the comparison between the $e\mu$ signals from the 1γ production of heavy lepton pairs and the 2γ production of μ pairs. It is already known that 2γ signals characterize themselves by usually being coplanar but not collinear, while lepton produced $e\mu$ pairs are both coplanar and collinear (within limits depending on s and the mass of the heavy lepton). Since theorists like to look ahead I calculated the collinearity distribution at a beam energy of 70 GeV for various masses: $m_L = m_\tau$; $m_L = 20$ GeV/c^2 and $m_L = 50$ GeV/c^2. The normalization of the corresponding 2γ signal is very sensitive to the cut on the electron angle and to whether the second muon is required to be invisible by being in the beam pipe. Assuming only one μ to be visible I obtain:

θ_e^{min} (deg.)	σ (pb)
2	99 ± 4
5	14.2 ± 1.6
15	1.14 ± 0.10
30	0.20 ± 0.01

The cut on the μ is taken to be $\theta_\mu > 30°$ and $E_\mu > 1$ GeV. The numbers in the table shou be compared with about 0.3 pb for τ produced $e\mu$ pairs and 0.1 pb for heavier leptons.

The 2γ collinearity angle distribution is rather insensitive to the exact value of θ_e^{min} (see Fig. 3). Considering the smallness of the 1γ cross-section it is clear that the 2γ cross-section has to be reduced even further so it will become necessary for the energy cuts on the electron and the muon to be much sharper. What values are needed depends, however, on the exact experimental acceptance, as one has to keep the real 1γ signal at a maximum.

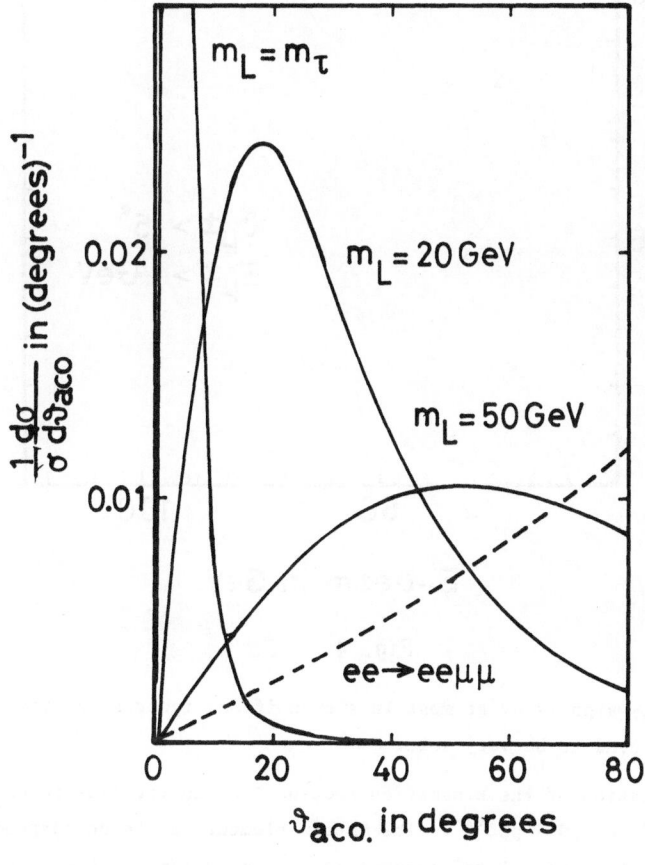

Fig. 3

The total cross-sections being this small it may be better to look for very heavy leptons in the 1 Z rather than in the 1γ channel.

The next interesting application is to study how the various 2γ signals behave as a function of s within a fixed experimental acceptance. For this fixed acceptance I took $E_\mu > 1$ GeV, $\theta_\mu > 15°$ and $\theta_e > 2°$ and calculated μμ and eμμ signals as a function of s (see Figs. 4 and 5). The asymptotic behaviour of the curves can be obtained by using a different scale. This yields that $\sigma_{\mu\mu} \sim \ln s$ and $\sigma(e\mu\mu) \sim \ln s/s$ once s is large enough. The main conclusion here is that untagged signals will keep increasing as s increases, while signals that involve one of the electrons will only gain over

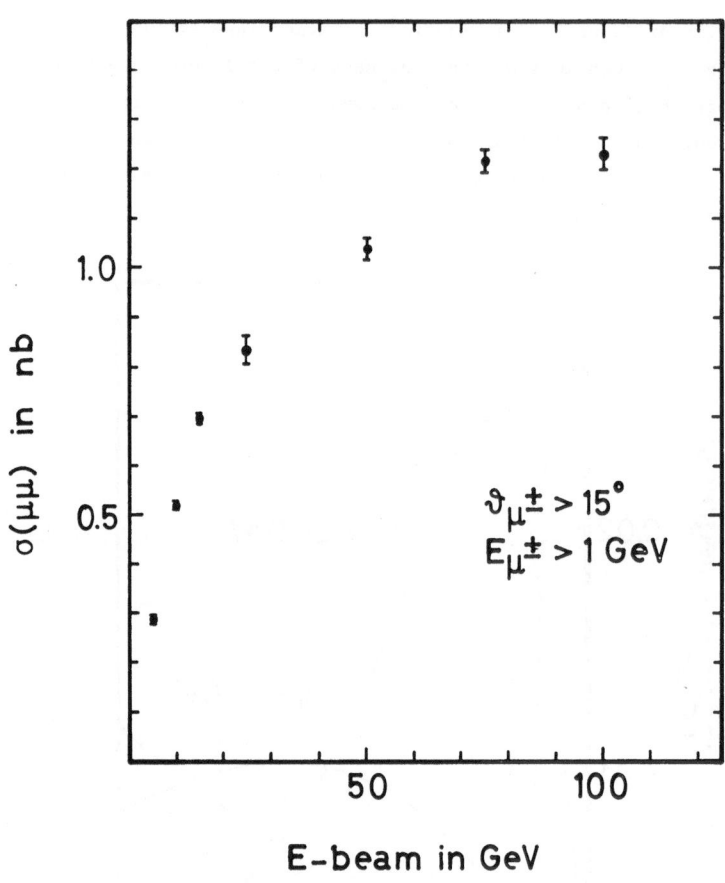

$$\vartheta_{\mu^\pm} > 15°$$
$$E_{\mu^\pm} > 1 \text{ GeV}$$

E-beam in GeV

Fig. 4

the corresponding 1γ signals by at most ln s even if one ignores the fact that the 2γ produced particles tend to be very soft.

Another application of the kinematics routine lies in the elastic channel of the reaction pp → pp$\ell^+\ell^-$ or p$\bar{\text{p}}$ → p$\bar{\text{p}}\ell^+\ell^-$. The matrix element has to be adapted to allow for the proton form factors. Total cross-sections and $d\sigma/dm_{\mu^+\mu^-}$ have been calculated already years ago[5]. Now that 2γ reactions have been observed at the ISR and the p$\bar{\text{p}}$ experiments are also expected to see them, it is clearly necessary to have a program for these reactions along the same lines as the eeμμ program.

The above-mentioned kinematics can also be used for the process $e^+e^- \to e^+e^-\pi^+\pi^-$, possibly via resonances like the f(1270) or an ε(1300?). A publication on this is in preparation[6].

4.   BACKGROUNDS TO 2γ PROCESSES DUE TO BHABHA SCATTERING

When doing a double-tag experiment as for instance proposed at PEP (PEP9) one would, in principle, like to trigger on each event where an electron and a positron

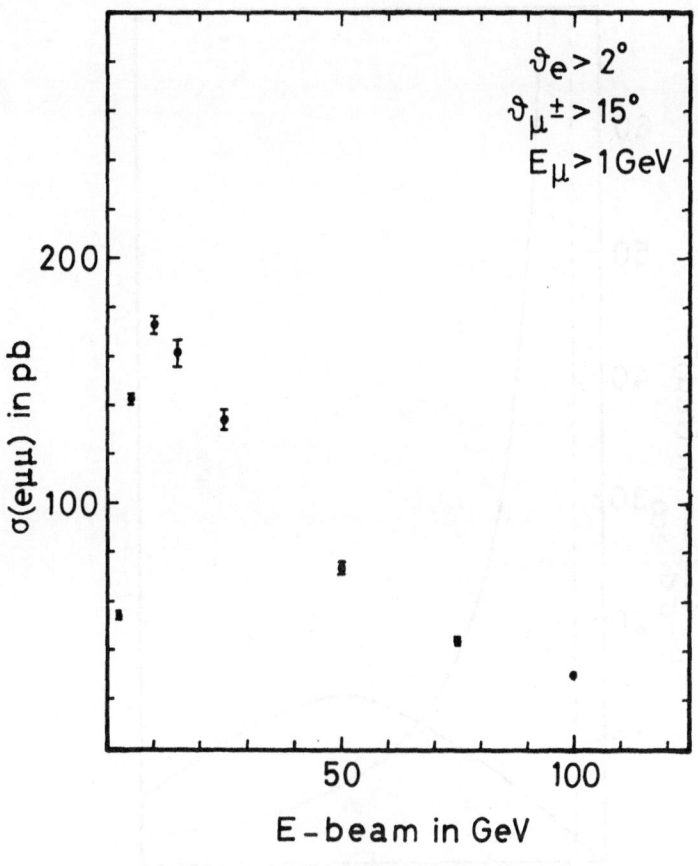

Fig. 5

are observed so as not to miss a single 2γ event. Their rate is, after all, not very high. This is, however, not very practical as the e^+e^- signals due to Bhabha scattering are so large that they will completely clog up the trigger system. Therefore, one has to find a way to get rid of them efficiently without hurting one's chances of seeing 2γ events. This can be done with an acollinearity cut, as Bhabha events are strictly collinear and 2γ events are not. This leaves, however, still the e^+e^- pairs due to the reaction $e^+e^- \rightarrow e^+e^-\gamma$: they are also not necessarily collinear and they come in copious quantities. With an acollinearity cut of 5 mrad one obtains 294 ± 4 nb if $\theta_{e^\pm} > 20$ mrad, while 10 mrad minimum acollinearity gives 190 ± 2 nb. This becomes rather problematic if luminosities should ever come close to 10^{32} cm^{-2}. Therefore one may need an acollinearity cut that is even larger than 10 mrad if the luminosity becomes rather high. To see what the effects are the differential cross-sections in $\theta_{acollinear}$ are given in Fig. 6. The scale is rather deceptive as the signal from eeγ is 300-500 × larger than the signal from eeμμ, even in the most favourable region around 40-60 mrad. A missing-mass cut during the analysis will of course take the events with missing γ's out, but due to the finite resolution this

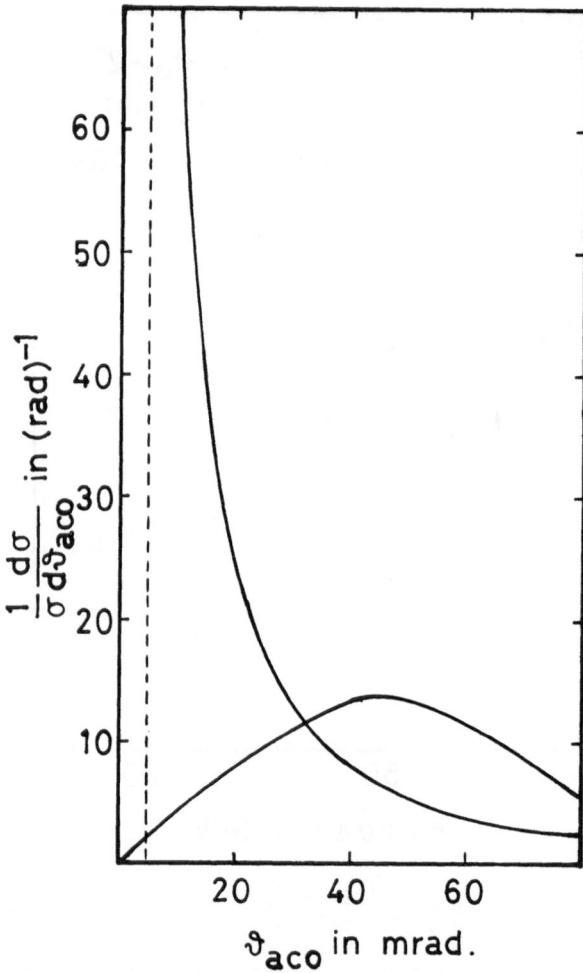

Fig. 6

means that the region with $M_{\mu\mu} < 1$ GeV/c^2 is completely inaccessible unless extra information from a central detector is used in the trigger.

5. NON-PERIPHERAL CONTRIBUTIONS

If one takes the first six diagrams of Fig. 1 into account, rather than the multiperipheral diagrams only, the total cross-section for ee → eeμμ increases by about 1%. Total cross-sections are, however, not indicative of what is observable when studying 2γ reactions as explained before. In the underlying case one would expect an electron that radiated a μ pair to be much more readily observable than an electron from a multiperipheral diagram. I studied therefore the effect on the double-tag experiment from the previous diagram. With $\theta_e > 2°$ I obtained

$$\sigma_{2 \text{ diagrams}} = 220 \pm 5 \text{ pb}$$
$$\sigma_{6 \text{ diagrams}} = 249 \pm 5 \text{ pb} \ .$$

This effect is of the order of 10%. The place where this excess can be found is revealed in Fig. 7. The excess is almost completely in the region where $M_{\mu\mu}$ is less than 1 GeV/c^2, while there is virtually no excess at masses larger than 2 GeV/c^2. This means that these diagrams can also be ignored when studying 2γ produced hadron systems with a high invariant mass.

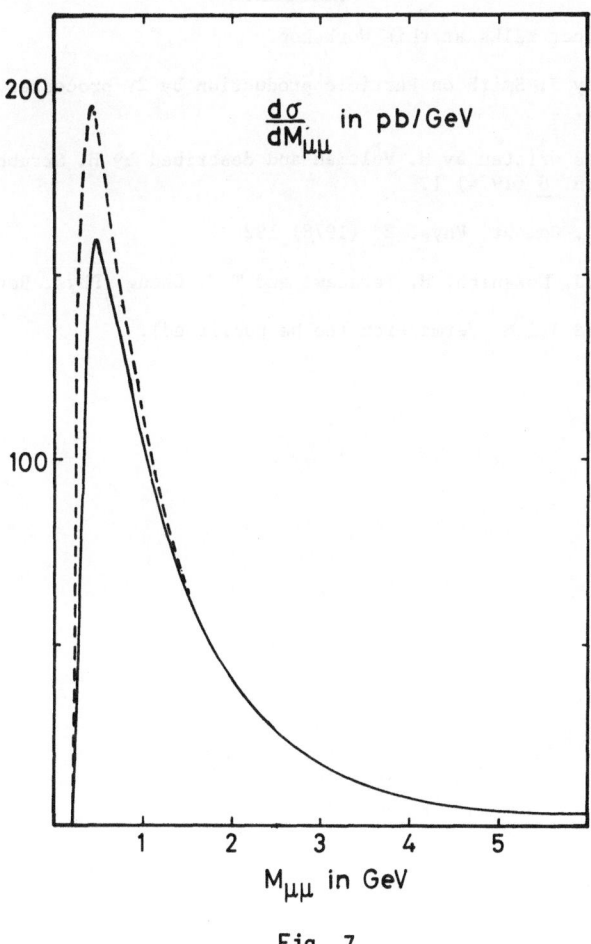

Fig. 7

If the electron angle cut is increased to larger values, it can even happen that the $1\frac{1}{2}\gamma$ diagrams get the upper hand. By that time cross-sections are, however, already of the order of single picobarns. The signature of these diagrams is anyway rather clear. While the 2γ produced μ pairs have usually a large opening angle the opposite holds for the μ pairs that are radiated: their opening angle peaks at $0°$.

Acknowledgements

Much of this work was done in collaboration with J. Smith of Stony Brook. I would also like to thank H. Paar for explaining to me the problems with the radiative Bhabha events.

REFERENCES

1) See various other talks at this Workshop.

2) See the talk by J. Smith on Particle production by 2γ processes, in these Proceedings.

3) SCHOONSCHIP was written by M. Veltman and described by H. Strubbe, Computer Phys. Commun. 8 (1974) 1.

4) G.P. Lepage, J. Comput. Phys. 27 (1978) 192.

5) M.-S. Chen, I.J. Muzenich, H. Terazawa and T.P. Cheng, Phys. Rev. 7 (1973) 3485.

6) H. Kraseman and J.A.M. Vermaseren (to be published).

RESULTS ON TWO-PHOTON INTERACTIONS FROM
MARK II AT SPEAR[*]

SLAC-LBL Mark II Collaboration[**]

Presented by:
P. Jenni[***]
Stanford Linear Accelerator Center
Stanford University
Stanford, California 94305, USA

ABSTRACT

Two-photon interactions have been studied with the SLAC-LBL Mark II magnetic detector at SPEAR. The cross-section for η' production by the reaction $e^+e^- \rightarrow e^+e^-\eta'$ has been measured over the beam energy range from 2 to 4 GeV. The radiative width $\Gamma\gamma\gamma(\eta')$ has been determined to be 5.8 ± 1.1 keV ($\pm 20\%$ systematic uncertainty). Upper limits on the radiative widths of the f(1270), A_2(1310), and f'(1515) mesons have been determined. The two-prong final state events show a signal for the f(1270) decaying into $\pi^+\pi^-$. The current status of this analysis has been discussed.

1. INTRODUCTION

In this talk some results on two-photon interactions from the Mark II experiment at the Stanford Linear Accelerator Center e^+e^- storage ring facility SPEAR are reported.

The observation of leptons and hadrons produced by two-photon interactions in electron-positron colliding-beam experiments has been a challenge ever since the importance of the two-photon mechanism was pointed out ten years ago[1-3]. The basic diagram for the two-photon process is shown in Fig. 1. Lepton pairs produced by the two-photon process have been observed in several experiments[3-7], but only very few events with hadrons in the final state have been observed until recently[7-9].

---

[*] Work supported by the Department of Energy under contract numbers DE-AC03-76SF00515 and W-7405-ENG-48.

[**] Members of the Stanford Linear Accelerator Center - Lawrence Berkeley Laboratory (SLAC-LBL) Collaboration: G.S. Abrams, M.S. Alam, C.A. Blocker, A.M. Boyarski, M. Breidenbach, D.L. Burke, W.C. Carithers, W. Chinowsky, M.W. Coles, S. Cooper, W.E. Dieterle, J.B. Dillon, J. Dorenbosch, J.M. Dorfan, M.W. Eaton, G.J. Feldman, M.E.B. Franklin, G. Gidal, G. Goldhaber, G. Hanson, K.G. Hayes, T. Himel, D.G. Hitlin, R.J. Hollebeek, W.R. Innes, J.A. Jaros, P. Jenni, A.D. Johnson, J.A. Kadyk, A.J. Lankford, R.R. Larsen, V. Lüth, R.E. Millikan, M.E. Nelson, C.Y. Pang, J.F. Patrick, M.L. Perl, B. Richter, A. Roussarie, D.L. Scharre, R.H. Schindler, R.F. Schwitters, J.L. Siegrist, J. Strait, H. Taureg, V.I. Telnov, M. Tonutti, G.H. Trilling, E.N. Vella, R.A. Vidal, I. Videau, J.M. Weiss, and H. Zaccone.

[***] Now at CERN, European Organization for Nuclear Research, Geneva, Switzerland.

The first evidence of a meson resonance produced by the two-photon interaction has been reported by the SLAC-LBL Mark II Collaboration[10]; the reaction

$$e^+e^- \rightarrow e^+e^-\eta'(958) \tag{1}$$

has been observed by detecting the $\eta' \rightarrow \pi^+\pi^-\gamma$ final state. Total cross-section measurements for photon-photon annihilation into hadrons have been obtained very recently by the PLUTO Collaboration[11].

In this talk the results of a search for meson resonances produced in two-photon processes are reported. The corresponding cross-sections are directly proportional to the radiative widths $\Gamma_{\gamma\gamma}(X)$ of the resonances X, as has been pointed out by Low[12]. We have used this relation to determine the radiative width $\Gamma_{\gamma\gamma}(\eta')$ of the $\eta'(958)$ meson to be 5.8 ± 1.1 keV ($\pm 20\%$ systematic uncertainty) from a measurement of reaction (1). This value is in good agreement with predictions from models with fractionally charged quarks[13-15] under assumptions[14] which will be mentioned later. The cross-section for reaction (1) has been measured to increase over the beam energy range covered by this experiment, 1.95 to 3.70 GeV, as expected for two-photon processes[1-3].

After a short description of the SLAC-LBL Mark II magnetic detector in Section 2, we will present the results on reaction (1) including all the data accumulated at the Stanford Linear Accelerator Center e^+e^- storage ring facility SPEAR from beam energies E_b above 1.95 GeV in Section 3. These data represent an increase in the integrated luminosity of about a factor of 3 over the previously published data sample[10]. We have also searched for the following final states from two-photon resonance production: $f(1270) \rightarrow \rho^0\gamma$, $A_2(1310) \rightarrow \rho^{\pm}\pi^{\mp}$ and $f(1270)$, $A_2(1310)$ or $f'(1515)$ decaying into K^+K^-. No signal has been found. Upper limits on the radiative width of these resonances have been obtained from these measurements which will be described in Section 4. A strong signal of $f(1270)$ decaying into $\pi^+\pi^-$ has been observed in the sample of two-prong final state events. The analysis of these events is not completed yet and we discuss briefly in Section 5 some of the preliminary results. The conclusions will be summarized in Section 6.

2. THE MARK II DETECTOR

A schematic view of the SLAC-LBL Mark II magnetic detector is shown in Fig. 2. Its configuration and performance have been described elsewhere[10-16]. The detector consists essentially of a large cylindrical drift chamber with 16 layers followed by time-of-flight (TOF) scintillation counters, both embedded in a solenoidal magnet, which in turn is surrounded by a liquid-argon (LA) electromagnetic shower calorimeter and a muon detection system. Additional shower counters cover both ends of the cylindrical detector.

The performance features may be summarized as follows. The azimuthal coordinates for charged tracks are measured in the drift chamber to a r.m.s. accuracy of about 210 μm per layer. The magnetic field is 4.1 kG, and when tracks are constrained to pass through the known beam position the momenta of charged particles are determined with a resolution of $\delta p/p = \pm \left[(0.005p)^2 + (0.0145)^2\right]^{\frac{1}{2}}$, where p is the momentum in GeV/c. The r.m.s. TOF resolution for hadrons is 300 ps. This provides a π versus K separation at the 1 standard deviation level at momenta of 1.35 GeV/c and K versus p separation at 2.0 GeV/c. The r.m.s. energy resolution for photons and electrons in the LA calorimeter has been measured to be $\delta E/E = 0.11/\sqrt{E}$ (E in GeV) at high energies (E ≳ 0.5 GeV) and slightly poorer $(0.13/\sqrt{E})$ at lower energies because of the increasing importance of the energy loss in the 1.36 radiation lengths of material (coil and supports) in front of the calorimeter. The r.m.s. angular resolution for low-energy photons is about 8 mrad both in azimuthal and dip angle. The measured photon detection efficiencies are 15% at 100 MeV, 50% at 200 MeV, and ≥ 90% above 400 MeV, exclusive of geometry. The LA detector is also used for electron-pion separation. Pion misidentification probabilities of less than 4% and electron efficiencies above 77% are achieved for particle momenta greater than 500 MeV/c. This performance improves at higher momenta. Finally, muons are detected above p ≃ 700 MeV/c with a segmented steel hadron absorber. The fraction of the full solid angle covered by the drift chamber and the TOF counters is 75%, by the LA detector is 65%, and by the muon detection system is 55%.

A two-stage hardware trigger[17] has been used to select, with efficiency ≥ 99%, all interactions that have at least one charged particle with transverse momentum p_T > 100 MeV/c, such that it traverses the entire drift chamber, and another particle which passes through at least the first five layers.

3. MEASUREMENT OF THE TWO-PHOTON PRODUCTION OF THE η'

First results on reaction (1) were reported recently from the SLAC-LBL Mark II experiment[10]. In the following analysis we used the same method with similar event selection criteria. All data accumulated at beam energies E_b above 1.95 GeV were used; the results presented here include the previously published data sample.

The events sought were $\eta' \to \pi^+\pi^-\gamma$ decays where no additional final-state particles were detected. The outgoing e^+ and e^- in reaction (1) were not detected. Therefore, events were selected which have only two oppositely charged tracks coming from the interaction region and one photon measured in the LA detector.

The charged particles were identified as pions if their TOF was within 3 standard deviations of the expected time, they deposited less energy in the LA detector than that expected for electrons, and there were no track-associated hits in the muon chambers behind the hadron absorber. Only those events with an invariant $\pi^+\pi^-$

pair mass of less than 1 GeV/c^2, with each pion momentum less than 1 GeV/c, and with a photon energy E_γ within $0.180 < E_\gamma < 1.0$ GeV were considered further. With the lower photon energy cut we removed background that is generated by electronic noise fluctuations (spurious photons).

Kinematical cuts were then applied to reduce the contributions from the two principal background sources. Possible background from one-photon e^+e^- annihilation events with some of the final-state particles not detected were decreased by requiring that the transverse momentum p_T of the $\pi^+\pi^-\gamma$ state be less than 250 MeV/c and that the acoplanarity angle $\Delta\phi$ between $\pi^+\pi^-$ and the γ momentum vectors projected into a plane perpendicular to the beam axis be less than 20° ($\Delta\phi = 0°$ for back-to-back decays). The background from lepton or pion pairs produced in two-photon inter-actions combined with noise-generated spurious photons was suppressed by requiring that the transverse momentum of the $\pi^+\pi^-$ state be larger than 50 MeV/c and that the acoplanarity angle between the two pions be larger than 3°.

The $\pi^+\pi^-\gamma$ invariant mass distribution $m_{\pi^+\pi^-\gamma}$ for the events which satisfy all the selection criteria is shown in Fig. 3a. There is a clear signal of events from the decay $\eta' \to \pi^+\pi^-\gamma$. The observed r.m.s. width of about 40 MeV/c for the η' mass signal is mostly due to the photon energy resolution and agrees well with a Monte Carlo calculation. The shift of the η' signal by ~ 25 MeV/c^2 towards higher masses is caused by the steep rise of the photon detection efficiency as a function of the deposited energy in the LA detector for energies below 400 MeV. This mass shift can be investigated experimentally in the following way. Resonances produced in two-photon interactions at SPEAR energies occur mainly at very low transverse momenta p_T with respect to the axis of the colliding electron beams. This fact can be exploited by constraining the events to zero net p_T and using a calculated photon energy in-stead of the measured one. This procedure reduces the expected mass resolution for the η' signal to about 15 MeV/c^2 (r.m.s.) and removes the mass shift. However, for events with an η' produced with non-zero p_T it gives an incorrect mass value. The mass distribution with the constraint $p_T = 0$ is shown in Fig. 3b for the η' region and displays the expected features.

The only explicit cut applied on the dipion mass was $m_{\pi\pi} < 1$ GeV/c^2. We find that the dipion mass distribution for the events in the η' mass region, defined as $900 < m_{\pi\pi\gamma} < 1050$ MeV/c^2, is compatible with the hypothesis that all pairs in the $\eta' \to \pi^+\pi^-\gamma$ signal come from ρ^0 decays.

The kinematics of two-photon reactions are very characteristic and different from other processes. For the η' produced by reaction (1) the kinematics are distinct, for example, from those of mesons in multi-hadrons events from one-photon e^+e^- annihilation reactions. The transverse momentum p_T distribution is shown in Fig. 4 for all the data (full histogram) and for the subsample of events lying in the η' mass region (shaded). The η' mesons have lower p_T than the background events.

The distribution of the total energy E is shown in Fig. 5. The energy of η' appears to be confined to low values, excluding the possibility that the η' is produced in a two-body annihilation reaction like $\eta'\gamma$ with the γ not detected. The angular distribution of the η' mesons is strongly peaked along the beams. Their rapidity (y) distribution, given in Fig. 6, is flat over the whole detector acceptance of about $-0.5 < y < 0.5$, whereas the background events tend to peak around $y = 0$.

The background was studied with two different methods. In the first method we analysed multihadron e^+e^- annihilation events. The same analysis cuts as for reaction (1) except for the topology selection were applied to events with three or more charged prongs and at least one photon. The resulting mass distribution is smooth and reaches a broad maximum over the range from 0.8 to 1.3 GeV/c^2. The transverse momentum distribution rises below about 125 MeV/c and stays approximately constant above. In the second method we used all the original selection criteria, including the exclusive two charged prongs and only one photon topology but then combined the dipion state from one event with the photon from the next event. This analysis reproduces the shape and the normalization of the observed background in the mass and transverse momentum distributions. Both these studies suggest a smooth background shape under the η' signal, and we have therefore made a direct subtraction using the adjacent mass regions (see Table 1). This subtraction makes no specific assumptions about the origin of the background. We notice that the background contribution is lower for the data taken at the higher beam energies than for the low-energy part of the data (see Fig. 3a).

The cross-section was calculated using the branching ratio[18] $B(\eta' \rightarrow \pi^+\pi^-\gamma) =$ $= 0.298 \pm 0.017$ and the detection efficiency ε (see Table 1) determined by a Monte Carlo simulation. Events were generated according to the cross-section calculation and angular distribution of Ref. 19. These events were then subjected to the same detector geometry and selection criteria as the real data except for the LA shower cuts on the charged tracks. Because of the difficulties of describing in detail the interaction of pions in the shower counter material, the efficiencies of the latter cuts (typically 85%) were determined experimentally with unambiguously identified pions from ψ decays. Furthermore, there is a small (5%) loss of events due to additional spurious photons which was also determined experimentally. The observed cross-section $\sigma(\eta')$ is given in Table 1. The cross-section $\sigma(\eta')$ is also displayed in Fig. 7 as a function of the beam energy and is found to be compatible with the expected slow rise with increasing energy. The errors shown are statistical only and do not include an estimated over-all systematic uncertainty of $\pm 20\%$. The principal contributions to this value come from the estimated uncertainties in the two-photon model dependence in the Monte Carlo calculation ($\pm 15\%$), the photon detection efficiency ($\pm 10\%$), the experimentally measured losses due to the LA cuts on the charged tracks ($\pm 5\%$), and the integrated luminosity ($\pm 6\%$). We did not correct the displayed cross-sections for initial-state radiation in reaction (1).

The two-photon production cross-section for a resonance X is directly proportional to the radiative width $\Gamma_{\gamma\gamma}(X)$ of the resonance[12]. For example, in the equivalent photon approximation calculation of Ref. 2

$$\sigma_{ee \rightarrow eeX} = 16\alpha^2 (2J+1) m_X^{-3} \Gamma_{\gamma\gamma}(X) \times \left(\ln \frac{E_b}{m_e} - \frac{1}{2} \right)^2 f\left(\frac{m_X}{2E_b} \right) \tag{2}$$

with

$$f(\xi) = (2+\xi^2)^2 \ln \frac{1}{\xi} - (1-\xi^2)(3+\xi^2) .$$

J denotes the spin and m_X the mass of the resonance. E_b is the beam energy and m_e the electron mass. From the measured cross-section for reaction (1) we determined $\Gamma_{\gamma\gamma}(\eta') = 5.8 \pm 1.1$ keV using the two-photon calculation of Ref. 19. In this calculation we included the correction for initial-state radiation effects. The error reflects the statistical accuracy of the measurement, but does not include the estimated over-all systematic uncertainty of ±20%. Upper limits on $\Gamma_{\gamma\gamma}(\eta')$ in the range 11 to 35 keV were obtained in other searches[6-8] for reaction (1). From the radiative width we deduced the total width $\Gamma_{tot}(\eta') = 293 \pm 67$ keV or equivalently, the mean life $\tau(\eta') = (2.2 \pm 0.5) \times 10^{-21}$ s (±20% systematic uncertainty) using[18] $B(\eta' \rightarrow \gamma\gamma) = 0.0197 \pm 0.0026$. This measurement of $\Gamma_{tot}(\eta')$ is in good agreement with the value 280 ± 100 keV reported in Ref. 20.

The radiative width for the η' was calculated by several authors[13-15]. Quark models with fractionally charged quarks lead, under the assumptions of a small pseudoscalar octet-singlet mixing angle and of equal singlet and octet decay constants, to the prediction of $\Gamma_{\gamma\gamma}(\eta') \simeq 6$ keV in very good agreement with our measurement. Applying the same assumptions, the data are not compatible with integral charge quark models, for which a radiative decay width of about 26 keV was predicted. However, the validity of these assumptions was questioned in Ref. 14, where alternative analyses are proposed.

4. UPPER LIMITS ON THE RADIATIVE WIDTHS OF THE TENSOR MESONS f, A₂, and f'

The radiative widths of the tensor mesons f(1270), A_2(1310), and f'(1515) are experimentally not known. From SU(3) symmetry they are expected to have values in the ratio 25:9:2 for the case of ideal mixing and with phase-space corrections neglected[*]. The $\Gamma_{\gamma\gamma}$ for the f(1270) meson was calculated by several authors, see for example Refs. 21 and 22, which predict $\Gamma_{\gamma\gamma}(f)$ to be in the range of about 5 to 20 keV.

---

[*] This result follows from the quark flavour assignment $(u\bar{u}+d\bar{d})/\sqrt{2}$ for the f^0(1270) $(u\bar{u}-d\bar{d})/\sqrt{2}$ for the A_2^0(1310), and $s\bar{s}$ for the f'(1515) with fractionally charged quarks.

We investigated exclusive final states for the following reactions

$$e^+e^- \to e^+e^- \ f(1270)$$
$$ \longrightarrow \rho^0\gamma \tag{3}$$

$$e^+e^- \to e^+e^- \ A_2(1310)$$
$$ \longrightarrow \rho^\pm\pi^\mp \tag{4}$$

$$e^+e^- \to e^+e^- \ f(1270)$$
$$ \longrightarrow K^+K^- \tag{5}$$

$$e^+e^- \to e^+e^- \ A_2(1310)$$
$$ \longrightarrow K^+K^- \tag{6}$$

$$e^+e^- \to e^+e^- \ f'(1515)$$
$$ \longrightarrow K^+K^- \tag{7}$$

by searching for signals in the invariant mass distributions of the respective final states which would occur at low transverse momenta. As in the measurement of reaction (1), the final-state e^+ and e^- remained undetected. No signal above background was detected for reactions (3) to (7). We used the data taken at beam energies above 2.25 GeV (14 pb^{-1} integrated luminosity) to determine upper limits on the radiative widths of these tensor mesons on the basis of reactions (3) to (7).

4.1 $\rho^0\gamma$ final state

It was recently proposed[22] that the f(1270) meson could have a relatively large branching fraction into $\rho^0\gamma$, namely 3 to 5%. We searched for reaction (3) using the same method as described in the preceding section for the measurement of reaction (1). The same cuts were applied, except that the lower photon energy cut was increased to 250 MeV in order to further suppress the background. The resulting $\pi^+\pi^-\gamma$ mass distribution is shown in Fig. 8a and the p_T distribution in Fig. 8b with the events from the mass interval $1.15 < m_{\pi\pi\gamma} < 1.40$ GeV/c^2 shaded. No signal at the f(1210) mass is present. The η' signal appears reduced owing to the more stringent photon energy cut. The background mass distribution can be well described with artificially generated events in which the dipion state has been combined with the photon from the next event of the same two charged prongs and only one photon topology. The over-all detection efficiency for reaction (3) was determined in the same way as for reaction (1) and found to be 0.020. The data therefore allow us to set a 95% confidence level (C.L.) upper limit on the product $\sigma(f) \times B(f \to \rho^0\gamma) < 0.14$ nb at the luminosity-weighted average beam energy of 2.85 GeV. With Eq. (2) this implies $\Gamma_{\gamma\gamma}(f) \times B(f \to \rho\gamma) < 0.8$ keV. Both limits include an estimated 20% systematic uncertainty.

4.2 $\rho^\pm\pi^\mp$ final state

To search for reaction (4), we used the event topology which contains two oppositely charged pions and two photons. All events with two photons in the energy range $0.1 < E_\gamma < 1.0$ GeV and with a two-photon invariant mass $m_{\gamma\gamma}$ within $0.075 < m_{\gamma\gamma} < 0.200$ GeV/c^2 were selected. For these events the photon energies were adjusted to constrain $m_{\gamma\gamma}$ to m_{π^0}. These π^0 were combined with each of the charged pions to form the invariant mass $m_{\pi^\pm\pi^0}$. Only those events with a $\rho^\pm$ candidate, defined as $0.5 < m_{\pi^\pm\pi^0} < 1.0$ GeV/c^2, were retained any further. Special cuts were necessary to suppress the background coming from τ lepton pair production with their subsequent decays into $\rho^\pm \nu_\tau$ and final states containing single pions or multipions which were measured in the same experiment[23]. Both the $\rho^\pm$ and the $\pi^\mp$ momenta were required to be less than 800 MeV/c. Finally, the invariant $\rho^\pm\pi^\mp$ mass was formed, and a transverse momentum cut of $p_T < 250$ MeV/c was applied to reduce the background from one-photon e^+e^- annihilation events. The $m_{\rho^\pm\pi^\mp}$ and p_T distributions are shown in Figs. 9a and 9b, with the events lying in the A$_2$ mass region $1.20 < m_{\rho^\pm\pi^\mp} < 1.45$ GeV/c^2 shaded in Fig. 9b. From the observed background in the $m_{\rho^\pm\pi^\mp}$ distribution, the known[18] $B(A_2 \to \rho^\pm\pi^\mp) = 0.703 \pm 0.021$ and the over-all detection efficiency for reaction (4), $\varepsilon = 0.0028$, which is greatly reduced owing to the low-energy photon efficiency, one deduces a 95% C.L. upper limit for $\sigma(A_2) < 0.36$ nb at $E_b = 2.85$ GeV for reaction (4). From Eq. (2) it follows that $\Gamma_{\gamma\gamma}(A_2) < 2.5$ keV. A possible systematic error of 25% was included in these limits.

4.3 K^+K^- final state

The K^+K^- decay modes of the f(1270), A$_2$(1310), and f'(1515) mesons can be used to study their two-photon production with little background from one photon e^+e^- annihilation processes. For this purpose we selected events with only two oppositely charged prongs and no detected photons. Both tracks were required to be unambiguously identified as kaons by the TOF measurement. This was achieved by a cut on the probability level[*] which was required to be larger than 0.65 for the two tracks to be kaons. In order to reduce sources other than two-photon production we applied two loose kinematical cuts: the acoplanarity angle $\Delta\phi$ between the two kaons had to be < 20° and the p_T of the K^+K^- state had to be < 250 MeV/c. The invariant mass $m_{K^+K^-}$ and the p_T distributions are given in Figs. 10a and 10b. The p_T of most of the events is seen to be very small as expected for reactions (5) to (7).

Non-resonant production can also contribute to K^+K^- final states from two-photon interactions. We estimated the expected background in the $m_{K^+K^-}$ distribution from the non-resonant process

$$e^+e^- \to e^+e^-K^+K^- \tag{8}$$

with the equivalent photon approximation of Refs. 2 and 24. The curve in Fig. 10a shows the result using a simple kaon form factor with a Φ pole of the form

---

*) Each track is assigned weights for being a pion, kaon, or proton. Each weight is proportional to the probability that if the particle has the assumed identity, its flight time would have the measured value. The normalization is such that the sum of all weights for a given track is unity.

$|F_K(s)| = |(1-s/m_\phi^2)^{-1}|$, where s is m_{KK} squared and m_ϕ is the Φ mass. With this assumption for the form factor, the contribution of reaction (8) appears to be small (10% over the range of observed K^+K^- masses), and the excess of events is suggestive of resonance production by the reactions (5) to (7). The poor statistics of the data only allow us to extract upper limits on the production cross-sections for the f(1270) and A_2(1310) mesons in reactions (5) and (6), where we used the branching ratio[18] into K^+K^- of 0.0165 ± 0.0020 and 0.0235 ± 0.0025 and over-all detection efficiencies of 0.0167 and 0.0172, respectively, at the average beam energy of 2.85 GeV. The results as well as the upper limits on $\Gamma_{\gamma\gamma}$ from Eq. (2) are given in Table 2. The branching ratio of the f'(1515) meson into $K\bar{K}$ is not known, but expected to be dominant[18] [i.e. $B(f' \to K^+K^-)$ near 0.5]. The data, together with the calculated detection efficiency of 0.0195, provide therefore only upper limits on $B(f' \to K^+K^-) \times \sigma(f')$ and $B(f' \to K^+K^-) \times \Gamma_{\gamma\gamma}(f')$ for reaction (7). These upper limits are also listed in Table 2. A systematic uncertainty for the K^+K^- final state of 15% was included in the listed 95% C.L. upper limits.

We have summarized in Table 2 the upper limits on the radiative widths of the f(1270), A_2(1310), and f'(1515) mesons, obtained from reactions (3) to (7). All of these limits are consistent with the expected values mentioned at the beginning of this section.

5. CURRENT STATUS OF THE ANALYSIS OF THE FINAL-STATE PION PAIRS

The study of final-state pion pairs, from the reaction

$$e^+e^- \to e^+e^- \; f(1270)$$
$$\hookrightarrow \pi^+\pi^- \qquad\qquad (9)$$

and from the non-resonant two-photon process $e^+e^- \to e^+e^-\pi^+\pi^-$, is complicated by the problem of distinguishing pions unambiguously from muons and electrons in the momentum range of about 300 to 700 MeV/c in the Mark II detector. The threshold momentum imposed by the muon detection system is $p_T \gtrsim 700$ MeV/c. The data in and below the mass region of the f(1270) are obscured by the competing QED two-photon processes $e^+e^- \to e^+e^-\mu^+\mu^-$ and $e^+e^- \to e^+e^-e^+e^-$, which occur with large cross-sections[1-3].

We selected events with only two oppositely charged prongs and no detected photons. Both tracks were required to be compatible with being pions by the TOF measurement within 3 standard deviations. Tracks were only accepted if they deposited less energy in the LA detector than that expected for electrons and if there were no track-associated hits in the muon chambers. However, the rejection of electrons and muons in the data due to these cuts is strongly dependent on the particle momenta. The same two loose kinematical cuts as for the K^+K^- final states were then applied to reduce sources other than two-photon production: the acoplanarity angle $\Delta\phi$ between the two tracks had to be < 20° and the p_T of the candidate dipion state had to be < 250 MeV/c.

The preliminary invariant pair mass distribution of the same data sample as used for the kaon-pair analysis -- Section 4.3 -- is shown in Fig. 11a, assuming pion masses for both particles. The distribution has an enhancement in the f(1270) mass region even though no subtraction of the significant two-photon lepton pair background has been attempted yet. The p_T of the events peaks at low values, as expected for reaction (9), and is given in Fig. 11b.

The resonance behaviour in the f(1270) region is more evident after correcting the event distribution, Fig. 11a, for the mass dependence of the acceptance and of the two-photon cross-section[2,24], which are important in the case of a broad resonance. In Fig. 12 we show the corrected pair mass distribution for the same events as those of Fig. 11a.

The background subtraction of the two-prong lepton pairs and finally the determination of the radiative width of the f(1270) meson are not yet completed.

6. SUMMARY

We have reported evidence for meson resonance production by the two-photon interaction in a colliding e^+e^- beam experiment. We observed exclusive events containing no other particle in the detector than an $\eta'(958)$ detected in the decay mode $\eta' \rightarrow \pi^+\pi^-\gamma$. On the basis of their kinematics these events were interpreted as coming from the reaction $e^+e^- \rightarrow e^+e^-\eta'$, where the final-state e^+ and e^- were not detected. The cross-section for these events was measured to rise over the beam energy range 1.95 to 3.7 GeV as expected for two-photon processes. From the cross-section we determined the radiative width of the η' meson $\Gamma_{\gamma\gamma}(\eta')$ to be 5.8 ± 1.1 keV (±20% systematic uncertainty). It was argued that this value is in good agreement with (model-dependent) predictions from various quark models with fractionally charged quarks and excludes, within these models, the possibility of integral charge quarks.

We investigated the two-photon production of the tensor mesons f(1270), A_2(1310) and f'(1515) by searching for signals in the mass distributions of various exclusive final states. A signal for f(1270) decaying into $\pi^+\pi^-$ has been observed, and the problem of extracting a radiative width for this resonance, in the presence of important backgrounds from lepton pairs from QED two-photon reactions, is currently under study. Also, an excess of low transverse momentum K^+K^- events was observed. However, the poor statistics of these data did not allow us to separate the contributions from the f, A_2, f', and other possible resonances. Upper limits on the radiative widths of these resonances were determined.

REFERENCES

1) N. Arteaga-Romero, A. Jaccarini and P. Kessler, C.R. Acad. Sci. B269, 153 (1969)
   and B269, 1129 (1969);
   N. Arteaga-Romero, A. Jaccarini, P. Kessler and J. Parisi, Nuovo Cimento Lett. 4,
   933 (1970) and Phys. Rev. D3, 1569 (1971).

2) S.J. Brodsky, T. Kinoshita and H. Terazawa, Phys. Rev. Lett. 25, 972 (1970) and
   Phys. Rev. D4, 1532 (1971).

3) V.E. Balakin, V.M. Budnev and I.F. Ginsburg, Zh. Eksp. Teor. Fiz. Pis'ma v Red. 11,
   559 (1970) [Sov. Phys.-JETP Lett. 11, 388 (1970)].

4) V.E. Balakin et al., Phys. Lett. 34B, 663 (1971);
   C. Bacci et al., Nuovo Cimento Lett. 3, 709 (1972);
   G. Barbiellini et al., Phys. Rev. Lett. 32, 385 (1974).

5) A. Courau et al., Phys. Lett. 84B, 145 (1979).

6) H.J. Besch et al., Phys. Lett. 81B, 79 (1979).

7) R. Baldini Celio et al., Phys. Lett. 86B, 239 (1979).

8) S. Orito et al., Phys. Lett. 48B, 380 (1974);
   L. Paoluzi et al., Nuovo Cimento Lett. 10, 435 (1974).

9) Ch. Berger et al., Phys. Lett. 81B, 410 (1979).

10) G.S. Abrams et al., Phys. Rev. Lett. 43, 477 (1979).

11) Ch. Berger et al., Phys. Lett. 89B, 120 (1979).

12) F.E. Low, Phys. Rev. 120, 582 (1960).

13) S. Matsuda and S. Oneda, Phys. Rev. 187, 2107 (1969);
    S. Okubo *in* Symmetries and quark models (ed. R. Chand), (Gordon and Breach,
    New York, 1970);
    H. Suura, T.F. Walsh and B.-L. Young, Lett. Nuovo Cimento 4, 505 (1972).

14) M.S. Chanowitz, Phys. Rev. Lett. 35, 977 (1975) and Phys. Rev. Lett. 44, 55 (1980).

15) E. Etim and M. Greco, Nuovo Cimento 42, 124 (1977);
    N.M. Chase and M.T. Vaughn, Z. Phys. C 2, 23 (1979).

16) W. Davies-White et al., Nucl. Instrum. Methods 160, 227 (1979);
    G.S. Abrams et al., IEEE Trans. Nucl. Sci. NS-25, 309 (1978).

17) H. Brafman et al., IEEE Trans. Nucl. Sci. NS-25, 692 (1978).

18) Particle Data Group, Phys. Lett. 75B, 1 (1978).

19) V.M. Budnev and I.F. Ginzburg, Phys. Lett. 37B, 320 (1971);
    V.N. Baier and V.S. Fadin, Lett. Nuovo Cimento 1, 481 (1971) and private communi-
    cation.

20) D.M. Binnie et al., Phys. Lett. 83B, 141 (1979).

21) B. Renner, Nucl. Phys. B30, 634 (1971);
    B. Schrempp-Otto, F. Schrempp and T.F. Walsh, Phys. Lett. 36B, 463 (1971).

22) W.N. Cottingham and I.H. Dunbar, J. Phys. G 5, L155 (1979).

23) G.S. Abrams et al., Phys. Rev. Lett. <u>43</u>, 1555 (1979).

24) S.J. Brodsky, Suppl. J. Phys. <u>35</u>, C2-69 (1974).

<u>Table 1</u>

Summary of the cross-section calculation. E_b is the beam energy, $\int \mathcal{L}\,dt$ is the integrated luminosity, ε is the detection efficiency not including $B(\eta' \to \pi\pi\gamma)$, $n_{\eta'}$ is the background subtracted number of η' events, and $\sigma(\eta')$ is the observed cross-section. Only statistical errors are shown.

E_b (GeV)	$\int \mathcal{L}\,dt$ (nb^{-1})	ε	$n_{\eta'}$	$\sigma(\eta')$ (nb)
1.95-2.21	4199	0.0231	7.7 ± 5.1	0.27 ± 0.18
2.25-2.50	2131	0.0224	4.3 ± 2.6	0.30 ± 0.18
2.50-3.00	6655	0.0211	25.9 ± 7.1	0.62 ± 0.17
3.00-3.35	4009	0.0177	20.0 ± 5.9	0.94 ± 0.28
3.70	984	0.0125	3.1 ± 2.2	0.84 ± 0.60

<u>Table 2</u>

Upper limits (95% C.L.) on the two-photon production cross-section σ and the radiative width $\Gamma_{\gamma\gamma}$ of the tensor mesons at the luminosity weighted average beam energy 2.85 GeV. The over-all detection efficiency is listed under ε, and B stands for branching ratio.

Final state	Meson	ε	σ (nb)	$\Gamma_{\gamma\gamma}$ (keV)
$\rho^0\gamma$	f(1270)	0.0200	$\sigma \times B(f \to \rho\gamma)$ < 0.14	$\Gamma_{\gamma\gamma} \times B(f \to \rho\gamma)$ < 0.8
$\rho^\pm\pi^\mp$	A_2(1310)	0.0028	< 0.36	< 2.5
K^+K^-	f(1270)	0.0167	< 4.2	< 24.0
K^+K^-	A_2(1310)	0.0172	< 2.6	< 17.0
K^+K^-	f'(1515)	0.0195	$\sigma \times B(f' \to K^+K^-)$ < 0.052	$\Gamma_{\gamma\gamma} \times B(f' \to K^+K^-)$ < 0.6

Figure captions

Fig. 1   :   Diagram for the two-photon production of the state x.

Fig. 2   :   Schematic view of the Mark II detector. A = vacuum chamber,
              B = pipe counter,   C = drift chamber,   D = time-of-flight counters,
              E = solenoid coil,   F = liquid-argon shower counters,   G = iron ab-
              sorber, and   H = muon proportional tubes.

Fig. 3   :   a) $\pi^+\pi^-\gamma$ invariant mass distribution. Events from beam energies
              equal to or above 2.6 GeV are shown shaded. b) $\pi^+\pi^-\gamma$ invariant
              mass distribution with p_T = 0 constraint.

Fig. 4   :   Transverse momentum distribution. Events in the η' peak
              $0.90 < m_{\pi^+\pi^-\gamma} < 1.05$ GeV/c^2 are shaded.

Fig. 5   :   Total energy distribution. Events in the η' peak $0.90 < m_{\pi^+\pi^-\gamma} <$
              1.05 GeV/c^2 are shaded.

Fig. 6   :   Rapidity distribution for the $\pi^+\pi^-\gamma$ states, with the events in the η'
              peak shaded.

Fig. 7   :   Cross-section for $e^+e^- \rightarrow e^+e^-\eta'$ as a function of the beam energy.
              The curve is the result of Eq. (2) with $\Gamma_{\gamma\gamma}(\eta')$ = 6 keV.

Fig. 8   :   a) $\pi^+\pi^-\gamma$ invariant mass distribution for the f(1270) $\rightarrow \rho^0\gamma$ search.
              b) Transverse momentum distribution; events from the f(1270) mass
              region $1.15 < m_{\pi^+\pi^-\gamma} < 1.40$ GeV/c^2 are shaded.

Fig. 9   :   a) $\rho^\pm\pi^\mp$ invariant mass distribution. b) Transverse momentum distri-
              bution; events from the A$_2$(1310) mass region $1.20 < m_{\rho^\pm\rho^\mp} < 1.45$ GeV/c^2
              are shaded.

Fig. 10   :   a) Invariant K$^+$K$^-$ mass distribution. The curve shows the expected
              contribution from non-resonant two-photon interactions (see text).
              b) Transverse momentum distribution of the K$^+$K$^-$ system.

Fig. 11   :   a) Preliminary invariant $\pi^+\pi^-$ mass distribution (see text for comments
              on the lepton-pair background).
              b) Preliminary transverse momentum distribution for the pion-pair can-
              didate events.

Fig. 12   :   Preliminary differential cross-section dσ/dm^2 for pion-pair candidate
              events corrected for the mass dependence of the two-photon cross-
              section.

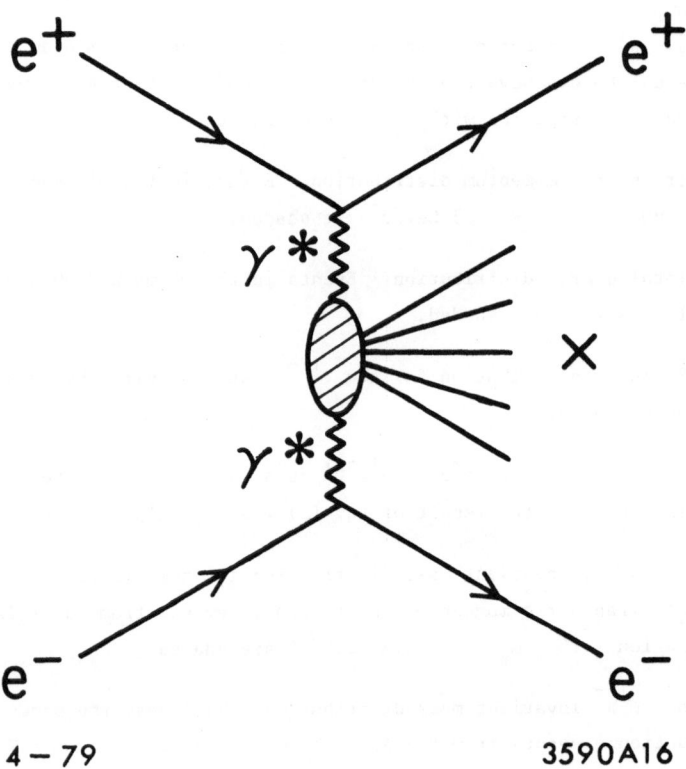

4 – 79   3590A16

Fig. 1

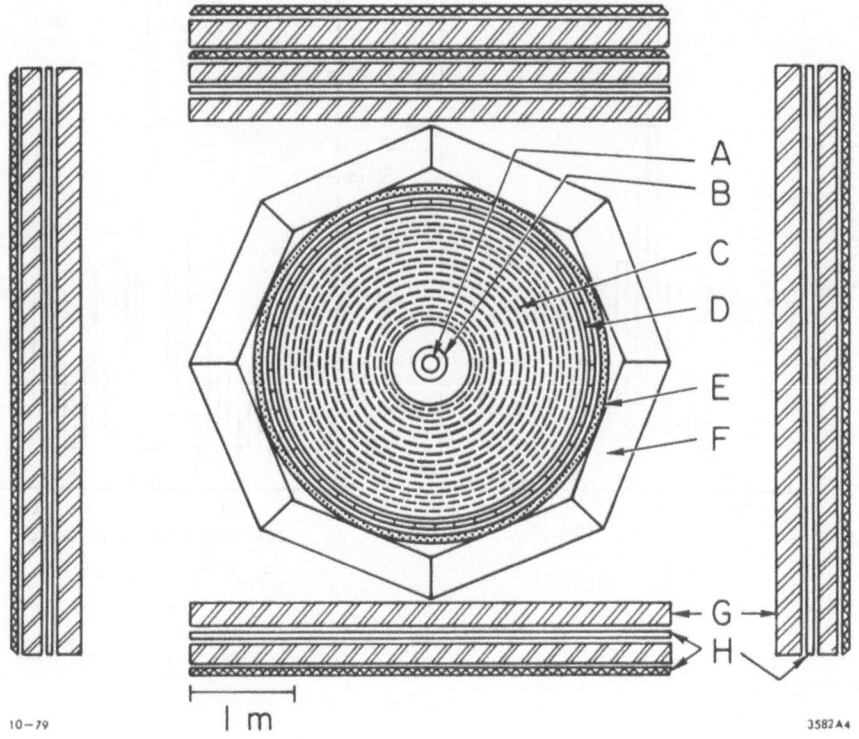

A
B
C
D
E
F

G

H

1 m

10–79

3582 A4

Fig. 2

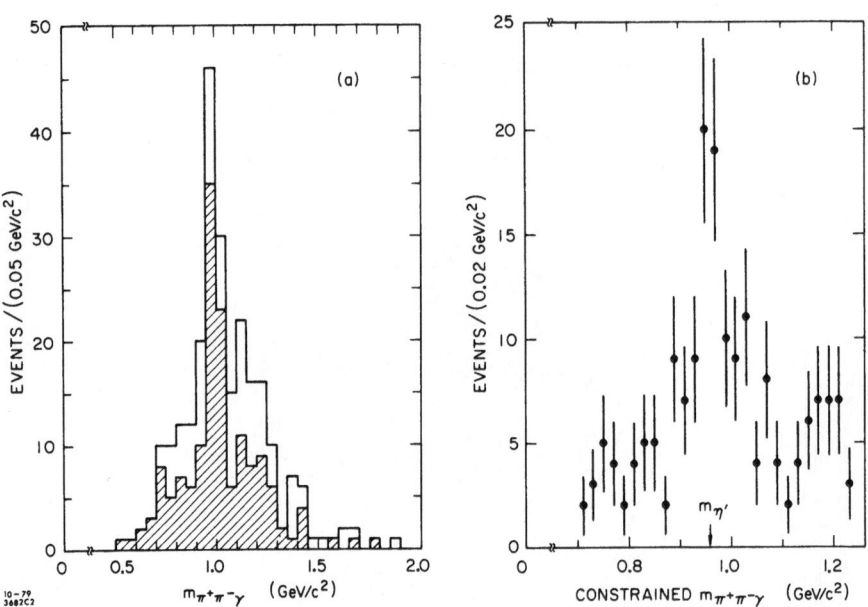

Fig. 3

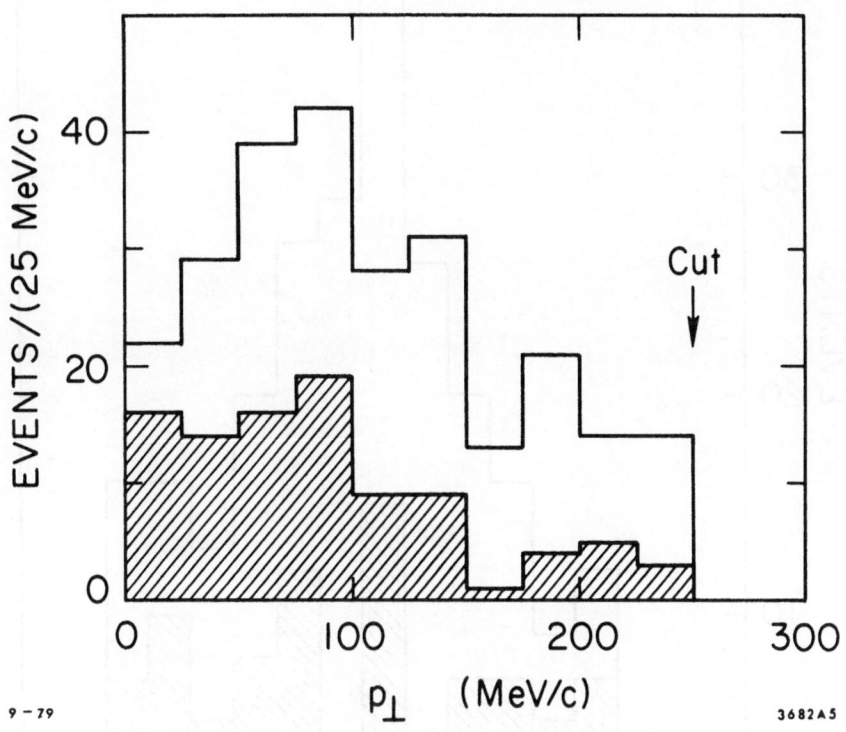

3682A5

Fig. 4

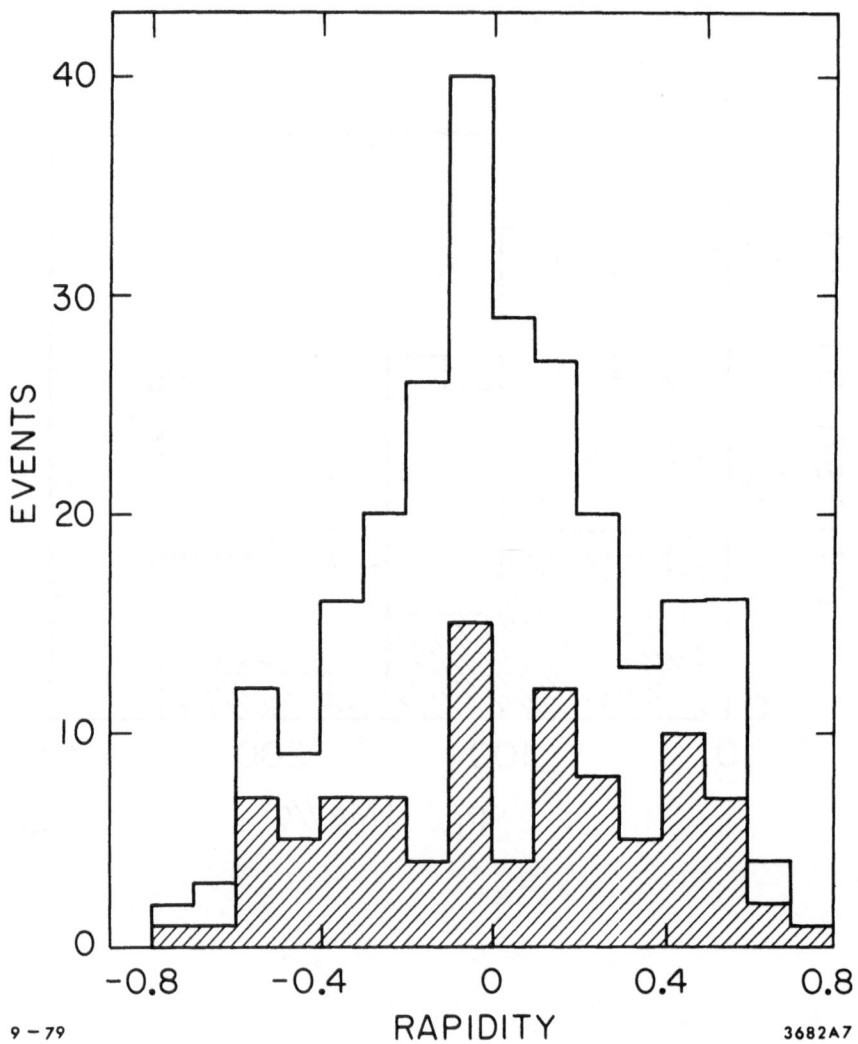

Fig. 6

9 – 79   3682A7

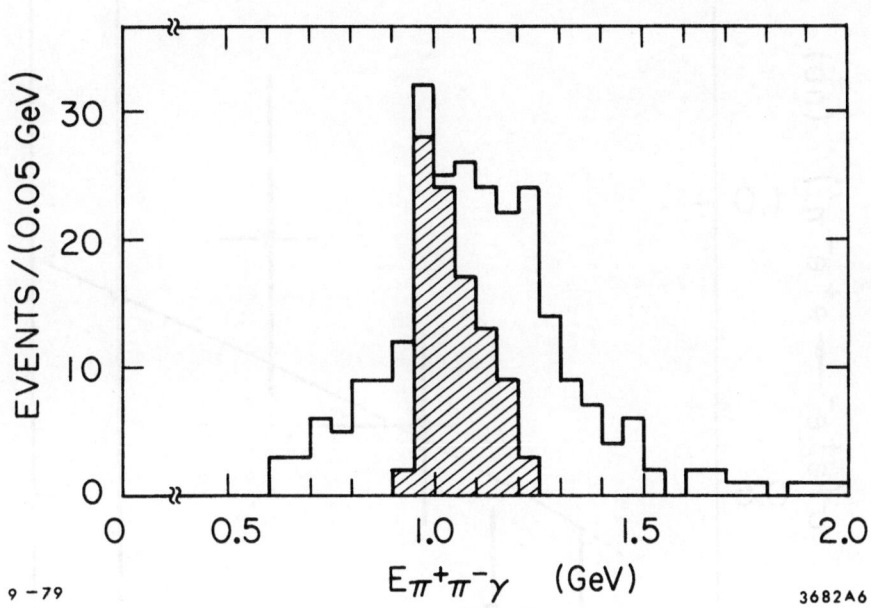

9 -79

3682A6

Fig. 5

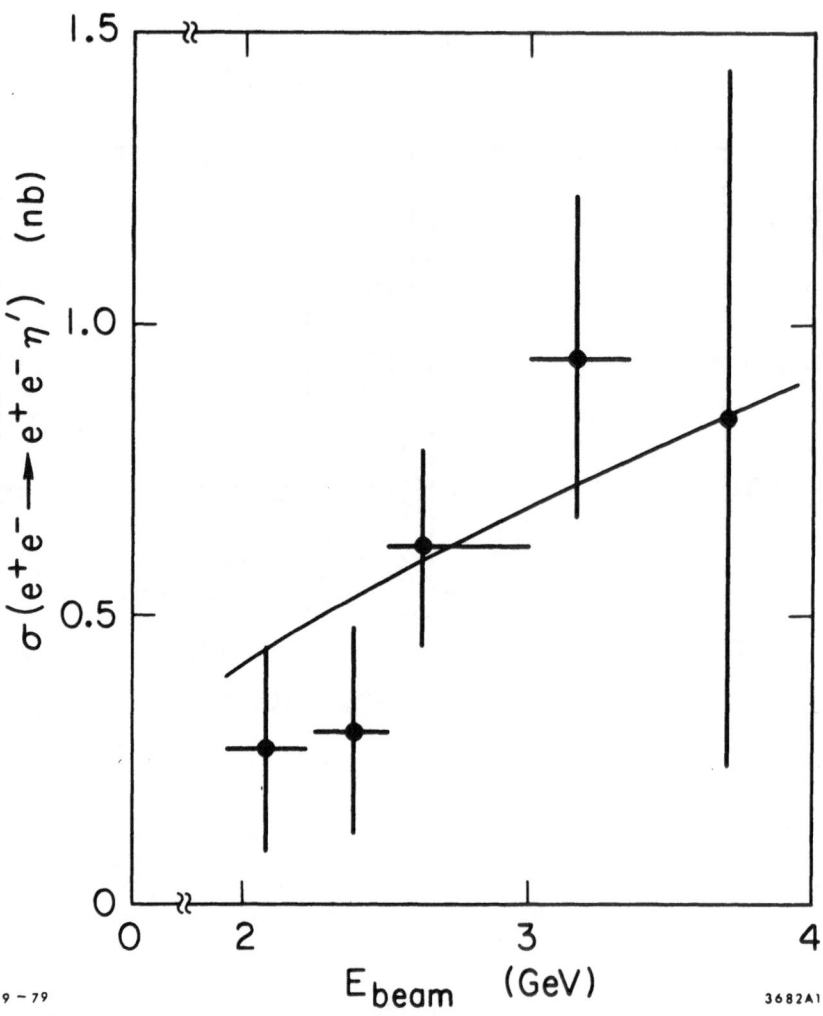

Fig. 7

9 – 79

3682A1

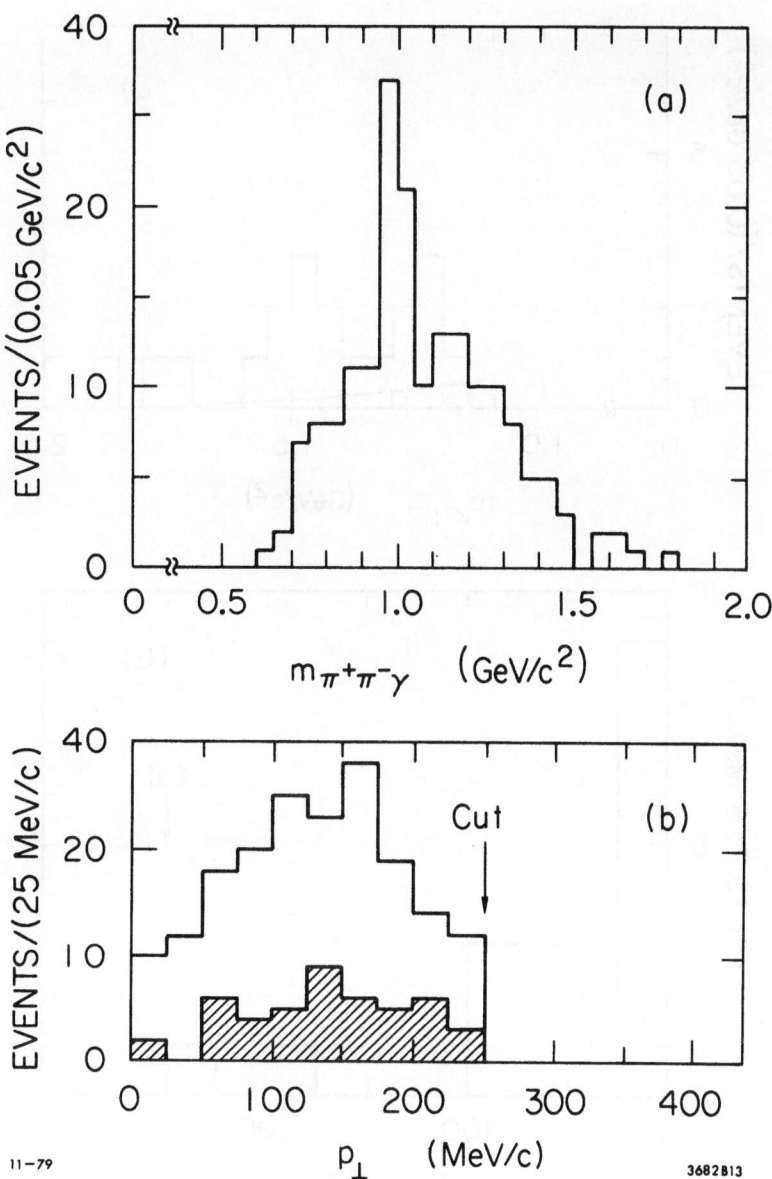

11-79

3682813

Fig. 8

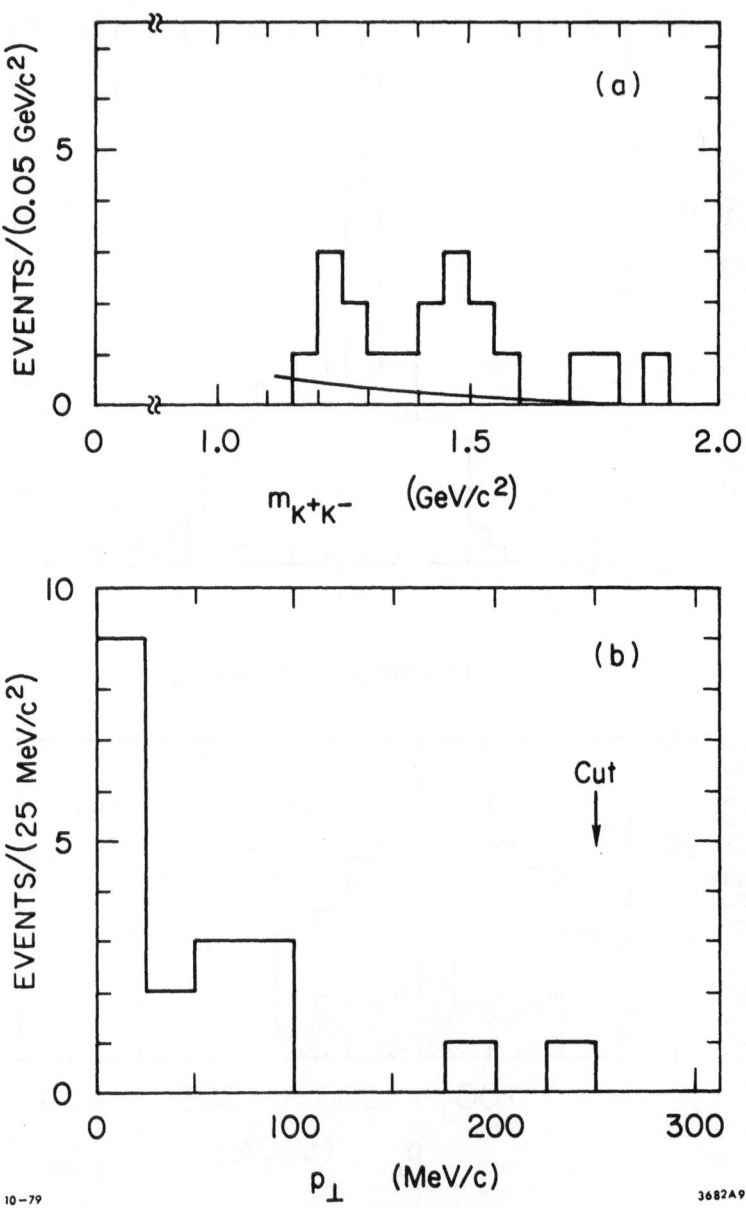

Fig. 10

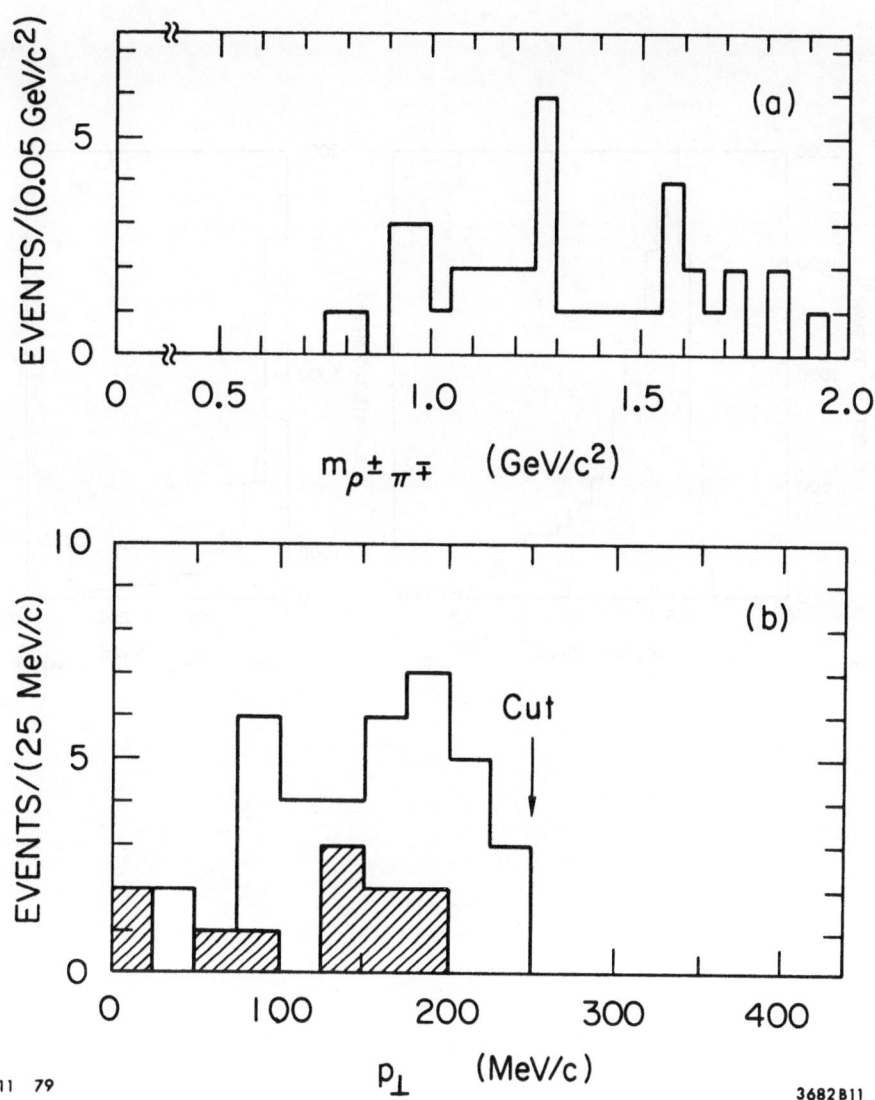

11 79   3682B11

Fig. 9

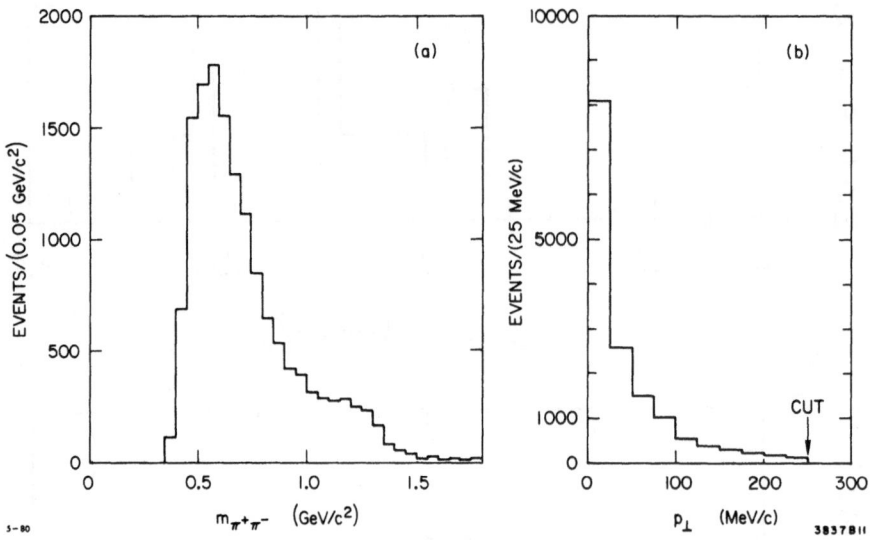

Fig. 11

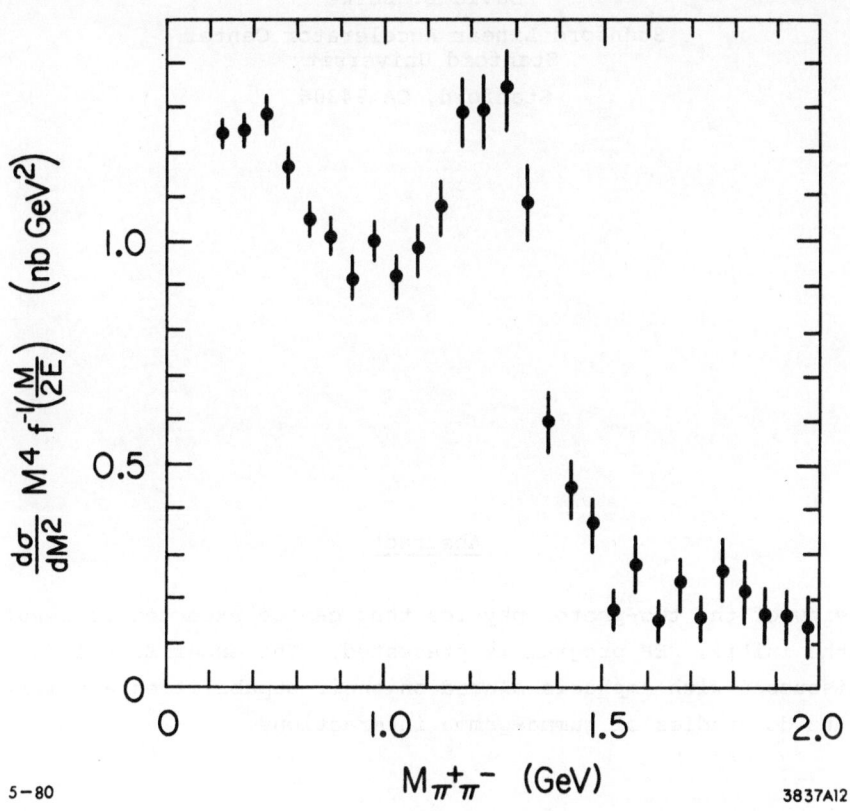

Fig. 12

5-80

3837A12

OUTLOOK FOR STUDIES OF GAMMA-GAMMA COLLISIONS AT PEP[*]

David L. Burke

Stanford Linear Accelerator Center
Stanford University

Stanford, CA 94306

Abstract

A preview of the two-photon physics that can be expected to result
from the initial PEP program is presented. The experimental detectors
are discussed with emphasis placed on their capabilities and limitations to do studies of gamma-gamma interactions.

---

[*]Work supported by Department of Energy, Contract DE-AC03-76SF00515
Invited talk presented at the International Workshop on Gamma-Gamma
Collisions, Amiens, France, April 8-12, 1980.

Introduction

The detectors that comprise the initial experimental program at PEP are outlined in Table I. These detectors have been primarily designed to study the annihilation process $e^+e^- \rightarrow \gamma^* \rightarrow X$, with only the Two-Gamma forward spectrometer specifically intended to study the two-photon process. The remaining detectors offer a wide range of capabilities and limitations, but for the purposes of this preview they can be summarized as follows. The Mark II, TPC, and HRS are magnetic solenoids. Each offers essentially complete particle identification and precise momentum and energy measurements in the central polar angle region, but has generally worse to nonexistent capabilities in the more forward regions that are covered by endcaps. The Magnetic Calorimeter (MAC) and the Direct Electron Counter (DELCO) are highly specialized detectors. MAC is designed to provide hadron calorimetry and muon identification over nearly 4π solid angle, while the DELCO detector is capable of electron identification with a high degree of background rejection.

Table I. PEP Detectors. The approximate acceptance of the central and endcap parts of each detector is separately given.

Detector	Type	Acceptance
TPC/Two Gamma	Time Projection Chamber	Central > 45°
		Endcap > 15°
	Forward Spectrometer	21 mr - 180 mr
MAC	Magnetic Iron Calorimeter	Central > 15°
		Endcap > 10°
HRS	High Resolution Spectrometer	Central > 35°
		Endcap > 17°
DELCO	Direct Electron Counter	Central > 45°
		Endcap > 12°
MARK II	Solenoid Spectrometer	Central > 45°
		Endcap > 20°
	Small Angle Tagger	21 mr - 80 mr

Since the Mark II, TPC, and HRS detectors are quite similar to each other, I will choose one of them (the Mark II) and discuss it in some detail. A brief section will be devoted to the MAC detector, and finally the Two-Gamma detector will be presented and its unique capabilities discussed. For the remainder of this discussion I will assume that the average luminosity of the PEP machine is 10^{31} cm^{-2} sec^{-1} (or 10^{-2} nb^{-1} sec^{-1}), and a typical experiment will be assumed to consist of an integrated luminosity of $\sim$ 50,000 nb^{-1}.

A Solenoidal Detector: Mark II

For polar angles greater than θ ≃ 45°, the Mark II detector, shown in
Fig. 1, provides charged particle identification, momentum measurement,
and photon detection and calorimetry. The properties of the central
part of this detector have been given elsewhere in these proceedings.[1]
At the angles between 20° ≤ θ ≤ 45°, the endcaps of the detector and
the inner drift chamber provide charged particle tracking and photon
detection, but with reduced efficiency and poorer resolution. The
small angle tagger (SAT) offers charged particle tracking and electron
calorimetry (no magnetic field) in the region 21 mr < θ < 82 mr. The
probability that a two-photon event will produce an electron in the
SAT ranges from 7% to 12% per side, depending upon the minimum photon
energy allowed by the constraints of the physics and the remainder of
the detector.

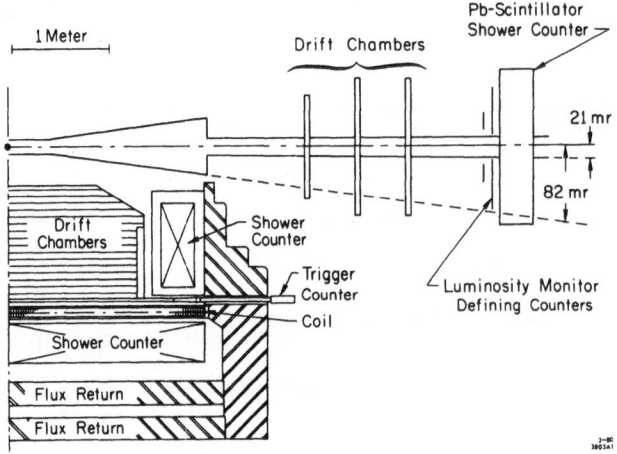

Fig. 1   Sliced view of the Mark II detector.
The muon detection system is not
shown, nor is the inner tracker drift
chamber.

The Mark II trigger is expected to be a logical combination of information acquired from the separate subsystems of the detector. On each
beam crossing the drift chamber electronics produces a fast count of
the number of charged tracks. The liquid argon signals are summed to
give the total energy deposited in each of the modules, and a count of
the number of modules with energy exceeding a fixed threshold is determined. The SAT shower counter signals are also discriminated and made
available to the trigger logic. The trigger logic itself is fully

programmable and can produce a trigger on any combination of subsystem signals. This situation is typical of the large detectors used at e^+e^- storage rings.

Most of the two-photon processes that produce hadrons in the final state can be normalized to the QED process ee → ee$\mu\mu$. The visible cross section for this is calculated[2] to be 4.4nb at a beam energy of 15 GeV. "Visible" events are those that result in two final state muons, both at θ > 45° and both with transverse momenta greater than 100 MeV. To properly separate the μ-pair final state from other two-prong events it is necessary to require that the mass of the pair exceed approximately 1.4 GeV. This effectively reduces the detectable cross section to about .2nb, or a rate of .002 Hz. It may prove impossible to operate the detector with a two-prong trigger due to large background rates. If this is the case then the SAT trigger will provide a sample of two-prong events but with a corresponding reduction in event rate.

Detection of resonance production by γγ collisions can, in principle, be done in two ways. Either the decay products of the resonance can be detected, or the existence of the resonance can be deduced by detecting both of the final state electrons. The resolution in the γγ center of mass energy provided by the Mark II SAT is ∿700 Mev, which implies that the detection of a narrow resonance must be done by reconstructing the decay products. The kinematics of the two-gamma process are perhaps best thought of in terms of the longitudinal rapidity variable y (≡ ½ ln (E+p_L) / (E−p_L)). The length of the rapidity spectrum of hadrons produced in a γγ collision is given by ln (s/m^2), where s = 4E$_b$^{2} is the maximum γγ c.m.s. energy, and m is the effective mass of the produced final state. A central detector like the Mark II is limited to a fixed rapidity interval ΔY (∿1 unit for the Mark II), and so the acceptance of the detector scales as (ΔY / ln (s/m)). This effectively nullifies the energy dependence of the production cross section σ ∿ (ln (E/me))^2 (ln s/m^2). For example the Mark II has been used at SPEAR to detect the production of the η' (958) by two photons. An integrated luminosity of 18 pb^{-1} resulted in ≃ 70 detected η's. At a beam energy of 15 GeV an integrated luminosity of 18 pb^{-1} will yield ≈100 detected η's.

The need to reconstruct a decay mode of a given resonance makes it difficult to detect higher mass states with a central detector. If, for example, the charmed pseudoscaler η$_c$ has a 10% branching fraction into a decay mode that, in turn, is detected 1% of the time, then

50 pb^{-1} will result in only 2-3 detected events. The decay mode $\eta_c \rightarrow 4\pi$ will have a detection efficiency of about 5% at a beam energy of 15 GeV, and substantial improvement cannot be made without increasing the solid angle acceptance of the detector.

Gamma-gamma collisions that proceed through a hard-scattering subprocess are expected to provide interesting tests of models of hadronic interactions, and perhaps ultimately, tests of QCD.[3] These processes are experimentally accessible through the detection of high transverse momentum hadrons, either single particles or multiparticle jets. The simple point coupling of the photon to quark lines, shown in Fig. 2a, will produce hadronic jets with a power law P_T^{-4} behavior. Fig. 2b is

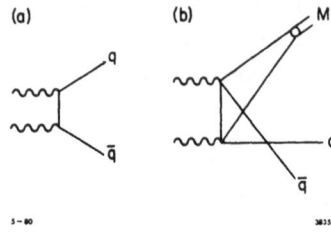

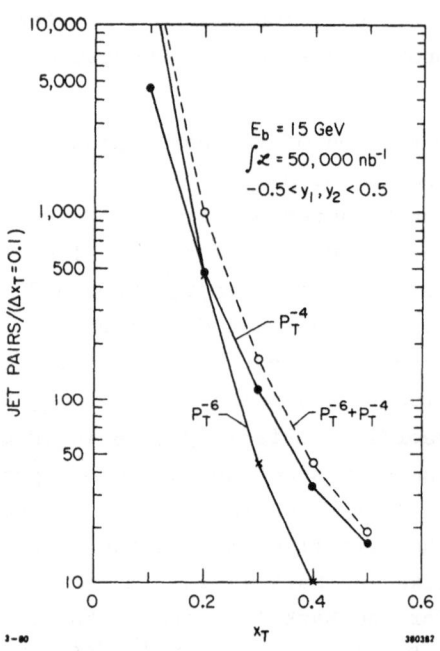

Fig. 2 a) Point coupling of photons to fermions. b) Constituent interchange diagram that results in a single meson at high transverse momentum balanced by an opposing quark jet.

an example of a proposed[3a] constituent-interchange process (CIM) that would result in a cross section characterized by P_T^{-6}. Shown in Fig. 3 is the number of untagged events that produce jet pairs within the acceptance of the Mark II as calculated with the model of Ref. 3a. Experimentally it will almost cer-

Fig. 3 Jet production by gamma-gamma collisions. The number of untagged events that produce a pair of jets within the acceptance of the Mark II is shown as a function of scaled transverse momentum.

tainly be necessary to work with tagged events in order to eliminate backgrounds from radiative annihilation events. This will result in a reduction in the number of events by a factor of four to five, and thus, only a few events are expected at $x_T > 0.4$. Reconstruction of individual jets produced at x_T below 0.4 will become increasingly difficult, and so, the analysis will probably proceed by measuring the distribution of a sphericity-like variable, and then making comparisons to Monte Carlo calculations based on various models. This is similar to the problem of finding jet signatures in e^+e^- annihilation events at SPEAR.

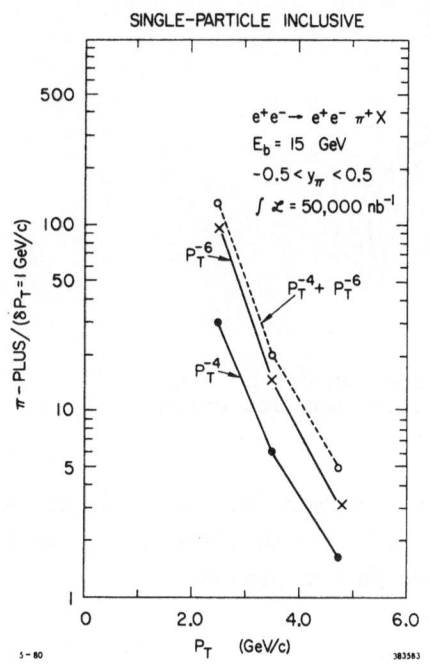

SINGLE-PARTICLE INCLUSIVE

$e^+e^- \rightarrow e^+e^- \pi^+ X$

$E_b = 15$ GeV

$-0.5 < y_\pi < 0.5$

$\int \mathcal{L} = 50{,}000$ nb^{-1}

P_T^{-6}

$P_T^{-4} + P_T^{-6}$

P_T^{-4}

π – PLUS/($\delta P_T = 1$ GeV/c)

P_T (GeV/c)

5 – 80   383583

Fig. 4 Single pion production at high transverse momentum. Untagged event yields are shown for the production of π^+. Neutral pions and π^- will be produced at the same rate.

Detection of single particles at high transverse momentum is, of course, free of the reconstruction problems inherent to jet studies, but at a given P_T the event rates are lower. Shown in Fig. 4 is the single π^+ untagged yield within the Mark II acceptance as taken from Ref. 3a. The tagged fraction is approximately 20%, which means that events at $P_T > 3.0$ GeV will be scarce. Nevertheless, yields at lower P_T become quite large and studies of these inclusive cross sections should be a basic measurement made by the central detectors at PEP.

A Large Solid Angle Calorimeter: MAC

A section of the Magnetic Calorimeter (MAC) is shown in Fig. 5. This detector is capable of hadron calorimetry ($\sigma \simeq 55\%\sqrt{E}$) for polar angles larger than 10° and electromagnetic calorimetry ($\sigma \simeq 18\%\sqrt{E}$) at $\theta > 30°$. Muons with momenta greater than 1 GeV are identified by counters external to the calorimeter. Initially only a luminosity monitor will be available for detecting electrons near the beam pipe. The tagging fraction is ~1-2%.

MAC is a large-acceptance, relatively homogeneous detector that is probably best suited to measure the total hadronic $\gamma\gamma$ cross section. This cross section is a fundamental parameter of $\gamma\gamma$ physics, yet it is quite difficult to properly measure. MAC represents a significant

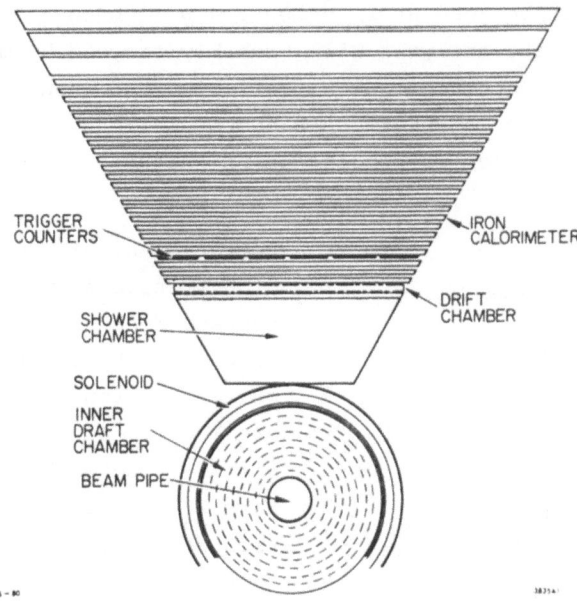

Fig. 5   Sliced view of one section of the MAC
         calorimeter. The outer muon detection
         system is not shown.

increase in total acceptance relative to the more conventional solenoidal detectors, and a significant improvement in calorimetry can be
expected even for the forward-peaked two photon process.

A Dedicated Detector: Two Gamma

The Two-Gamma small angle spectrometer[4], shown in Fig. 6, has been
designed to specifically study the two-photon process. The detector
includes the angular region from 21mr to 180mr in both the forward and
backward directions. Taken in combination with the TPC solenoid, the
solid angle coverage is essentially complete. (Neither detector
covers the range 180mr $\leq \theta \leq$ 240mr.)

Septum magnets with $\int$B·dl = 3 kg · m and drift chambers measure
charged particle momenta with δP/P $\lesssim$ 1% for momenta below 3 GeV/c.
The bulk of the hadrons that enter the spectrometer are expected to
have momenta in this range. Time-of-flight hodoscopes provide 3σ
separation of π's and K's up to momenta of 1.7 GeV/c, and K/p separation up to 2.8 GeV/c. At angles between 21mr and 100mr, arrays of
sodium iodide crystals are used to measure electron energies
($\sigma \simeq$ 1% / $E^{\frac{1}{4}}$); Pb-scintillator shower counters ($\sigma \simeq$ 12%$\sqrt{E}$) provide
electromagnetic calorimetry at larger angles (100mr-180mr).

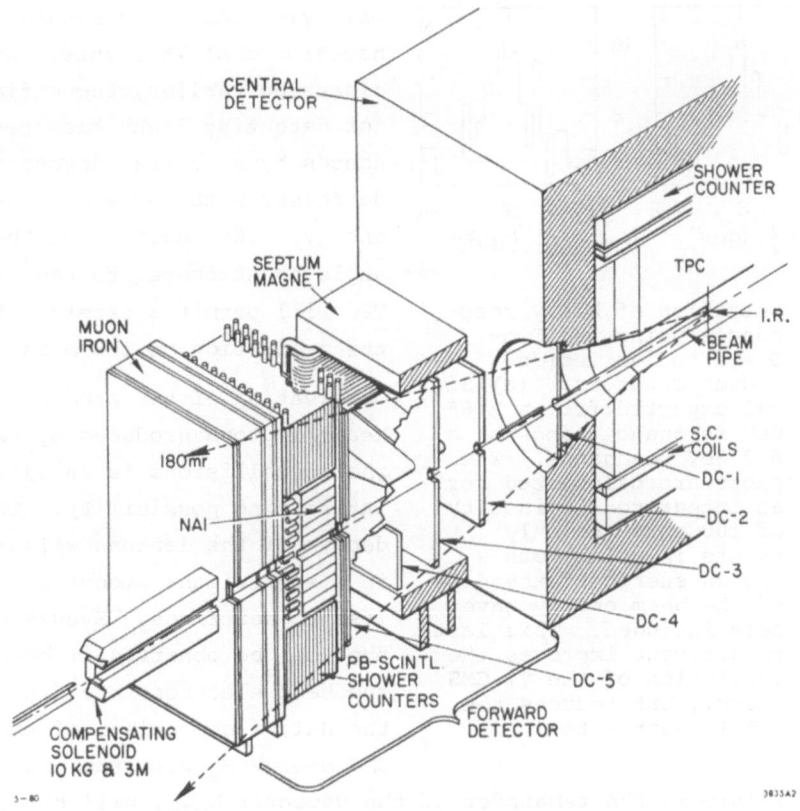

Fig. 6   The Two-Gamma forward spectrometer. Also
shown schematically is the TPC solenoidal
detector. There are two arms to the
spectrometer, one on each side of the
central detector.

Identification of muons with momenta greater than 1 GeV is done in
more or less standard fashion with drift chambers embedded in iron
absorber. The pion misidentification is 1% at momenta of 3 GeV/c.

Obviously the γγ physics that can be studied with these detectors
includes as a subset the physics that has been discussed in the previous sections. I will therefore limit the following discussion to
those features that are unique to the Two Gamma/TPC combination.
Particularly of interest is the resolution of the NaI and the detection of particles in the angular range below θ = 180mr.

The NaI measures the γγ c.m.s. energy of doubly tagged events with a
resolution $\sigma_{m_x} \sim 150$ MeV. This is sufficient to detect a heavy resonance by searching for a peak in the missing mass m_x. Figure 7 shows
the signal that can be expected[4] from a narrow resonance with mass of
either 2.85 GeV or 6.0 GeV and with a partial width Γm → γγ = 13 keV.
In the case of the 2.85 GeV state, an integrated luminosity of 50 pb^{-1}

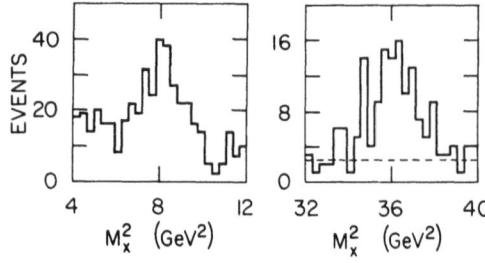

EVENTS

M_x^2 (GeV^2) M_x^2 (GeV^2)

3835A4

Fig. 7 Detection of heavy resonances using the Two-
Gamma sodium iodide
shower counters. (a) Sig-
nal expected from a 2.85
GeV resonance, and (b) a
6.0 GeV resonance. The
plots are calculated for
an integrated luminosity
of 100 pb^{-1} and only
events in which both
photon energies exceed 5%
of the beam energy have
been included. This last
requirement improves the
resolution of the $\gamma\gamma$ CMS
energy, but reduces the
total event rate.

will yield about 50 events over a background of 75 events. As was discussed earlier, the efficiency for detecting light mass reso- nances by a central detector decreases with increasing beam energy. The addition of the small angle spectrometer to the central TPC will permit a careful study of the production of low mass states.

Observation of the production of heavy leptons produced by two photon collisions is an extremely interesting possibility. The decays of the leptons will result in a significant amount of energy loss to neutrinos. Events of this type can be observed if both of the beam electrons are detected in the NaI; a comparison of the $\gamma\gamma$ CMS energy M_x with the total energy visible in the remainder of the detector M_{vis}, will show a neutrino signature if $M_x \gg M_{vis}$. The rate of tau production by $\gamma\gamma$ collisions, for example, is expected[2] to be equal to the corresponding rate in the annihilation process at a beam energy of 15 GeV. The nominal 50 pb^{-1} experiment would result in approximately 50 doubly tagged $\gamma\gamma \to \tau^+\tau^-$ events. Since most of the events are produced near threshold (because of the falling $\gamma\gamma$ energy spectrum), then a rela- tively clean signal should be seen. Searching for new particles in this manner has the feature that the entire energy range of the machine is scanned at one time. One is faced, however, with the need to detect the beam electrons and with the fact that nearly 4π calorim- etry is necessary to eliminate backgrounds.

The photon structure function F_2^{γ} (x,Q^2) has been calculated exactly to leading order in QCD, and estimates of higher order contributions have been carried out.[5] In contrast to hadronic structure functions, it is predicted that F_2^{γ} will be essentially constant over the range $.3 \lesssim x \lesssim .7$, for fixed Q^2. Furthermore, F_2^{γ} is expected to violate scaling with a factor of $\ln Q^2$ (i.e., $F_2^{\gamma} \sim \ln Q^2 \cdot f(x)$). Experi- mentally, however, measurement of this quantity is quite difficult. It is necessary to isolate events in which one of the photons is far

off the mass shell. The energy of the target photon must be reconstructed, and the scattered electron must be detected and identified.
To eliminate backgrounds from radiative Bhabha events it will also be
important to detect at least one other produced particle. A parton-
model estimate[4] indicates that with the Two Gamma/TPC detectors a
luminosity of 50 pb^{-1} will result in a hundred or so events at
$Q^2 > 3$ GeV and $x > 0.3$. Only a few events can be expected at $x > 0.7$.
These are clearly low yield experiments and will require optimal beam
conditions in order to be successful.

Comments

Two photon collisions will be a copious source of events visible to
the detectors at PEP. Experiments will be difficult, however, and in
many cases machine luminosities that are near to design specifications
will be required. Despite nearly ten years of simmering theoretical
discussion, experimentally the field is still largely unexplored. PEP
and PETRA are perhaps the ideal locations to begin serious studies of
gamma-gamma interactions.

References

[1] P. Jenni, invited talk to this conference.

[2] Computer routines contained in the library AXOLIB where used to
compute these rates. See:
R. Bhattacharya, J. Smith, and G. Grammer, Phys. Rev. D 15, 3267
(1977).
J. Smith, J.A.M. Vermaseren, and G. Grammer, Phys. Rev. D 15, 3220
(1977).

[3] (a) S. J. Brodsky, et al., Phys. Rev. D 19, 1418 (1979).
(b) K. Kajantie, Helsinki preprint, HU-TFT 79-5
    K. Kajantie and R. Raitio, Helsinki preprint HU-TFT 79-13

[4] PEP Proposal PEP-9 (1976)
Addendum to Proposal PEP-9 (1977)

[5] W. A. Bardeen and A. J. Buras, Phys. Rev. D 20, 166 (1979) and
references therein.

TWO PHOTON PHYSICS IN THE PLUTO DETECTOR

BY

CHRISTOPH BERGER

I. PHYSIKALISCHES INSTITUT, RWTH AACHEN, GERMANY

I. Introduction

The scattering of light by light has been investigated for the first
time in a 1935 theoretical paper by Euler and Kockel[1]. This process was
soon recognized to be very interesting, because it is characteristic
of quantized electrodynamics and cannot occur in the classical notion
of linear Maxwell equations. Although the basic motivation has changed
a lot, light light scattering is attracting much interest   in todays
physics.

Nowadays (inelastic) two photon reactions can be studied at high energy
e^+e^- storage rings. The basic diagram is given in fig.1. The incoming
leptons (e^+, e^-) radiate (virtual) photons, which subsequently scatter
into the final state X.

The spin averaged inclusive cross section for the reaction $e^+e^- \rightarrow e^+e^- X$
reads (the kinematical symbols used, are explained in fig.1)

$$\frac{d\sigma}{d\Omega_1 dE_1' d\Omega_2 dE_2'} = \sum_{i=1}^{8} A_i F_i \qquad (1)$$

The eight structure functions F_i depend on q_1^2, q_2^2, W, i.e. on the
virtual photon masses and the invariant mass of the system X. The A_i
are kinematical factors, which can be calculated from the electron
(positron) scattering angles and energies[2]. Fortunately for most practical cases, which have been investigated until now, the eight term formula (1) reduces to a two term or one term formula.

I.1 Electron photon scattering

One of the incoming leptons (say the positron) is scattered into very
small angles, with the virtual photon practically on mass shell
($q_2^2 \approx E E_2' \theta_2^2$). For most experiments this means, that the scattered
positron remains within the beam pipe and is either undetected ('no tag')
or detected with a special apparatus making use of the storage ring
magnets ('0 degree tagging'). Using the symbols explained in fig.1 we
introduce the invariants

$$Q^2 = -q_1^2 = 2EE_1'(1-\cos\theta_1) \qquad (2)$$

$$\nu = q_1 \cdot q_2 = 2E_\gamma \ (E - E_1'\cos^2\theta_1/2) \qquad (3)$$

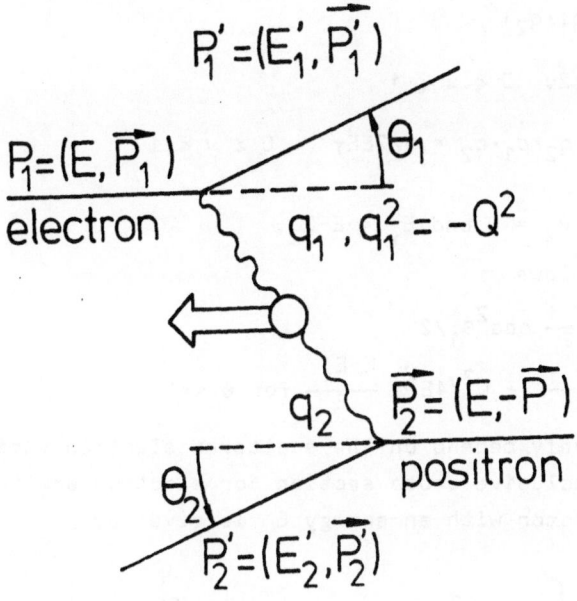

Fig.1 Basic two photon scattering diagram
with explanation of symbols used.

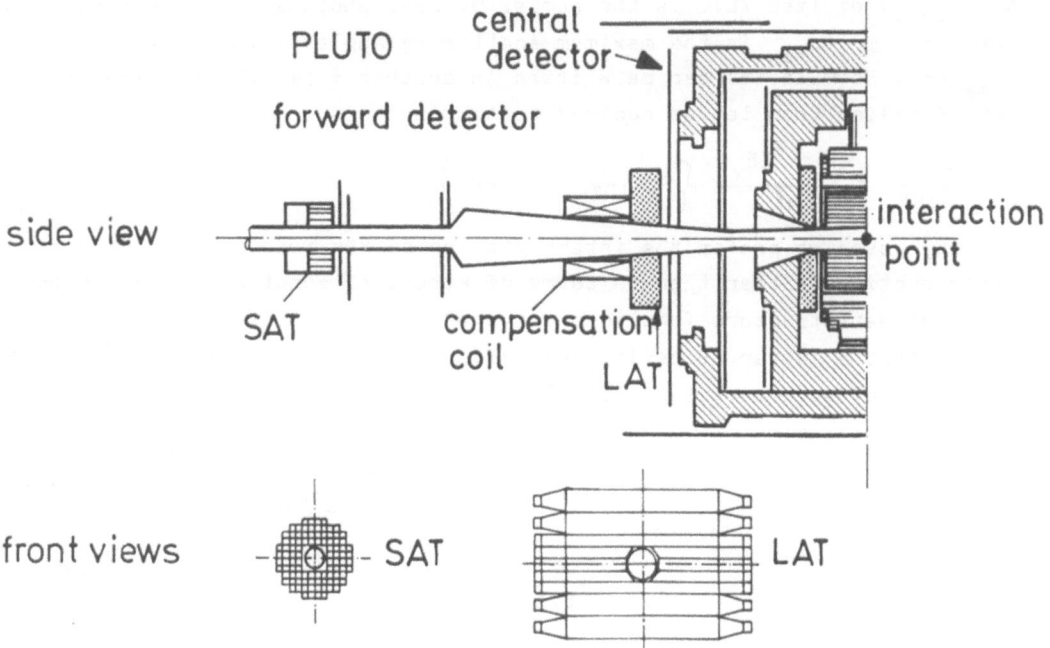

Fig.2 The PLUTO detector with forward spectrometers.

$$W^2 = (q_1 + q_2)^2 \tag{3a}$$

$$x = Q^2/2\nu \quad 0 \leqslant x \leqslant 1 \tag{4}$$

$$y = q_1 \cdot q_2 / p_1 \cdot q_2 = \nu/2EE\gamma \quad 0 < y < 1 \tag{5}$$

Herein we have used $\theta_2 = 0$ and $E_\gamma = E - E_2$.

From (3) and (5) follows

$$y = 1 - \frac{E_1'}{E} \cos^2 \theta_1/2$$

or

$$y = \frac{E - E_1'}{E} + Q^2/4E^2 \approx \frac{E - E_1'}{E} \text{ for } \theta_1 \ll 1$$

Note that Q^2 and y only depend on the scattered electron variables. The spin averaged inclusive cross section for electron scattering off a real transverse photon with an energy E_γ is given by[3]

$$\frac{d\sigma}{d\Omega_1 dE_1'} = \frac{4\alpha^2 E_1'}{Q^4 y} \left\{ F_2(x, Q^2)(1-y) + F_1(x, Q^2)xy^2 \right\} \tag{6}$$

and therefore

$$d\sigma_{e^+ e^- \to e^+ e^- X} = \frac{4\alpha^2 E_1'}{Q^4 y} \left\{ F_2(1-y) + F_1 xy^2 \right\} N(z, \theta_{2max}) dz d\Omega_1 dE_1' \tag{7}$$

$N(z, \theta_{2max}) dz$ ($z = E_\gamma/E$), is the number of real photons in the energy interval dz, θ_{2max} is the maximum scattering angle of the positron, $\theta_{2max} \ll 1$. Formula (6) can be written in another form, which again is very familiar from lepton nucleon scattering

$$\frac{d\sigma}{dxdy} = \frac{16\pi\alpha^2 EE_\gamma}{Q^4} \left\{ F_2(1-y) + F_1 xy^2 \right\} \tag{8}$$

In many cases we prefer the interpretation of the results in terms of cross sections rather than in terms of structure functions. The connection between structure functions and photon photon cross sections is to some extent a matter of definition. After introducing transverse and longitudinal structure functions F_t, F_1[3]

$$F_t = F_1 \tag{9}$$

one may define[2]

$$F_1 = \frac{1}{2x} F_2 - F_1$$

$$\sigma_{t,1} = \frac{4\pi^2\alpha}{\nu} F_{t,1} \tag{10}$$

and thus

$$\frac{d\sigma}{d\Omega_1 dE_1'} = \Gamma_t \left\{ \sigma_t(Q^2, W) + \epsilon \sigma_1(Q^2, W) \right\} \tag{11}$$

and

$$d\sigma_{e^+e^-\rightarrow e^+e^-X} = \Gamma_t\left\{\sigma_t + \varepsilon\sigma_1\right\} N(z,\theta_{2max})dz \ d\Omega_1 E_1'$$ (12)

Γ_t is the flux factor for the incoming virtual photons (radiated from the electron) and ε is a polarization parameter for the virtual photons. σ_t is the cross section for the scattering of virtual transverse unpolarized photons off real photons and σ_1 the respective cross section for virtual longitudinal photons. Γ_t and ε are respectively given by

$$\Gamma_t = \frac{\alpha \ E_1' \ (1+(1-y)^2)}{2\pi^2 \ Q^2 \ y}$$ (13)

$$\varepsilon = \frac{2(1-y)}{1+(1-y)^2}$$ (13a)

A technical problem is the calculation of the photon flux factor $N(z,\theta_{max})$. This is usually done in the so called equivalent photon approximation (EPA) or Weizsaecker Williams approximation (WWA). There are three classical papers[4] giving different analytical expressions for $N(E,\theta_{max})$. Numerically the difference is less than 5 percent for the experimental conditions accessible to the PLUTO detector. A criterium for the validity of an EPA has been given by Carimalo, Kessler and Parisi[5]:

$$2|q_2|/W \ll 1$$

This requirement is satisfied in our detector for W>1 GeV, because with $0<\theta_2<23$mrad the average $|q_2^2|$ is very small.

Using the standard expression[6] (m is the electron mass)

$$N(z,\theta_{max}) = \frac{\alpha}{\pi z} \left\{(1+(1-z)^2) \ \ln \ (\frac{E(1-z)}{m \ z}\theta_{max}) \ -1+z\right\}$$ (14)

the product $\Gamma_t \cdot N(E,\theta_{max})$ exactly reproduces the single tag luminosity function given in formula (11) of ref.[7].

If one instead of formula (10) defines the $\gamma\gamma$ cross sections analogous to electroproduction

$$\sigma_{t,1} = \frac{4\pi^2\alpha}{\nu - Q^2/2} \ F_{t,1}$$ (15)

the expression for Γ_t[6,8] is depending an electron and positron variables. Up to now we use definition (15), but I prefer in the future to use definition (10), which leads to completely factorizing luminosity functions for single tag experiments, a fact which has not been realized before. The data presented in this talk are still at small enough Q^2 values to make the numerical difference of both approaches neglegible except for the smallest W bin. (see chapter IV).

I.2 Photon Photon scattering

If also the incoming electron is scattered into angles close to 0°
(no tagging or 0° tagging) formula (12) reduces to the well known one
term formula

$$d\sigma_{e^+e^- \to e^+e^-X} = N(z, \epsilon_{1max})dz_1 \cdot N(z, \theta_{2max})dz_2 \sigma_{\gamma\gamma}(W) \qquad (16)$$

with $z_i = (E-E_i)/E$

It is this expression, which has been used in most early discussions
of two photon reactions. If one has no 0° tagging system both electrons
remain undetected and travel down the beam pipe. Thus formula (16)
establishes a 'background' to the usual annihilation process $e^+e^- \to X$.
In the following chapter we will emphasize, that even in this case a
separation between 1γ (annihilation) and 2γ reactions is possible. In
any case $\sigma_{\gamma\gamma}(W)$ is giving the $Q^2 = 0$ limit of $\sigma_t(Q^2,W)$ and thus a careful and accurate measurement of this quantity (e.g. using a high level
0° double tagging system) is of great importance.
I have put such an emphasis on the discussion of the $\gamma\gamma$ kinematics,
because there is no 'standard' analysis around and because it is very
important to define the assumptions used as precise as possible.

II.   Short description of the apparatus

The PLUTO detector is operated by a collaboration of physicists from
DESY, 4 german universities, the university of Bergen in Norway and
the university of Maryland, USA :

PLUTO Collaboration

Ch. Berger, H. Genzel, R. Grigull, W. Lackas, F. Raupach
I. Physikalisches Institut der RWTH Aachen, Germany
A. Klovning, E. Lillestöl, E. Lillethun, J.A. Skard
University of Bergen, Norway
H. Ackermann, G. Alexander, F. Barreiro, J. Bürger, L. Criegee, H.C.
Dehne, R. Devenish, A. Eskreys, G. Flügge, G. Franke, W. Gabriel,
Ch. Gerke, G. Knies, E. Lehmann, H.D. Mertiens, U. Michelsen, K.H. Pape,
H.D. Reich, M. Scarr, B. Stelle, T.N. Ranga Swamy, U. Timm, W. Wagner,
P. Waloschek, G.G. Winter, W. Zimmermann
Deutsches Elektronen Synchrotron DESY, Hamburg, Germany
O. Achterberg, V. Blobel, L. Boesten, V. Hepp, H. Kapitza, B. Koppitz,
B. Lewendel, W. Lührsen, R. van Staa, H. Spitzer
II. Institut für Experimentalphysik der Universität Hamburg, Germany
C.Y. Chang, R.G. Glasser, R.G. Kellogg, K.H. Lau, R.O. Polvado, B.Sechi-
Zorn, A. Skuja, G. Welch, G.T. Zorn
University of Maryland, College Park, USA
A. Bäcker, S. Brandt, K. Derikum, A. Diekmann, C. Grupen, H.J. Meyer,
B. Neumann, M. Rost, G. Zech
Gesamthochschule Siegen, Germany
T. Azemoon, H.J. Daum, H. Meyer, O. Meyer, M. Rössler, D. Schmidt,
K. Wacker
Gesamthochschule Wuppertal, Germany

The main components of PLUTO are (fig. 2)

(a)   a central detector with 13 cylindrical proportional chambers
      operating in a magnetic field of 1.65 T. The momentum resolution
      for charged tracks is $\sigma_p/p = 3\%p$ (p in GeV) for p>3 GeV.

(b)   barrel and endcap shower counters with proportional tubes for
      position measurement of the showers. The energy resolution for
      electrons and photons with energy E > 1 GeV is $\sigma_E/E \sim 35\%/\sqrt{E}$
      (E in GeV) in the barrel and $\sim 19\%/\sqrt{E}$ in the endcaps. The geometri-
      cal acceptance of (a) and (b) is 87% and 94% of 4π sterad.

(c)   a muon identifier with a 1m iron absorber for hadrons. The tracks
      are sampled at two depths within the absorber by a set of pro-
      portional and drift chambers.

(d)   Forward spectrometers on each side of the detector for luminosity
      measurements and for selection of reactions coming from two photon

interactions. Because these spectrometers are relatively new
and some understanding of their operation is essential for the
results I will describe them in more detail.

Each arm of the forward spectrometers consists of a 'large angle tagger'
(LAT) and a 'small angle tagger' (SAT). The LAT covers the polar angle
region between 70 and 260 mrad. The energy of electrons and photons is
determined with a lead scintillator shower counter of 14.5 radiation
length thickness. The position of charged particles is determined by
six planes of proportional tube chambers with a wire spacing of 1cm.
The SAT covers the angular region between 23 and 70 mrad. Energy information of electrons and photons is obtained from a lead glass shower
counter matrix. It consists of 96 blocks (each with a front area of
6.6×6.6 cm^2),in a concentric arrangement around the beam pipe. The
thickness of this counter is 12.5 radiation length. Tracking of charged
particles is achieved by a set of four planar porportional wire chambers
(wire distance 0.3 cm). In a test beam the energy resolution of the LAT
was measured to be 11%/$\sqrt{E}$ (rms) and of the SAT 8.5%/$\sqrt{E}$ (rms), E in GeV.
These values have been reproduced by analyzing small angle Bhabha
scattering.

All results discussed in the following two chapters are derived from
the single tag or no-tag data sample. The tagging system is thus only
used to select a '2γ event' and to determine the Q^2 of the virtual
photon. The mass W of the hadronic system or of the QED pairs is
measured in the central detector. This procedure is quite different,
from what has been dicussed by many people years ago, but obviously
very successful.

III. Results on 2 prong final states

Two-photon production of lepton and hadron pairs $\gamma\gamma \to ee, \mu\mu, \pi\pi$ can be
studied at high energy electron-positron storage rings via the reaction
$e^+e^- \to e^+e^-X$, where X is a pair of charged particles (see fig.3a). These
processes are of considerable interest, because lepton-pair production
and the pointlike production of pion pairs can be calculated in QED.
By comparing the measured rate of e.g. lepton pairs with that predicted,
one is thus testing QED in processes, the amplitude of which is proportional to the fourth power of the coupling constant e, compared to e^2
in annihilation processes . On the other hand the production of C = +1
resonances, which decay into pion pairs, should show up as a deviation
from the calculated QED two-prong cross section.

The outgoing electrons and positrons (outer lepton lines in fig. 3a)are
not identified for the bulk of our data ('no-tag condition', θ_{max}=23mrad).
I report, however, also on a small subsample of the data, where either
the electron or the positron is measured in one of the forward spectrometers ('single-tag condition').

The produced pairs (inner lines in fig.3a) are identified using the
central track detector of PLUTO. For the no-tag sample we require two
oppositely charged tracks subject to the following conditions:

        a) $|\cos\theta|<0.56$ for each track
        b) $p_\perp>400$ MeV for each track
        c) acolinearity angle between the track $>15^o$

The polar angle θ and the transverse momentum $p_\perp$, are measured with
respect to the beam axis. Events with neutral energy not related to
the tracks are rejected. The rather strong conditions (a) and (b) are
used to ensure uniform efficiency of the central detector. Condition (c)
is needed for rejecting cosmic rays and Bhabha events. The track-selec-
tion criteria for tagged events are less stringent and are identical to
the ones described in ref.8.

The vertex distribution of the no-tag events is shown in fig.4. There
is a very clear peak around the interaction point with only a small
contamination (~5%) from beam-gas scattering. In fig.5a,b we plot for
the background subtracted no-tag events, the total energy in the central
detector and the vector sum of the transverse momenta, $p_\perp^{sum}=|\vec{p}_\perp^1+\vec{p}_\perp^2|$.
The energy distribution peaks below 2 GeV and decreases steeply toward
higher energies. There are only 5% of all events above a total energy
of 10 GeV. They are attributed to radiative Bhabha events and are
eliminated. The $p_\perp^{sum}$ distribution demonstrates that the transverse
momentum of the two tracks is balanced, i.e. on the average the missing
transverse momentum is compatible with zero. Both features strongly

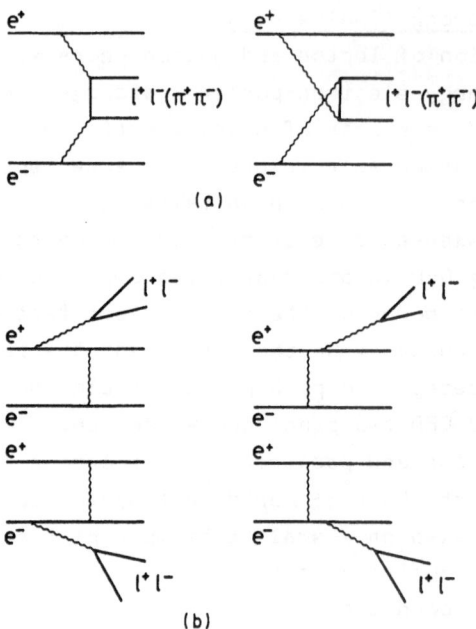

Fig. 3

Feynman diagram for pair production via the two-photon
exchange process (a) and the bremsstrahlungprocess (b),
$(1^+1^- = e^+e^-, \mu^+\mu^-)$.

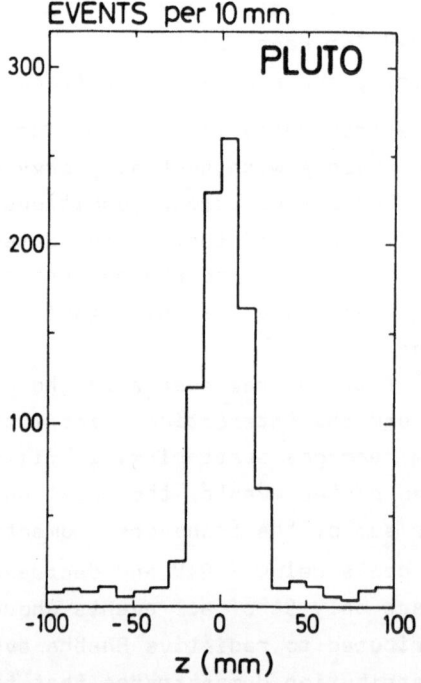

Fig. 4

Distribution of the reconstructed event vertices along
the beam line.

support the conclusion that these events originate from two-photon
interactions. The energy distribution results from the bremsstrahlungs-
spectrum of the photons and the transverse momentum is balanced because
most of the photons are radiated along the direction of the incoming
beams. In fact the estimated contribution from annihilation processes
which lead to two-prong events with similar characteristics is completely negligible.

In fig. 6 the number of events is plotted versus the invariant mass of
the pairs W which is obtained by assigning pion masses to all particles.
The solid line is the result of an absolute QED prediction for the production of lepton pairs (ee,μμ). It was calculated using a computer
program written by J. Vermaseren[9]. This program includes an exact calculation of the diagrams in fig.3a plus their interference with bremsstrahlung terms (fig. 3b). Pion-pair production via the two-photon
interaction is normally assumed to be small. On the other hand, the
acceptance cuts favour pions relative to leptons. We have estimated
this contribution by taking the production cross section from ref.10
and calculating the photon fluxes via the equivalent photon approximation discussed above. The predicted rate is ~15% of the lepton signal
in fig. 6.

A background to the two-prong spectra could result from two photon initiated multi-hadron production with missing particles in the central
detector. Using the experimental results and the Monte-Carlo method
described in ref. 8, we have estimated this background to be smaller
than 2% in our data sample.

The agreement of the QED calculation with the data in the bins below
1 GeV and above 1.5 GeV is very good, taking into account the systematic error of our data (15%) and the fact that radiative corrections
have not been included. There have been very few attempts to calculate
radiative corrections to two-photon processes[11]. These corrections are
strongly dependent on specific detector cuts. We expect them to reduce
the cross section by a few percent[12] and to have a smooth W dependence,
in accordance with the good agreement of our data with the shape of
the QED curve above 1.5 GeV.

For 1 GeV<W<1.5 GeV there is a clear excess of the data above the QED
prediction. The insert in fig.4 (128±27 events) shows the difference
between the data and the QED prediction. It has a typical resonance
behaviour. A very good candidate for such a resonance is the f^0 (1270).
Another possible candidate in the same mass region is the ε (1300). It
was rejected mainly because it is believed to have a much larger width
(Γ_{tot}>300 MeV)[13]. In principle one could distinguish the f^0 from the ε

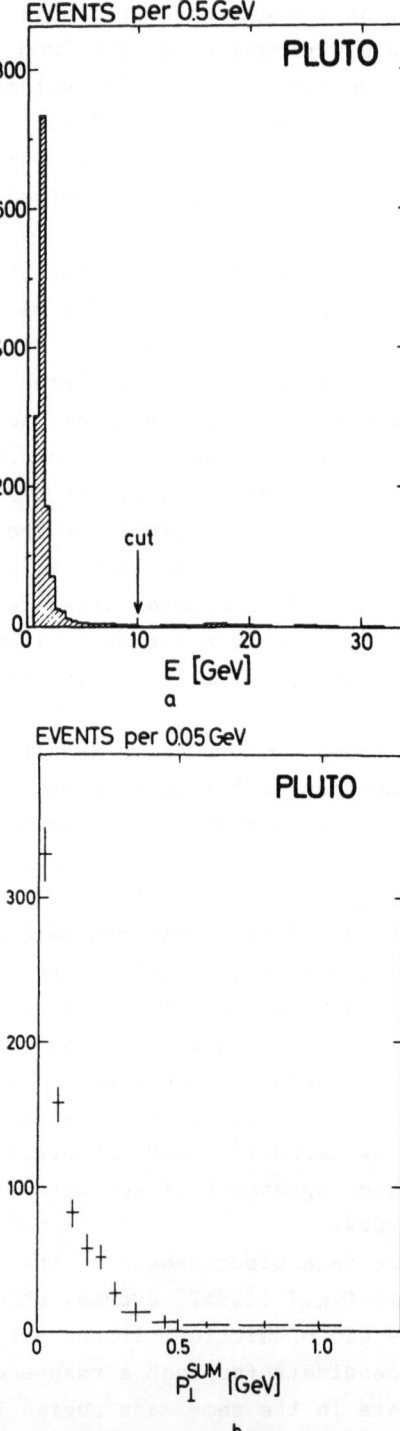

Fig. 5

Background subtracted distribution of the total energy in
the central detector (a) and the vector sum of the trans-
verse momenta $p^{SUM} = |\vec{p}^1 + \vec{p}^2|$ (b).

hypothesis via the angular distribution of the decay pions. In practice this is not conclusive because of our limited angular acceptance and the large QED background.

For a quantitative comparison we calculate the expected number of events (N^i_{exp}) in each mass bin i from a superposition of lepton-pair production (N^i_{QED}) and f^o production (N^i_{fo})

$$N^i_{exp} = a\ N^i_{QED} + b\ N^i_{fo}$$

In order to determine N^i_{fo} we simulate production and decay of f^o mesons (ee $\rightarrow$ e e f^o $\rightarrow$ e e π π) in a Monte-Carlo program using

$$d\sigma = N(z,\theta_{1max})dz_1 \cdot N(z,\theta_{2max})\ dz_2 \sigma_{\gamma\gamma}(W^2)P(\theta^{CM})\ d\Omega^{CM} \qquad (17)$$

with a Breit-Wigner ansatz for the total cross section $\sigma_{\gamma\gamma}$ in the reaction $\gamma\ \gamma \rightarrow f^o \rightarrow \pi^+\pi^-$:

$$\sigma_{\gamma\gamma}(W^2) = 40\pi \quad \frac{\Gamma_{\gamma\gamma}\ \Gamma_{\pi^+\pi^-}}{(W^2-M^2_{fo})^2+\Gamma^2 M^2_{fo}} \qquad (18)$$

$\Gamma_{\pi^+\pi^-}$, Γ and M_{fo} were taken from the standard-data compilation[13]. The decay width $\Gamma_{\gamma\gamma}$ is arbitrarily set to 1 keV, thus b will be the experimental width in units of keV. $N(E,\theta_{max})$ dz is the number of photons with energies between E_γ and $E_\gamma + dE_\gamma$ (Formula 14) radiated from the electron (positron). Being a J = 2 state, the f^o can be produced via two helicity amplitudes $|\lambda| = 2$, $\lambda = 0$ in photon-photon reactions. These amplitudes lead to different decay angular distribution $P(\theta^{CM})$ in the pion pair center of mass system.

$$|\lambda| = 2 \quad P(\theta^{CM}) = \frac{15}{32\ \pi}\ \sin^4\theta^{CM}$$

$$\lambda = 0 \quad P(\theta^{CM}) = \frac{5}{16\ \pi}\ (3\cos^2\theta^{CM}-1)^2$$

We have chosen $\lambda = 2$, because the dominance of this amplitude is predicted from widely differing theoretical approaches to the problem of radiative f^o decay[14]. The results of our fit are given by a = 0.97±0.05 and b = 2.3±0.5 (χ^2 per degree of freedom = 0.6). The value of a confirms QED within the errors and limitations discussed above. The fit value for the $\gamma\gamma$ decay width of the f^o meson is $\Gamma_{\gamma\gamma}$ = 2.3±0.5 keV with an additional systematic error of ±15%. The virtual photon mass squared is very low, $Q^2 = 0.007$ GeV^2 on the average.

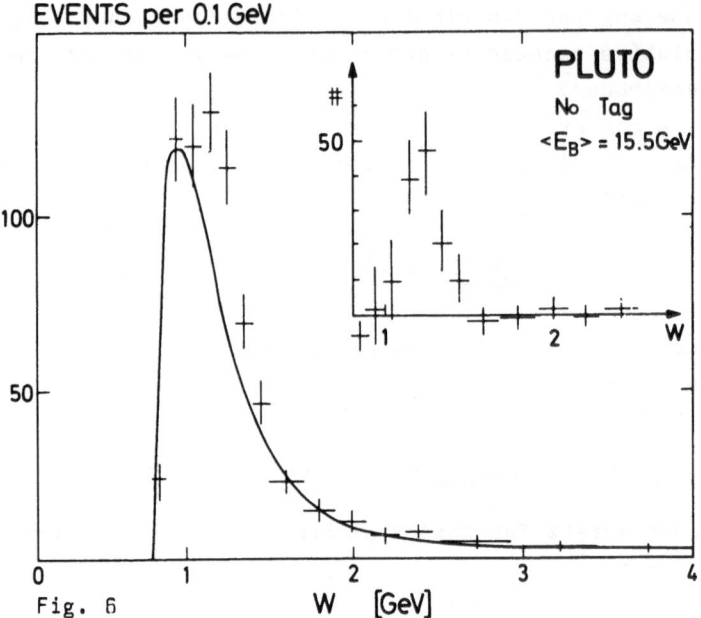

Fig. 6

Invariant mass distribution of the two-prong events. The
solid line shows the absolute QED prediction for the production of lepton pairs (ee,µµ). The insert shows the
difference between the data and the QED prediction.

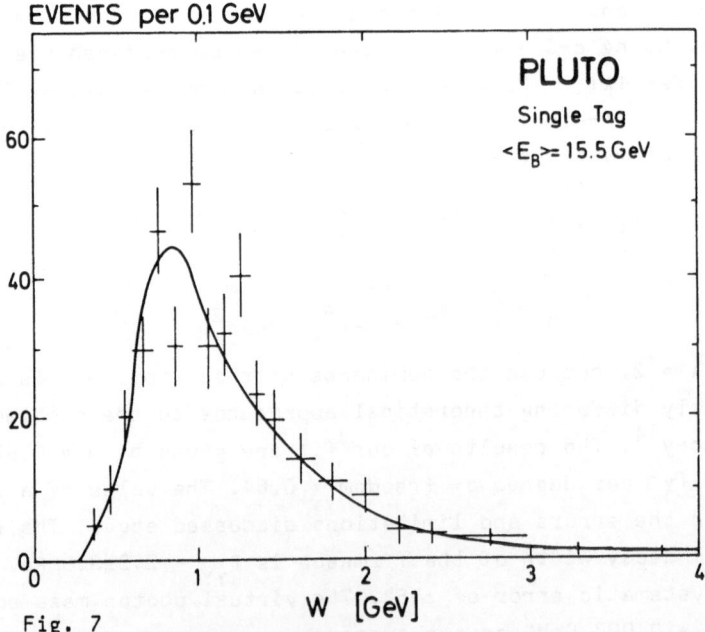

Fig. 7

Invariant mass distribution of the two-prong events with
a single tag in the SAT.

The experimental result is to be compared with the numerous theoretical predictions, which are listed in table 1. Our result is close to the value obtained in calculations[17,18] using the non relativistic quark model with an oscillator potential. With the exception of ref. 15 all methods based on finite energy sum rules, tensor meson dominance etc. lead to larger values for $\Gamma_{\gamma\gamma}$. It should be noted also that tentatively assuming the $\lambda = 0$ hypothesis we find $\Gamma_{\gamma\gamma} = 5.7\pm1.3$ keV.

Finally we have analyzed the much smaller data sample with the single-tag condition, i.e. an electron (positron) scattered into one of the small-angle taggers, SAT, of the forward spectrometers[8]. The invariant mass distribution is shown in fig.7 along with the QED prediction. The small enhancement in the f^0 mass region can be attributed to f^0 production via one almost real (as above) and one virtual ($Q^2 = 0.28$ GeV^2) photon. Due to the limited statistics we only give an upper limit of $\Gamma_{\gamma\gamma} < 2.6$ keV (95% confidence level) again using the $\lambda = 2$ hypothesis. For extracting this radiative width the flux factor for the virtual photons radiated from the electron scattered into the SAT was taken from the $e\gamma$ scattering formalism discussed in chapter I.

<u>Table 1</u>   Theoretical predictions for $\Gamma(f^0 \rightarrow \gamma\gamma)$

Reference	$\Gamma_{\gamma\gamma}$ (keV)
15	0.8
16	>1
17	1.2-2.3
18	2.6
19	5.07
20	7
21	5.7
22	8
14	8
23	9.2
24	11.3
25	21±6
26	28
this experiment	2.3±0.5

IV. Results on multihadron production

IV.1 The total cross section

We have measured for the first time 2γ induced multihadron production
in a wide range of center of mass energies (1<W<8 GeV). For this study
hadronic events were defined by

a) three or more tracks in the central detector, or

b) two tracks in the central detector and at least one shower which
   is not associated with the tracks ($E_{neutral}$ >350 MeV, $|\cos\theta^n|$<0.997).
   The two tracks are defined by the following conditions:
   $|\cos\theta_1^h|$ < 0.743, $p_{\perp_1}$ > 300 MeV; $|\cos\theta_2^h|$ < 0.820, $p_{\perp_2}$ > 80 MeV,
   where $p_{\perp}$ is the transverse momentum, θ^n and θ^h are the polar angles
   of showers and hadrons relative to the beam axis.

The minimum energy required in the SAT was 4GeV. We have taken data at
low beam energies (43nb^{-1} at E = 6.5 GeV and 88nb^{-1} at E = 8.5 GeV) and
high beam energies (2700nb^{-1} at $\bar{E}$ = 15 GeV). The vertex distribution
(fig. 8) of the central detector tracks for our high energy data sample
shows a very clear peak, thus proving that the beam gas background in
these data is small ($\leq$20%). The distribution of the total energy (fig.
9) of the background subtracted tagged events is again (cf fig.5a)
characteristic of 2γ reactions. None of these events survives our
standard cuts [27]   for e^+e^- annihilation into multihadrons.
Following the formalism described in chapter I.1 we are studying electron photon scattering at moderate Q^2 values. Our low energy data
(52 $\pm$ 9 events) correspond to an average Q^2 of .1 GeV^2 and our high
energy data (565 $\pm$ 26) to an average Q^2 of .25 GeV^2.
We have calculated the detection efficiency via a Monte Carlo program
simulating the electron photon scattering process in the PLUTO detector.
For the hadronic part of the cross section (σ_t + $\epsilon\sigma_l$) we have used a
multipion phase space model with limited transverse momentum and constant production rate, independent of Q^2 and W^2. One should note, that
the polarization parameter ϵ is practically constant ($\epsilon \approx 0.9$) in the
kinematical range discussed in this report. The transverse momentum
distribution has been taken from the data (fig.10) which can be very
well described by a superposition of two exponentials

$$\frac{dN}{dp_{\perp}^2} \sim e^{-5p_{\perp}^2} \quad \text{and} \quad \frac{dN}{dp_{\perp}^2} \sim e^{-1p_{\perp}^2}$$

The charged multiplicity has been taken from low energy e^+e^- annihilation experiments, n_{ch} = 2+0.7 ln W^2 and the ratio $n_{ch}/n_{neutral}$ has
been assumed to be 2:1.
The measured charged multiplicity distribution (fig.11) n_{vis}^{ch} versus

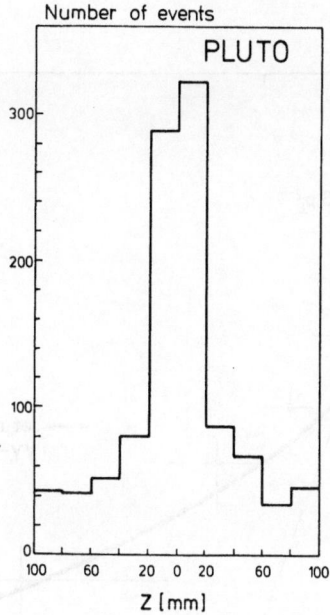

Fig.8 Vertex distribution of
multi-hadron events.

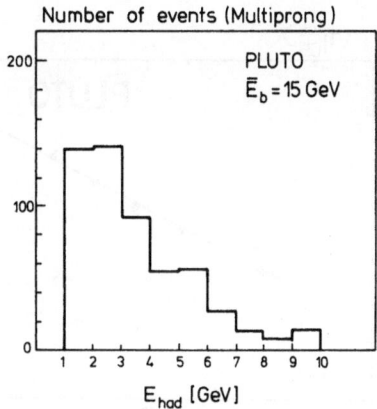

Fig.9 Total energy distribution
of multi-hadron events in
the central detector.

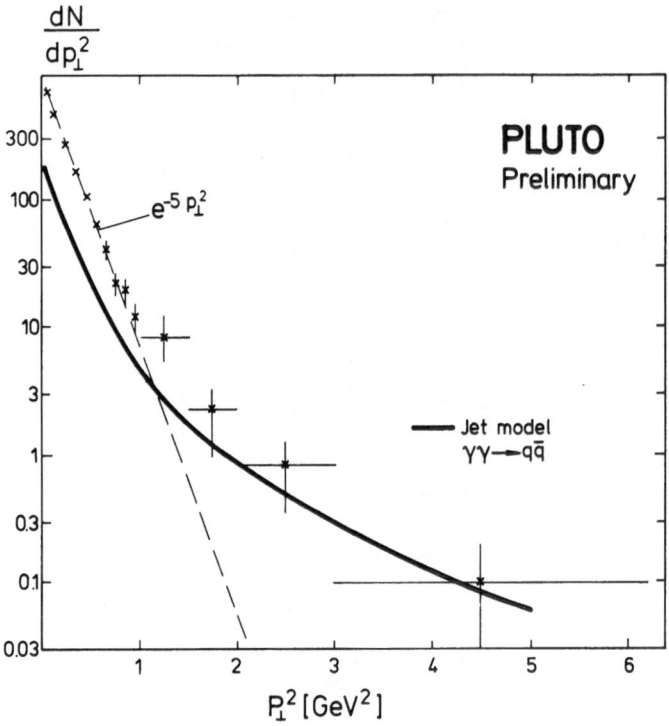

Fig.10   Inclusive $p_\perp^2$ distribution of hadronic tracks.

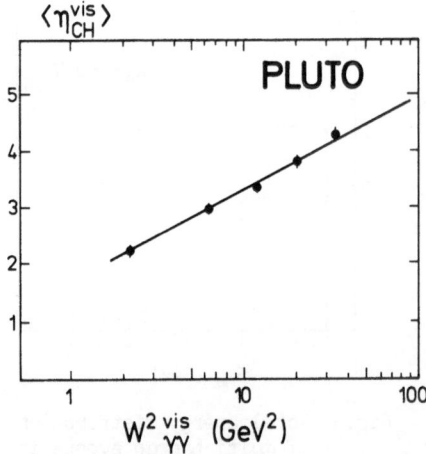

Fig.11   Visible charged multiplicity
versus the visible invariant
mass squared.

W_{vis} is well described by the Monte Carlo simulation (solid line in
fig.11), indicating again, that multiplicities are rather independent of the production mechanism. W_{vis} is the invariant mass of the
hadronic system as determined in the central detector assuming pion
masses for all particles.

The cross section $\sigma_t + \epsilon\sigma_1$ versus W_{vis} at $Q^2 = 0.1$ GeV^2 (E = 6.5,
8.5 GeV) and at $Q^2 = .25$ GeV^2 ($\bar{E}$ = 15 GeV) is shown in figs. 12 and
13. Due to detection inefficiencies W_{vis} is smaller than W, especially
for the lower bins. The range of W that contributes to a certain W_{vis}
bin has been calculated in the Monte Carlo program. If this range
is larger than the bin width of W_{vis}, this is indicated by the dashed
horizontal bars in the figs. 12 and 13. Besides the statistical error,
we estimate an overall systematic error in the measured cross section
of   25%, mainly coming from the uncertainty in the acceptance calculation.

Towards higher values of W the cross section seems to reach an 'asymptotic' value of 170 nb at $Q^2 = .1$ GeV^2 and 100 nb at $Q^2 = .25$ GeV^2. This
is in agreement with the expectation based on factorization alone,

$$\sigma_{\gamma\gamma} = \sigma_t(0,W) = (\sigma_{\gamma p})^2/\sigma_{pp} = 240\text{nb}$$

if one takes into account a ρ form factor ansatz for the Q^2 dependence

$$\sigma_t(Q^2,W) = \sigma_{\gamma\gamma} \left(\frac{1}{1+Q^2/m_\rho 2} \right)^2 \tag{18}$$

Regge models[29] predict an increase of the cross section at lower invariant masses W

$$\sigma_{\gamma\gamma} = 240 \text{ nb} + 270 \text{ nb GeV/W} \tag{19}$$

The solid line in fig. 13 is calculated with this ansatz including the
detector resolution, the 'ρ like' form factor (18) and setting σ_1=0.
The observed increase in the cross section is steeper than this prediction in the $Q^2 = .1$ GeV^2 data as well as in the $Q^2 = .25$ GeV^2 data.
In principle this excess could be attributed to contributions from σ_1,
which we cannot separate experimentally at present. On the other hand
we know from electroproduction, that the ratio σ_1/σ_t is rather small
(15-20%). It has been argued[30], that one should include the quark box
diagram (fig. 14) in the evaluation of $\sigma_{\gamma\gamma}$. Its contribution can be
estimated very roughly by

$$\sigma_{\gamma\gamma} = \frac{4\pi\alpha^2}{W^2} \sum_{\substack{\text{color}\\\text{flavor}}} q_i^4 \ln(W^2/m_q^2)$$

$\sigma_t + \varepsilon\sigma_l$ (nb)

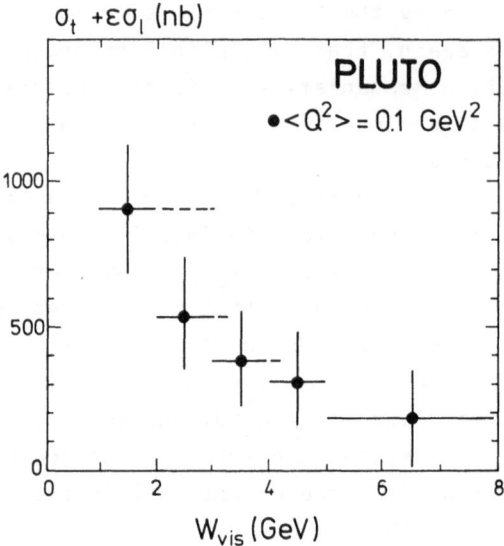

PLUTO

$\bullet <Q^2> = 0.1$ GeV^2

W_{vis} (GeV)

Fig.12 The total cross section taken
at small beam energies.

$\sigma_t + \varepsilon\sigma_l$ (nb)

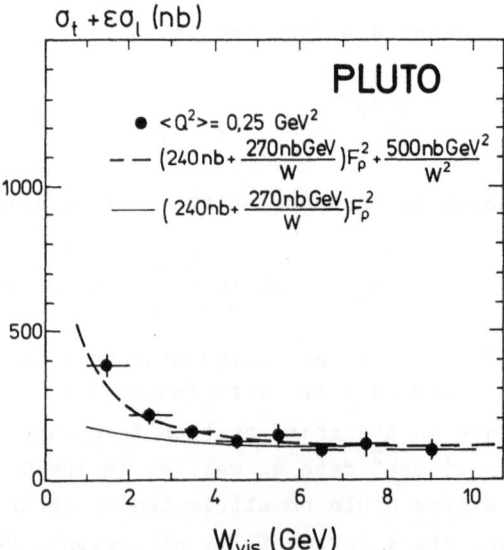

PLUTO

$\bullet \quad <Q^2> = 0,25$ GeV^2

$-- \left(240\,nb + \dfrac{270\,nb\,GeV}{W}\right)F_\rho^2 + \dfrac{500\,nb\,GeV^2}{W^2}$

$- \left(240\,nb + \dfrac{270\,nb\,GeV}{W}\right)F_\rho^2$

W_{vis} (GeV)

Fig.13 The total cross section
taken at high beam energies.

The dashed line in fig.13 has been calculated using m_q = 300 MeV. The agreement of the data with the theoretical prediction is improved considerably by the addition of the box diagram term. Because of the principle significance of the isolation of a pointlike contribution to $\gamma\gamma$ scattering, an accurate measurement of the cross section at low values of W is very important. On the other hand this is extremely difficult, because the detection efficiency is very small and crucially depending on model assumptions, especially at high beam energies. Our data taken at E = 15 GeV are still preliminary and the low Q^2 data suffer from the limited rate (52 events). Thus much more work is needed to establish the effect.

IV.2 Hard scattering subprocesses

Another way of isolating the pointlike structure of $\gamma\gamma$ reactions is to search for specific hard scattering subprocesses. Basically we have two classes of those processes

a) $\gamma\gamma \to q\bar{q}$ (Fig. 15a)

The photons are almost real (say Q^2 < .3 GeV^2 in a single tag experiment), but the internal quark propagator is far from mass shell. Experimentally this process should be observed as two photon initiated quark antiquark jets,with a rather large $p_\perp$ (say > 2 GeV) with respect to the beam pipe. The ratio of these jet events to 2γ initiated μ pair production is given by QCD [31] to

$$R_{\gamma\gamma} = \frac{\sigma_{\gamma\gamma \to q\bar{q}}}{\sigma_{\gamma\gamma \to \mu\mu}} = \sum e_q^4 (1+O(\alpha_s)) \qquad (20)$$
color
flavor

Neglecting the second term in the bracket and including charmed quarks we have

$$R = \frac{34}{27} \sim 1.2$$

Hadron production via large $p_\perp$ jets should show up as a tail ($\sim \frac{1}{p_\perp^4}$) in the inclusive $p_\perp^2$ distribution $\frac{dN}{dp_\perp^2}$ (fig.10).
Below $p_\perp^2 \approx 1$ GeV^2 this distribution is very well described by an exponential $e^{-5p_\perp^2}$ (dashed line). Above $p_\perp^2$ = 1.5 GeV^2 the data are not at all represented by an extrapolation of the low $p_\perp^2$ fit. The solid line is the result of a Monte-Carlo simulation of the process $\gamma\gamma \to q\bar{q}$ including the quark fragmentation into hadrons and detector resolution effects. The agreement between this absolute model prediction and the data (within the statistical and systematic limitations) is remarkable. There is probably room for the contribution of the $p_\perp^{-6}$ subprocess (fig. 15b) $\gamma\gamma \to q\bar{q}$ meson, which is considered to be the dominating process in our $p_\perp$ range by the authors of ref. 31.

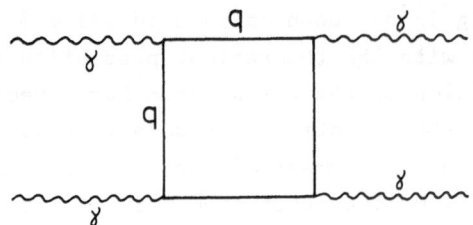

Fig.14 The quark box scattering diagram

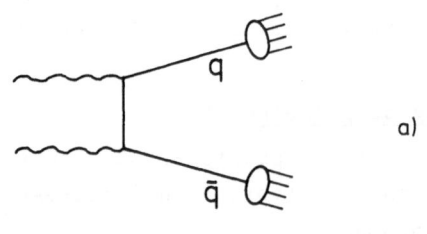

a)

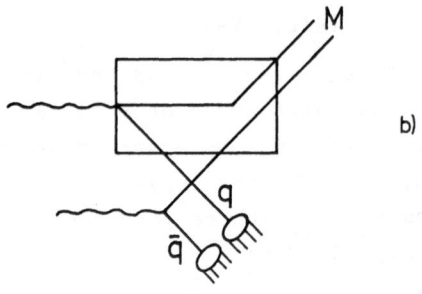

b)

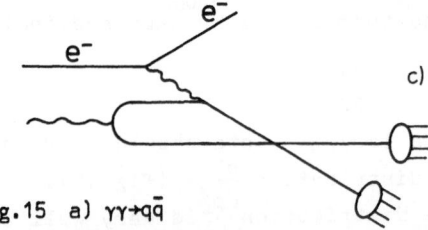

c)

Fig.15 a) $\gamma\gamma \rightarrow q\bar{q}$

b) $\gamma\gamma \rightarrow q\bar{q}$ meson

c) deep inelastic electron photon
scattering

We have also searched for single 2 jet events, i.e. for two jet
events, which are coplanar (with the incoming beams) but not collinear
because of the Lorentz boost. We have found a couple of those events
consistent in rate with the prediction from eq. (20).A particularly
nice example is shown in fig. 16

b) Deep inelastic electron photon scattering (fig. 15c). One photon
is almost real, but the other photon has a rather large mass (say
$Q^2 > 1$ GeV^2). There is no restriction on the $p_\perp$ of the outgoing hadrons.
As can be seen from fig. 15c this process is analogous to the well
known deep inelastic lepton nucleon scattering, the target nucleon
beeing replaced by a real photon! Using the Callan Gross relation

$$F_1 = F_2/2 \ x$$

formula (8) reduces to

$$\frac{d\sigma}{dxdy} = \frac{8\pi\alpha^2}{Q^4} \ E E_\gamma \ (1+(1-y)^2) \ F_2 \ (x,Q^2)$$

There are two quite orthogonal predictions for F_2. If the real photon
is hadronlike (Vector meson dominance) F_2 should drop to zero at large
x values, typically

$$F_2^\gamma = \frac{4\pi\alpha}{f\rho^2} \cdot \frac{2}{3} F_{2,proton}$$

If on the other hand the photon dissociates into a free quark pair
on can calculate the structure function in the Born approximation of
QCD, i.e. the quark model[31].

$$F_2^{\gamma \ Born} = \frac{3\alpha}{\pi} \sum_{flavor} e_q^4 \cdot x \left\{ x^2+(1-x)^2 \right\} \ln \frac{W^2}{4m_q^2}$$

Note that $F_2^{\gamma Born}$ is at maximum for x → 1!
The higher order QCD corrections to the quark model are quite large[32]
and a precise measurement of F_2 will be very interesting in the future.
It is easy to show[3] that for scaling structure functions $F_2 = F_2(x)$
the Q^4 weighted total cross section should scale with the beam energy
squared.

$$\int_{\Delta x \Delta y} Q^4 \ \frac{d\sigma}{dxdy} \ dxdy \sim E^2$$

In a first attempt to analyse our LAT data sample (110 single tag
events with $1 < Q^2 < 15$ GeV^2) we have plotted this quantity versus E^2
(fig. 17). The almost linear increase of these very preliminary data
is striking. Note in addition that the data are well described by the
quark model curve. Thus we hope very much to provide in the near
future a first measurement of the structure function of the photon.

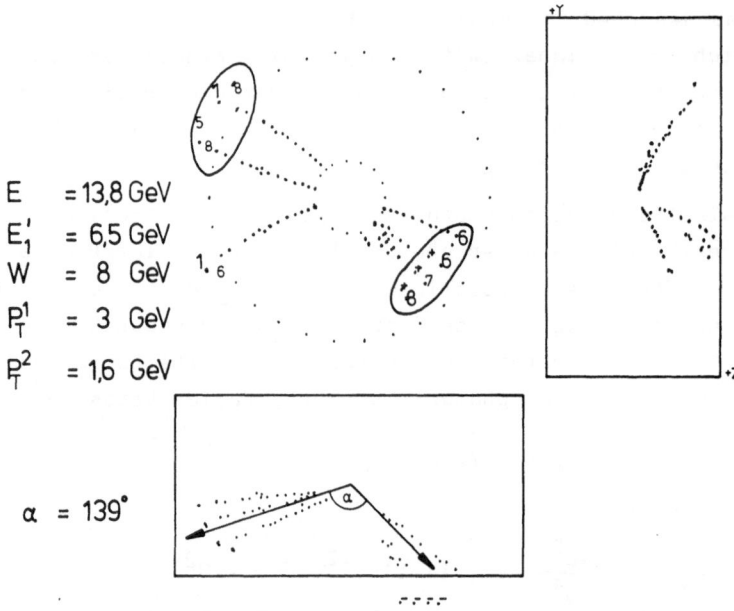

$E \quad = 13,8$ GeV

$E_1' \quad = 6,5$ GeV

$W \quad = 8$ GeV

$P_T^1 \quad = 3$ GeV

$P_T^2 \quad = 1,6$ GeV

$\alpha \quad = 139°$

Fig.16 Example of a two photon initiated
two jet event.

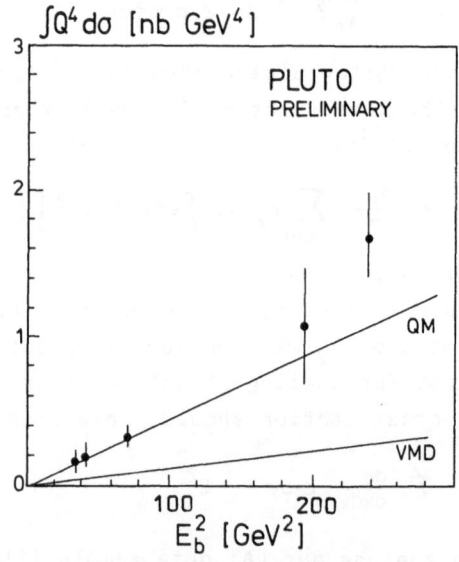

Fig.17 The Q^4 weighted total cross section
versus the beam energy squared.

Conclusions

Two photon experiments can be analyzed in the no-tag and single-tag
mode. Single tag data are described in terms of electron photon scattering with completely factorizing flux factors if a proper definition of
the cross section is used and the scattering angle of one lepton is
kept small (antitagging!)
The mass spectrum of charged particle pairs above W = 1.5 GeV agrees
fully with QED expectations.
The radiative width of the f^o (1270) meson has been determined for the
first time. The total cross section for hadron production in two photon
reactions agrees essentially with expectations from Regge like models.
Deviations at small invariant masses can be interpreted in terms of
a pointlike contribution.
The tail in the inclusive $p_\perp$ distribution agrees well with the prediction from the hard scattering subprocess $\gamma\gamma \rightarrow q\bar{q}$. A couple of two
photon initiated jet like events has been observed for the first time.

We have about 100 event with a rather large Q^2. Analyzed in terms of
deep inelastic electron photon scattering the data are consistent with
scaling structure functions.

It is a pleasure to thank my PLUTO colleagues for many very helpful
discussions. I have enjoyed the nice atmosphere at the Conference and
want to thank the organizers especially Prof. P. Kessler for all the
work they did.

REFERENCES

1. H. Euler and B. Kockel, Naturwiss. 23 (1935) 246

2. G. Bonneau, M. Gourdin and F. Martin, Nucl. Phys. B54 (1973)573
    and V.M. Budnev, I.F. Ginzburg, G.V. Meledin, V.G. Serbo,
    Phys. Reports 15C (1975) 181

3. P. Zerwas, Phys. Rev. D10 (1974) 1485

4. N. Arteago Romero, A. Jaccarini and P. Kessler
    C.R. Acad. Sci. B129 (1969) 153 and C.R. Acad. Sci. B269 (1969)
    1129, V.E. Balakin, V.M. Budnev and I.F. Ginzburg,
    Zh. E.T.F. Pisma 11 (1970) 559 (JETP Lett. 11, 388),
    S.J. Brodsky, T. Kinoshita and H. Terazawa, Phys. Rev. Lett. 25
    (1970) 972

5. C. Carimalo, P. Kessler and J. Parisi, Phys. Rev. D20 (1979)1057

6. C. Carimalo, P. Kessler and J. Parisi, Phys. Rev. D21 (1980) 669

7. J.H. Field, Luminosity functions for two photon processes in
    e^+e^- collisions, DESY 79/78

8. PLUTO Collaboration Ch. Berger et.al., Phys. Lett. 89B (1979) 120

9. J.A.M. Vermaseren, private communication and R. Bhattacharya,
    J. Smith, G. Grammer, Phys. Rev. D15 (1977) 3267

10. S.J. Brodsky, T. Kinoshita and H. Terazawa, Phys. Rev. D4 (1971)
    1532

11. G. Cochard, S. Ong, Phys. Rev. D19 (1979) 810

12. De Fries, private communication and discussion session at the
    International Workshop on γγ Interactions, Amiens (1980)

13. Particle data group C. Bricman et al., Phys. Lett. 75B (1978)

14. J. Babcock, J.L. Rosner, Phys. Rev. D14 (1976) 1286

15. A. Bramon, M. Greco, Lett. Nuovo Cimento 2 (1971) 522

16. D. Faiman, H.J. Lipkin, H.R. Rubinstein, Phys. Lett. 59B (1979)
    269

17. S.B. Berger, B.T. Feld, Phys. Rev. D8 (1973) 3875

18. V.M. Budnev, A.E. Kaloshin, Phys. Lett. 86B (1979) 351

19. N. Levy, P. Singer, S. Toaff, Phys. Rev. D13 (1976) 2662

20. H. Kleinert, L.P. Staunton, P.H. Weisz, Nucl. Phys. B38 (1972)87

21. B. Schrempp-Otto, F. Schrempp, T.F. Walsh, Phys. Lett. B36 (1971)
    463

22. B. Renner, Nucl. Phys. B30 (1971) 634

23. M. Greco, Y. Srivastava, Nuovo Cimento 43A (1978) 88

24. G. Schierholz, K. Sundermeyer, Nucl. Phys. B40 (1972) 125

25. V.N. Novikov, S.I. Eidelman, Soviet Journal of Nuclear Physics 21 (1975) 529

26. G.M. Radutzkij, Soviet Journal of Nuclear Physics 8 (1969) 65

27. PLUTO Collaboration, Ch. Berger et al. Phys. Lett. 86B (1979) 413

28. See G. Wolf, Proc. 1979 EPS Int. Conf. on High En. Phys. Geneva 1979 rapporteur talk and DESY 79/41

29. T. Walsh, J. Physique C2 Suppl.3 (1974) and J.L. Rosner in ISABELLE Physics Prospects, BNL Report 17522 (1972) 316

30. M. Greco, Y. Srivastava, Nuovo Cim. 43A (1978) 88
    See also the earlier work of S.J. Brodsky, F.E. Close and J.F. Gunion Phys. Rev. D6 (1972) 177, and T.F. Walsh and P. Zerwas Phys. Lett. 44B (1973) 195

31. S.J. Brodsky, T. De Grand, J. Gunion and J. Weis, Phys. Rev. D19 (1979) 1418

32. E. Witten, Nucl. Phys. B120 (1977) 189

TWO PHOTON RESULTS FROM TASSO

TASSO Collaboration

Aachen, Bonn, DESY, Hamburg, Imperial College, Rutherford, Oxford, Weizmann, Wisconsin [*]

presented by E. Hilger,
Physikalisches Institut der Universität Bonn, Nussallee 12, D-5300 Bonn

Summary

Two photon annihilation processes were observed with the TASSO detector at the e^+e^- storage ring PETRA. We report on the analyses of untagged and single tag events.

We select untagged events with a two prong final state seen in the central detector and compare these data to the prediction for two photon QED processes. We observe an excess of events in the expected invariant mass range for the f^o. The single tag events with two prongs seen in the central detector show a very good agreement to the QED prediction in all measured distributions. Multihadronic events with a single tag are interpreted in terms of electron photon scattering. The observed transverse momentum distribution of the hadrons exhibits an exponential fall-off at lower momenta up to about 1 GeV/c; beyond this the distribution flattens off. This may point to two different couplings of the photon to hadrons, a VDM-like and a pointlike coupling. From an overall fit to the measured distributions of these data we derive mean charged multiplicities and cross sections for hadron production as a function of the total energy of the two photon system.

After the early work on photon-photon scattering in the mid thirties /1/ the relevance of two photon physics was emphasized in 1960 /2/ and taken up again about ten years later by many people /3/.
Now a rising interest in two photon processes can be observed at the high energy e^+e^- storage rings.
At PETRA the TASSO experiment, for which the study of two photon collisions so far was not among the top priority issues on the physics menu, lately has increased its efforts in this field.

This is a report on the status of the investigations in two photon physics at TASSO. It is based on data mainly taken in the fall of 1979 at beam energies around 15 GeV. Analysis is still in progress and most of the results reported here are preliminary.

The two photon mechanism at e^+e^- machines is illustrated in fig.1. The process observed in the experiment is of the type

$$e^+e^- \rightarrow e^+e^- + X, \tag{1}$$

which is related in a non-trivial way /2/ to the genuine two photon reaction

$$\gamma\gamma \rightarrow X, \tag{2}$$

where X is any produced system of invariant mass M_X.
A specific signature of reaction (1) is the occurrence of an electron and a positron in the final state, both peaked at high energies and extreme forward angles.
Also distinctly different from the one photon annihilation processes is the distribution of the total energy of the produced system X. For two photon events this distribution rises steeply toward zero energy while for annihilation events it peaks in principle at twice the beam energy. Both characteristic features are used in TASSO to select $\gamma\gamma$ reactions.

Fig. 2 shows a cut through the TASSO detector along the beam axis. The central part consists of a large magnetic solenoid, the volume of which is filled with tracking chambers, a cylindrical proportional chamber surrounded by a cylindrical drift chamber /5/, and scintillation counters used in the trigger and to measure time of flight. Only this inner part, details of which can be found in /6/, was used in the present study of particle systems X produced in two photon collisions.
On either side of the central part further out along the beam pipe there are forward detectors. They measure the luminosity and allow to detect the scattered electrons and positrons from two photon processes.

Fig. 3 is a view at one of the forward detectors along the beam axis. A hodoscope for charged particles consists of 16 scintillation counters covering forward angles between 25 mrad and 60 mrad.

In TASSO the detection of tag-particles is restricted to this angular range by the hole in the end cap yoke of the main detector and by the size of the beam pipe. An array of 36 lead glass shower counters behind the hodoscope measures the energy of electrons and photons. The thickness of these counters is about 12 radiation lengths.

At present the study of two photon processes with TASSO follows two separate roads.

In one analysis the events with no tag-particles observed are used and two photon candidates are selected mainly on the basis of their measured total energy. These events correspond to scattering of almost real photons.

The other analysis considers the events with a single tag in one of the forward detectors. Due to the restricted angular acceptance the tagged photon has a rather small but finite four momentum Q^2 while the untagged photon again is almost real.

The analysis of the untagged events is not completed yet. Its status will be described first.

No special hardware trigger exists for untagged two photon events. But the standard triggers for 1γ annihilation events also accept two photon induced events.

One of these triggers, used for the analysis presented here, was a pair trigger. It demanded two or three charged tracks coplanar with the beam axis to within $27°$ having a minimum momentum transverse to the beam of 0.32 GeV/c. The pair trigger covered azimuthal angles between 0 and 2π and polar angles between $35°$ and $145°$.

Two photon induced two prong events were selected in two steps requiring at first two tracks, which originated from the interaction region, were oppositely charged, and had a time of flight difference of less than 3 ns. These cuts effectively removed all background from cosmic rays and beam gas scattering.

Then in a second step the 1γ annihilation events being mainly lepton pairs were rejected by requiring the two tracks to be non-collinear by more than $10°$ and the sum of their momenta to be less than 6 GeV/c, i.e. less than about 20% of the total energy available in the e^+e^- collision.

A total of 3049 events passing these criteria were found for an integrated luminosity of 3466 nb^{-1} taken at beam energies ranging from 13.7 GeV to 17.6 GeV [+)].

In fig. 4 these data are plotted versus the invariant mass M_X of the 2 prongs assuming they are muons.

Lepton pair production in two photon processes

$$e^+e^- \rightarrow e^+e^- + L^+L^- \tag{3}$$

was simulated in the detector using the event generator written by J.A.M. Vermaseren /7/. For the measured total luminosity electron pair and muon pair production were computed separately and then combined. The resulting M_X distribution, again assuming muon pairs, is also shown in fig. 4.

Above $M_X = 1.5$ GeV the agreement between the data and the Monte Carlo expectation is good. We observe an excess of events in the M_X region below 1.5 GeV, which may indicate a considerable contribution from the f° meson in the sample.

This may also be seen from angular distributions plotted separately for different M_X regions in fig. 5. The angle plotted on the abscissae is the angle between the beam axis and the direction of the observed pairs in their center-of-mass system.

Again, for $M_X > 1.5$ GeV the data are well represented by the 2γ-QED prediction. But for M_X between 1.0 GeV and 1.5 GeV the data by far exceed the expectation from two photon production of lepton pairs.

Fig. 6 is a very preliminary plot showing the difference between the data and the QED expectation from fig. 4. In the mass range of the f° a clear signal is observed. However, at present we do not want to draw any quantitative conclusions since several checks have still to be made as well as refinements to the Monte Carlo. These studies are in progress and should be completed soon.

We now turn to the analysis of the events with a single tag in the forward detector.

Candidates for two photon induced events were selected by a special trigger which demanded an energy of more than 3.5 GeV deposited by a charged particle in one of the forward detectors and at least one track with a momentum transverse to the beam exceeding 0.32 GeV/c in the

central detector. The off-line selection required at least one track
fully reconstructed in space and at least one more track reconstructed
in two dimensions (r,∅). That reduced the roughly 180 000 triggers to
2125 events, which then were scanned by inspection on a graphics termi-
nal. Track pairs from converted photons were removed and obvious background events rejected.

The analysis procedure now was split up in two branches.
In order to isolate events from two photon production of lepton pairs
(3) events with just two tracks in the central detector were separated.
The two tracks were required to balance in charge and their common
vertex had to lie within $\Delta z = \pm$ 7.5 cm off the nominal interaction
point, where z is the coordinate along the beam. The two tracks had to
be coplanar to within 45°.
These cuts reduced the background from beam gas scattering and from
multihadronic final states to less than 8 % each.

After background subtraction the final data sample contained 649 candidate events for reaction (3), collected for a total luminosity of
2040 nb^{-1} at beam energies between 14.95 GeV and 15.73 GeV.
Fig. 7 shows an especially remarkable event from this sample since it
is one of the rare specimens with a double tag.
The upper half of the figure shows the event in the central detector in
three views, the lower half shows the hits and deposited energies in
the forward detectors. From the insets one can tell that the two track
system is boosted toward +z (EAST), where the tag particle had a lower
energy than the one found in the WEST detector.

In the following three figures the data are compared to the absolute
expectation for two photon production of charged lepton pairs. The
predicted distributions were calculated using again the exact treatment
of the matrix element by J.A.M. Vermaseren /7/ in a Monte Carlo simulation of the acceptance. In fig. 8a the distribution of p_T is shown,
where p_T is the momentum of a track transverse to the beam axis, while
fig. 8b shows this transverse momentum squared. Fig. 9 is a plot of
Σp_z, the sum of the momentum components of the two tracks along the
beam axis.
Fig. 10 finally gives the invariant mass distribution of the events
assuming pion masses.

In all four plots the data agree well to the absolute expectation from two photon QED processes proving that event selection and acceptance calculation are properly understood.

The invariant mass plot for the single tag two prongs does not show much evidence for the f^o meson. This we view as a consequence of the mean Q^2 of the tagged photon being 0.26 GeV^2. This finite Q^2 may result in a form factor suppression for f^o production. It also also causes a different helicity structure, which may lead to an additional suppression factor.

The second branch of the analysis of single tag events concerned the production of multihadronic final states in two photon scattering.

Out of the 2125 preselected single tag events mentioned above candidates for the reaction

$$e^+e^- \rightarrow e^+e^- + \text{hadrons} \tag{4}$$

were selected requiring at least three tracks in the central detector. Two of these had to be fully reconstructed in space and one more reconstructed in two dimensions $(r, \emptyset)$, and not all three tracks were allowed to have the same charge sign.

Fig. 11 shows a vertex distribution of the events satisfying these conditions. A cut at $z = \pm 7.5$ cm left a background from beam gas events of about 12% in the sample which had to be subtracted statistically.

Fig. 12 presents one event from the final data sample containing 397 events collected for an integrated luminosity of 2280 nb^{-1} at beam energies between 13.7 and 15.73 GeV.

Fig. 13 shows the observed distribution of p_T^2, the squared momentum transverse to the beam axis, on a logarithmic plot. For small p_T^2 up to about 1 (GeV/c)^2 the data show an exact exponential fall-off. Above 1 (GeV/c)^2 we observe an excess over this exponential fall-off, which may indicate a contribution to process (4) from a pointlike coupling of the photon.

We have simulated the process (4) in the TASSO detector with Monte Carlo methods.

In order to extract cross sections for the reaction

$$\gamma\gamma \rightarrow \text{hadrons} \tag{5}$$

from the measured data we followed the approach recently discussed by Carimalo, Kessler and Parisi /4/ and used flux factors, or Weizsäcker-Williams approximation formulae, different for the tagged and the untagged photon. The cross section for reaction (5) was parameterized as in Regge-like models /8/ including a form factor ansatz for the virtual photon:

$$\sigma_{\gamma\gamma}(W_{\gamma\gamma}; Q_1^2, Q_2^2 = 0) = (A + B/W_{\gamma\gamma}) \cdot \left[\frac{1}{1 + Q_1^2/m_\rho^2} \right]^2 \tag{6}$$

The hadronic system was generated according to a multipion phase space model with limited transverse momentum relative to the photon-photon axis,

$$< p_T^{\gamma\gamma} > = C, \tag{7}$$

and a ratio of charged to neutral particles of 2:1. The mean charged multiplicity was assumed to be linear in $\ln W_{\gamma\gamma}$ and parameterized as

$$< n_{ch} > = D + E \cdot \ln W_{\gamma\gamma}. \tag{8}$$

The events generated according to this model were subjected to the acceptance of the detector, the trigger conditions, the efficiencies of the trigger and the track reconstruction, and the offline cuts.
The resulting distributions of observable quantities were compared to the data. The five parameters A to E were simultaneously fitted to three measured distributions, the squared momentum transverse to the beam axis [but so far only up to 0.9 (GeV/c)^2] , the charged multiplicity, and the visible energy.
In an iterative procedure the χ^2 was minimized in the five parameter space. A minimum was found for the following set of parameters

$$< p_T^{\gamma\gamma} > = 0.295 \text{ GeV/c} \tag{9}$$
$$< n_{ch} > = 1.9 + 1.5 \ln \left[W_{\gamma\gamma}(\text{GeV}) \right]$$
$$\sigma_{\gamma\gamma \to had}(W_{\gamma\gamma}; q_1^2 = q_2^2 = 0) = 300 + 840/\left[W_{\gamma\gamma}(\text{GeV}) \right] \text{ nb}.$$

The corresponding χ^2 values, both individual and total, were reasonably small

$$\chi^2{}_{p_T^2}/\text{d.o.fr.} = 1.2$$
$$\chi^2{}_{n_{ch}}/\text{d.o.fr.} = 1.2 \tag{10}$$
$$\chi^2{}_{W_{vis}}/\text{d.o.fr.} = 0.6$$
$$\chi^2{}_{total}/\text{d.o.fr.} = 1.0.$$

The result of the Monte Carlo simulation with the optimized set of parameters is also shown in fig. 13 and in the following figures. The excess at p_T^2 larger than 1 (GeV/c)^2 is not yet accounted for in our present model. This may give rise to appreciable modifications in the set of parameters (9).

The charged multiplicity distribution is presented in fig. 14. The agreement between data and simulation is very good. The same statement can be made for the distribution of visible energies plotted in fig. 15. We conclude from fig. 13 to 17 that the model described above with the set of parameters given in (9) represents our data quite well.

We then take one step further and use our optimized set of parameters to extract information directly on the two photon production process of hadrons.
In fig. 16 we have plotted the mean charged multiplicity for $\gamma\gamma \rightarrow$ hadrons (5). The central line is the result of the best fit $\langle n_{ch} \rangle = 1.9 + 1.5 \ln W_{\gamma\gamma}$, and the hatched area indicates the range up to ± 1 standard deviation as obtained by a careful analysis of the statistical errors in the five parameter space. There is an additional systematic error estimated to be less than $\pm 15\%$.
For comparison we have indicated also the mean charged multiplicities in other processes of hadron production, for one photon annihilation taken from a recent fit /9/ to the low energy region of center-of-mass energies between 1.4 and 7 GeV, as well as the corresponding curves for $\bar{p}p$ annihilation /10/ and pp scattering /11/.

Finally, fig. 17 shows the hadronic cross section for the scattering of two real photons at center-of-mass energies between about 1 and 6 GeV, as obtained by the fit to our data under the assumptions stated above. The central line corresponds to a functional dependence of the cross section on the two photon energy of

$$\sigma_{\gamma\gamma}(W_{\gamma\gamma}; 0,0) = 300 + \frac{840}{W_{\gamma\gamma}(GeV)} \text{ nb.} \tag{10}$$

In addition to the statistical error indicated in the figure there is a systematic uncertainty at present estimated to be less than $\pm 25\%$.

We believe we have demonstrated that our approach of an overall fit to the measured distributions in reactions $e^+e^- \rightarrow e^+e^- +$ hadrons allows to extract information on hadron production in two photon scattering within reasonable errors.
This analysis represents the first attempt to obtain detector independent results on two photon production of hadrons.

Acknowledgement

I am grateful to my colleagues in the TASSO collaboration for numerous discussions, in particular to W. Hillen, P. Leu, H.W. Martyn, M. Wollstadt, and R.J. Wedemeyer. I want to thank Prof. P. Kessler for the invitation and Prof. G. Cochard and the organizers of this Workshop for a most enjoyable meeting at Amiens.

References

/*/ The present members of the TASSO collaboration are
R. Brandelik, W. Braunschweig, K. Gather, V. Kadansky, K. Lübelsmeyer, P. Mättig, H.-U. Martyn, G. Peise, J. Rimkus, H.G. Sander, D. Schmitz, A. Schultz von Dratzig, D. Trines, W. Wallraff, I. Physikalisches Institut der RWTH Aachen, Germany;
H. Boerner, H.M. Fischer, H. Hartmann, E. Hilger, W. Hillen, L. Koepke, H. Kolanoski, G. Knop, P. Leu, B. Löhr, R. Wedemeyer, N. Wermes, M. Wollstadt, Physikalisches Institut der Universität Bonn, Germany;
D.G. Cassel, D. Heyland, H. Hultschig, P. Joos, W. Koch, P. Koehler, U. Kötz, H. Kowalski, A. Ladage, D. Lüke, H.L. Lynch, G. Mikenberg, D. Notz, J. Pyrlik, R. Riethmüller, M. Schliwa, P. Söding, B.H. Wiik, G. Wolf, Deutsches Elektronen-Synchrotron DESY, Hamburg, Germany;
R. Fohrmann, M. Holder, G. Poelz, O. Römer, R. Rüsch, P. Schmüser, II. Institut für Experimentalphysik der Univ. Hamburg, Germany;
D.M. Binnie, P.J. Dornan, N.A. Downie, D.A. Garbutt, W.G. Jones, S.L. Lloyd, D. Pandoulas, J. Sedgbeer, S. Yarker, C. Youngman, Department of Physics, Imperial College London, England;
R.J. Barlow, I. Brock, R.J. Cashmore, R. Devenish, P. Grossmann, J. Illingworth, M. Ogg, B. Roe, G.L. Salmon, T. Wyatt, Department of Nuclear Physics, Oxford University, England;
K.W. Bell, B. Foster, J.C. Hart, J. Proudfoot, D.R. Quarrie, D.H. Saxon, P.L. Woodworth, Rutherford Laboratory, Chilton, England;
Y. Eisenberg, U. Karshon, D. Revel, E. Ronat, A. Shapira, Weizmann Institut, Rehovot, Israel;
J. Freemann, P. Lecomte, T. Meyer, Sau Lan Wu, G. Zobernig, Department of Physics, University of Wisconsin, Madison, Wisconsin, USA;

/+/ This includes data meanwhile analyzed after the Workshop. Correspondingly, figs. 4 to 6 differ from those actually shown at the Workshop both in data and background simulation.

/1/   L.D. Landau, E.M. Lifschitz, Sov. Phys. 6 (1934) 244

/2/   F.E. Low, Phys. Rev. 120 (1960) 582; F. Calogero and C. Zemach,
       Phys. Rev. 120 (1960) 1860

/3/   A. Jaccarini, N. Arteaga-Romero, J. Parisi and P. Kessler, Compt.
       Rend. 269B (1969) 153, 1129; Nuovo Cimento 4 (1970) 933;
       V.E. Balakin, V.M. Budnev and I.F. Ginzburg,
       Zh.E.T.F. Pis'ma 11 (1970) 559 (JETP Lett.II, 388);
       S. Brodsky, T. Kinoshita and H. Terazawa, Phys.
       Rev. Lett.25 (1970) 972; Phys. Rev. D4 (1971) 1532;
       H. Terazawa, Rev. Mod. Phys. 45 (1973) 615

/4/   A. Carimalo, P. Kessler and J. Parisi,
       Phys. Rev. D20 (1979) 1957; Phys. Rev. D20 (1979) 2170; Phys.
       Rev. D21 (1980) 669;
       J. Field, DESY 79/78 (1979), to be published in Nucl. Phys.

/5/   H. Boerner et al, DESY 80/27 (1980), submitted to Nucl. Instr.
       and Meth.

/6/   TASSO Collaboration, R. Brandelik et al., Phys. Lett. 83B
       (1979) 20

/7/   J.A.M. Vermaseren, private communication

/8/   Reports of S. Brodsky and T. Walsh in the Proceedings of the
       Intern. Colloq. on Photon-Photon Collisions, Suppl. au Journal de
       Physique,
       Vol 35 C2 (1974)

/9/   TASSO Collaboration, R. Brandelik et al., Phys.Lett. 89B(1980)418

/10/   W. Thome et al., Nucl. Phys. B129 (1977) 365; see also the review
       by E. Albini, P. Capiluppi, G. Giacomelli, and R.M. Rossi, Nuovo
       Cim. 32A (1976) 101

/11/   Data given by R. Stenbacka et al., Nuovo Cimento 51A (1979) 63

Figure captions

Fig. 1: The two photon mechanism in e^+e^- interaction

Fig. 2: The TASSO detector (cut along the beam axis)

Fig. 3: The TASSO forward detector (view along the beam axis)

Fig. 4: Invariant mass distribution of the candidate events for two photon induced two prongs with no tag observed. This is compared to the expectation for lepton pair production in two photon processes.

Fig. 5: Angular distributions for different invariant mass regions (data and expectation as in fig. 4)

Fig. 6: Preliminary f^o signal obtained from fig. 4

Fig. 7: A double tag event with two charged particles seen in the central detector

Fig. 8: Distributions of p_T and p_T^2 for candidate events for two photon produced charged lepton pairs with a single tag observed (backgrounds subtracted). The curve is the absolute prediction obtained using the event generator written by J.A.M. Vermareren /7/.

Fig. 9: Distribution of the sum of the momentum components along the beam axis of the two tracks in the central detector for the single tag two prong events together with the absolute expectation.

Fig. 10: Invariant mass distribution for the single tag two prong events. The curve is the absolute expectation for two photon produced charged lepton pairs.

Fig. 11: Vertex distribution of the candidates for two photon produced multihadronic final states with a single tag observed.

Fig. 12: Candidate event for $e^+e^- \rightarrow e^+e^- +$ hadrons.

Fig. 13: Distribution of p_T^2 , the squared momentum transverse to the beam axis, for the two photon multihadronic events with a single tag. Also given is the Monte Carlo simulation according to our simplified model described in the text.

Fig. 14: Distribution of the charged multiplicities, observed and simulated

Fig. 15: Distribution of the visible energies, both observed and simulated

Fig. 16: Mean charged multiplicity for $\gamma\gamma \rightarrow$ hadrons vs the c.m. energy of the two photon system as obtained from the best fit. The hatched band indicates the range of $\pm 1\sigma$ in the statistical error. Also indicated are the mean charged multiplicities found in other reactions.

Fig. 17: The hadronic cross section for scattering of two real photons vs the c.m. energy of the two photon system as obtained from the fit to our data under the assumptions stated in the text. The hatched area indicates the range of the statistical error.

Note added in proof

After the Workshop the single tag multiprong events have been reana-
lyzed using the concept of two-photon luminosity functions as described
recently by J.Field /4/. This approach avoids the kinematical approximations of the Weizsäcker-Williams flux factors.

As a consequence especially for low z (z = $W_{\gamma\gamma}$/2E) the photon-photon
luminosity comes out considerably higher.

Correspondingly, a new fit to the data using luminosity functions but
otherwise as described in the report results in a somewhat different
set of parameters. The extrapolated total cross section is lowered,
especially at small $W_{\gamma\gamma}$. It may be described by

$$\sigma_{\gamma\gamma\to had} \ (W_{\gamma\gamma}; q_1^2 = q_2^2 = 0) = 380 + \frac{450}{W_{\gamma\gamma}(GeV)} \ \ nb,$$

while the mean charged multiplicity changes only slightly and the mean
transverse momentum remains unchanged. This latest analysis is also
still preliminary, and the errors are as given in the report.

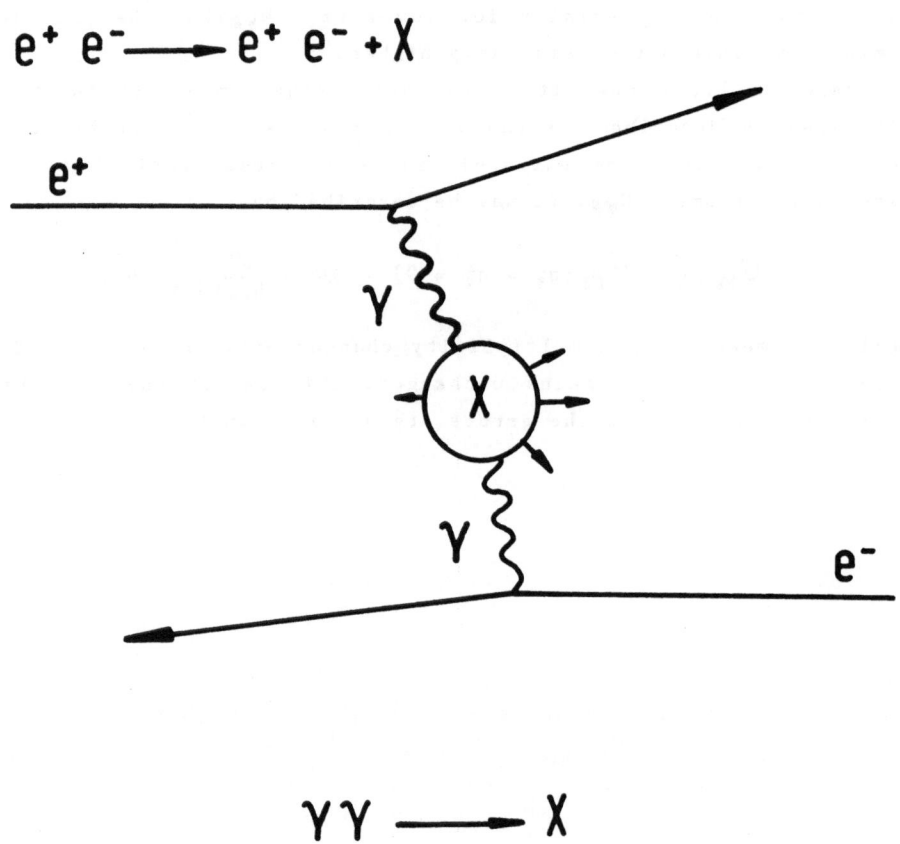

Fig.1

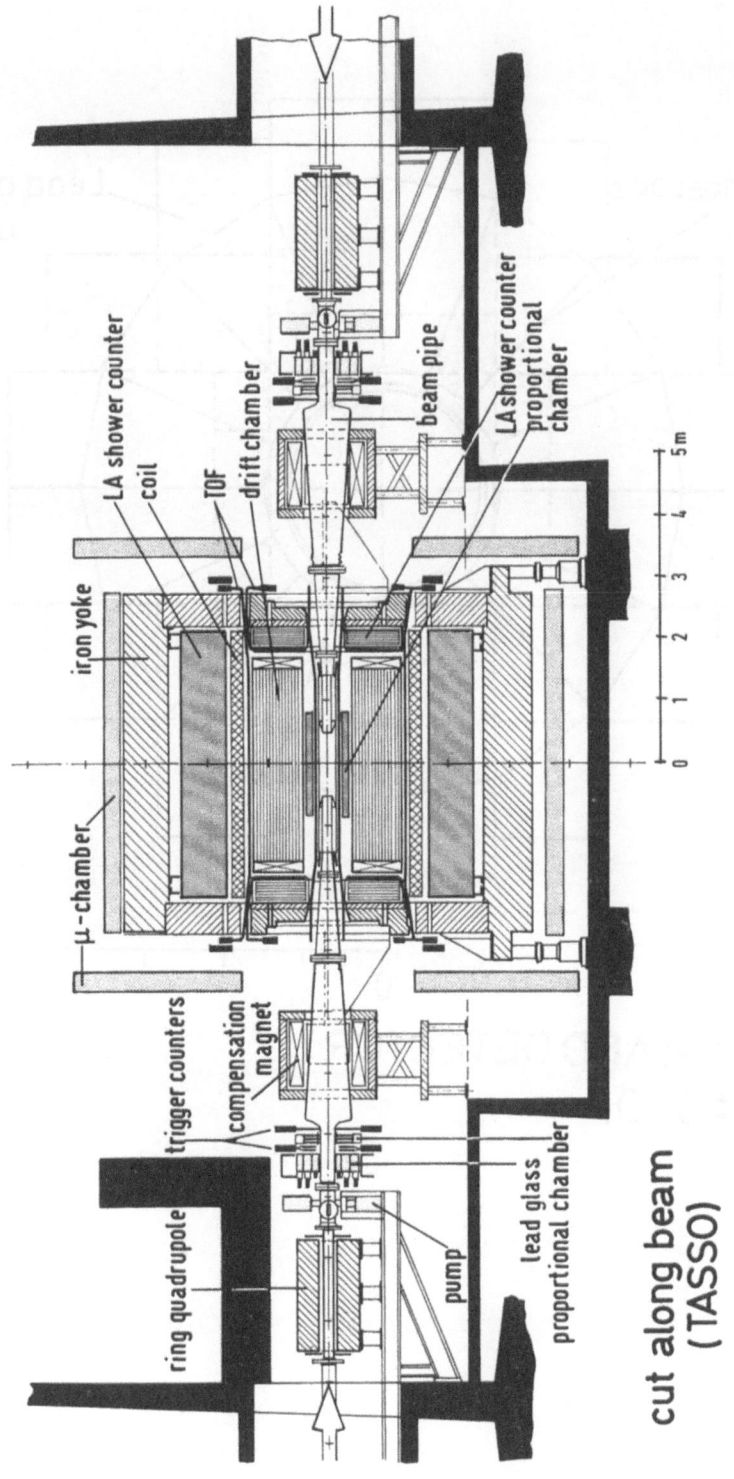

cut along beam
(TASSO)

Fig. 2

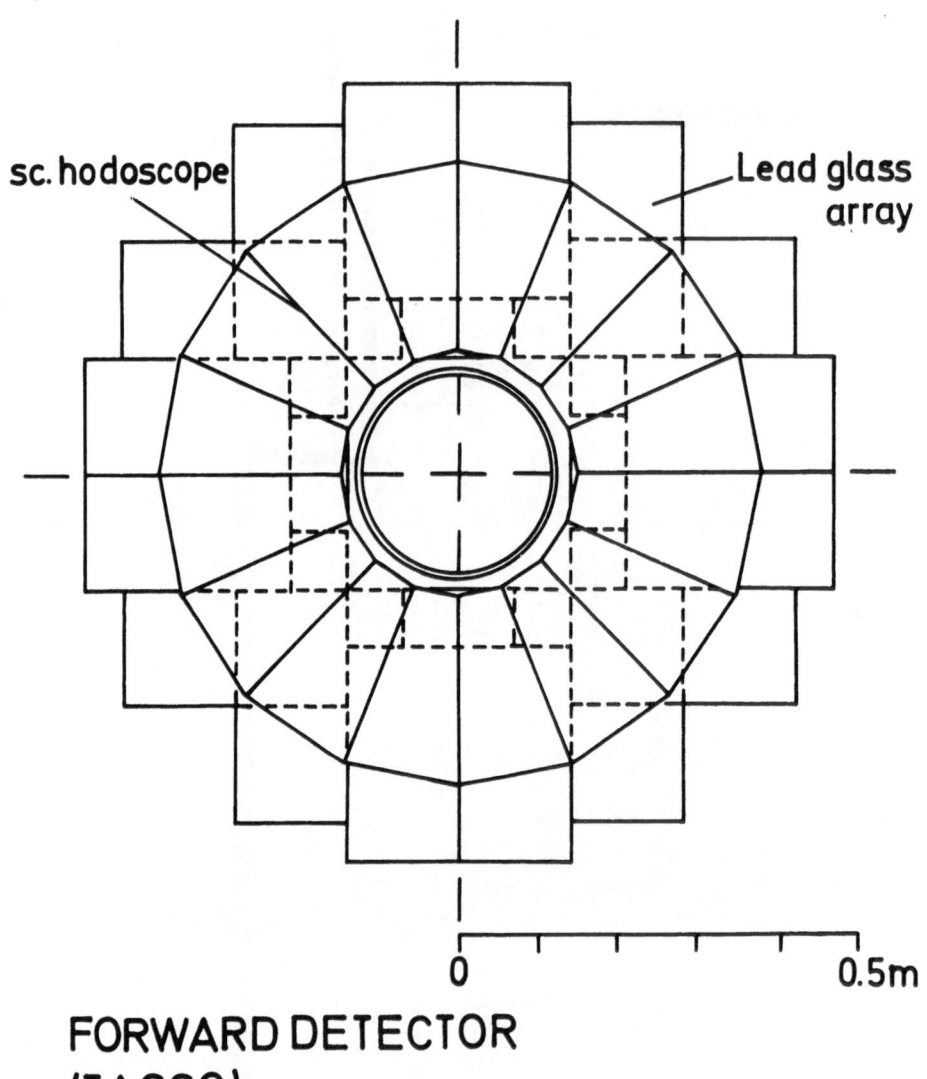

FORWARD DETECTOR
(TASSO)

Fig.3

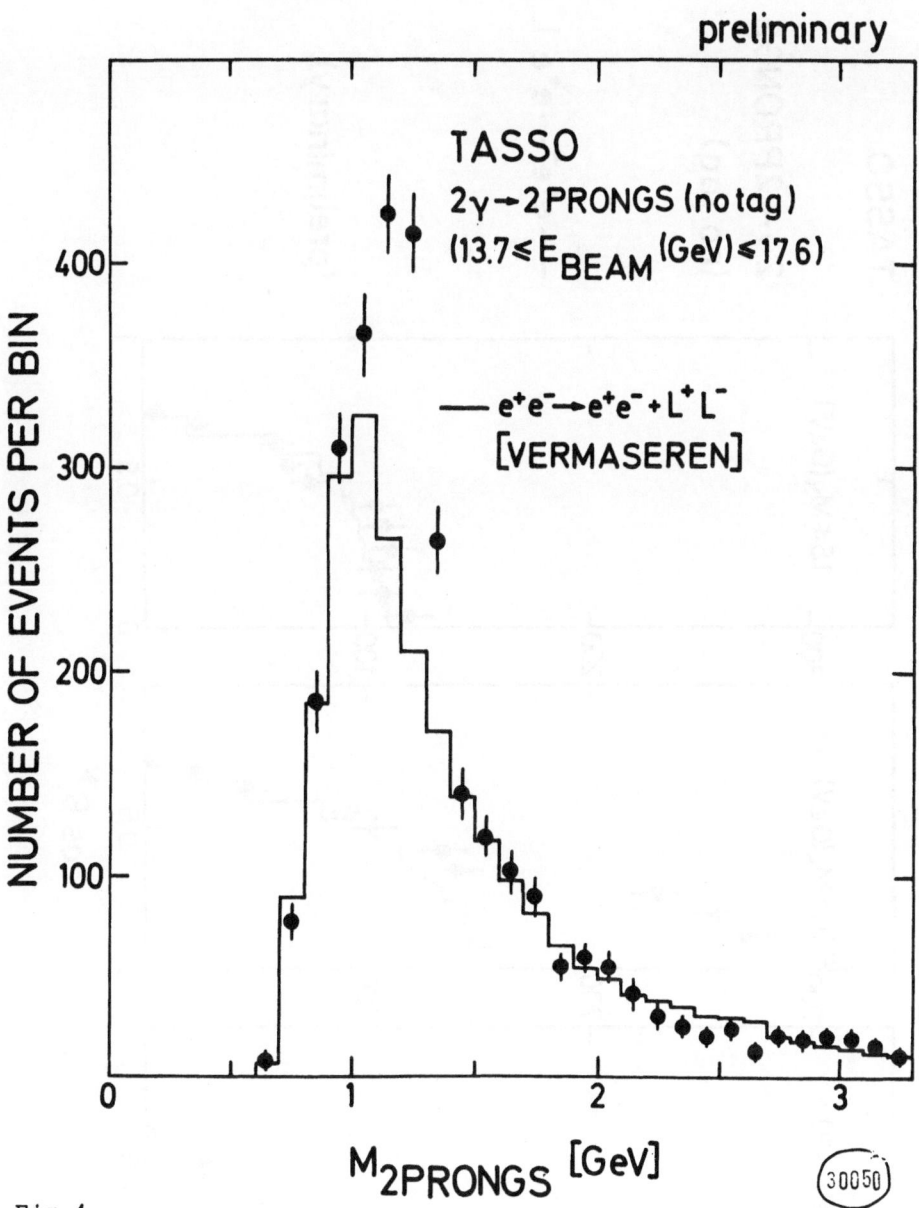

Fig.4

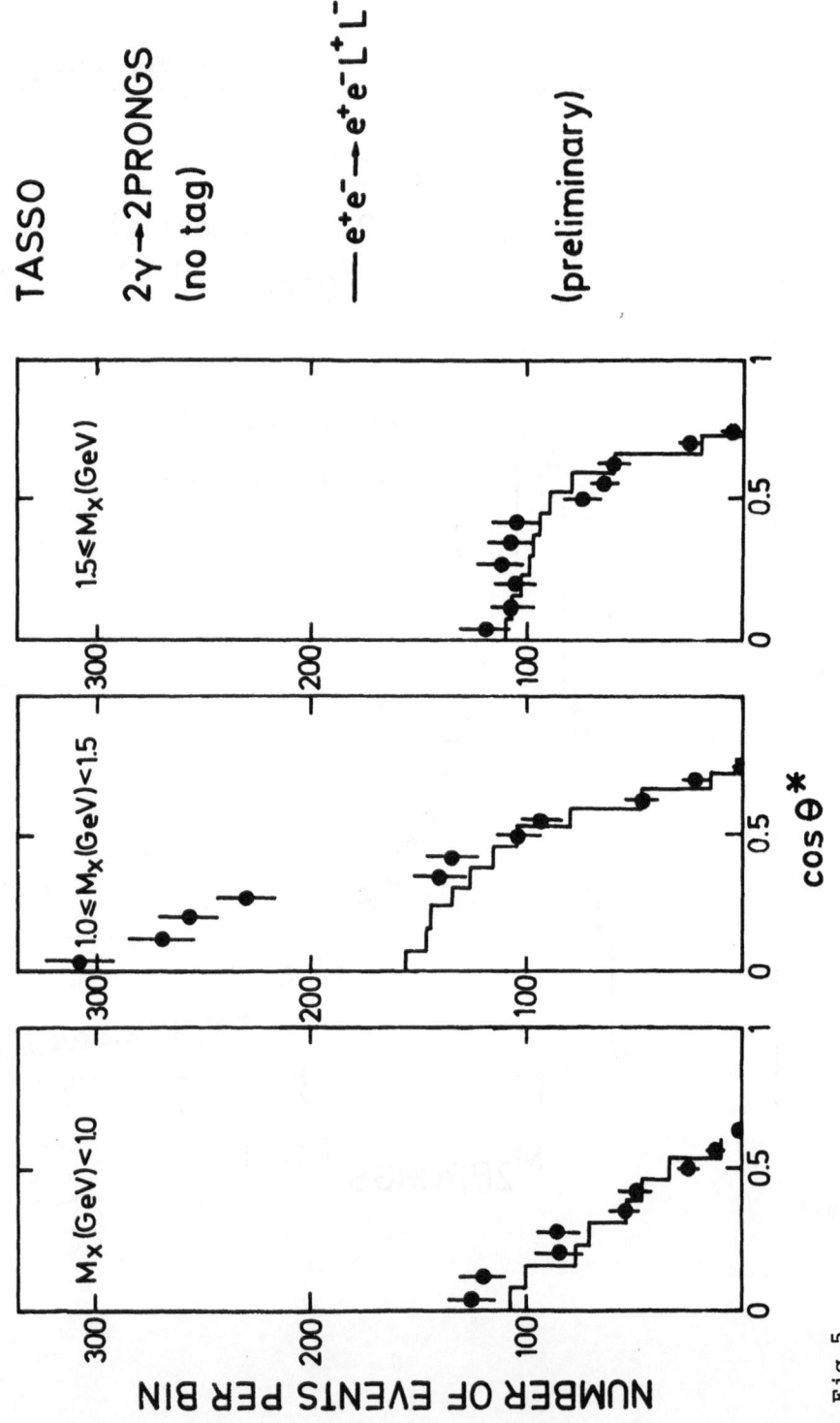

Fig. 5

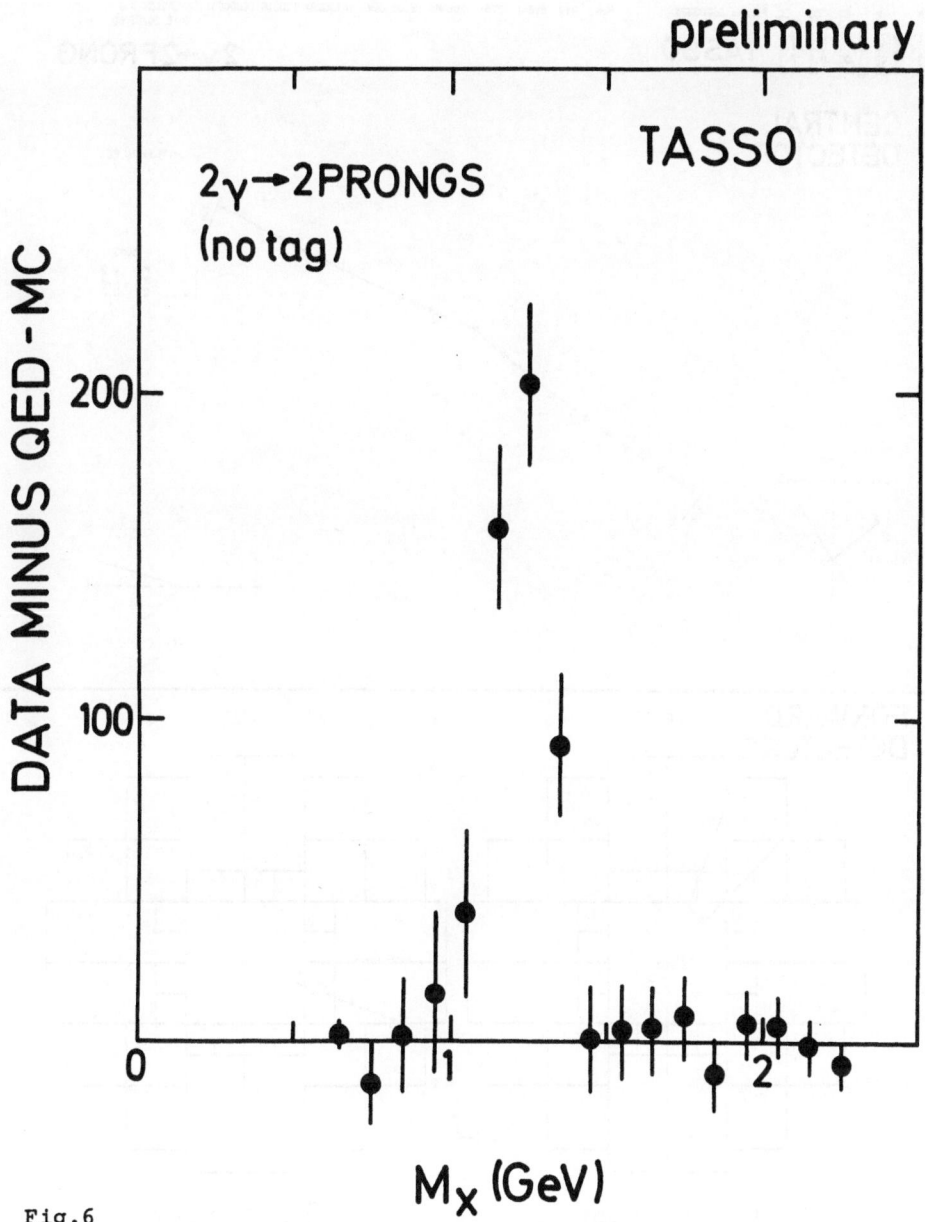

Fig.6

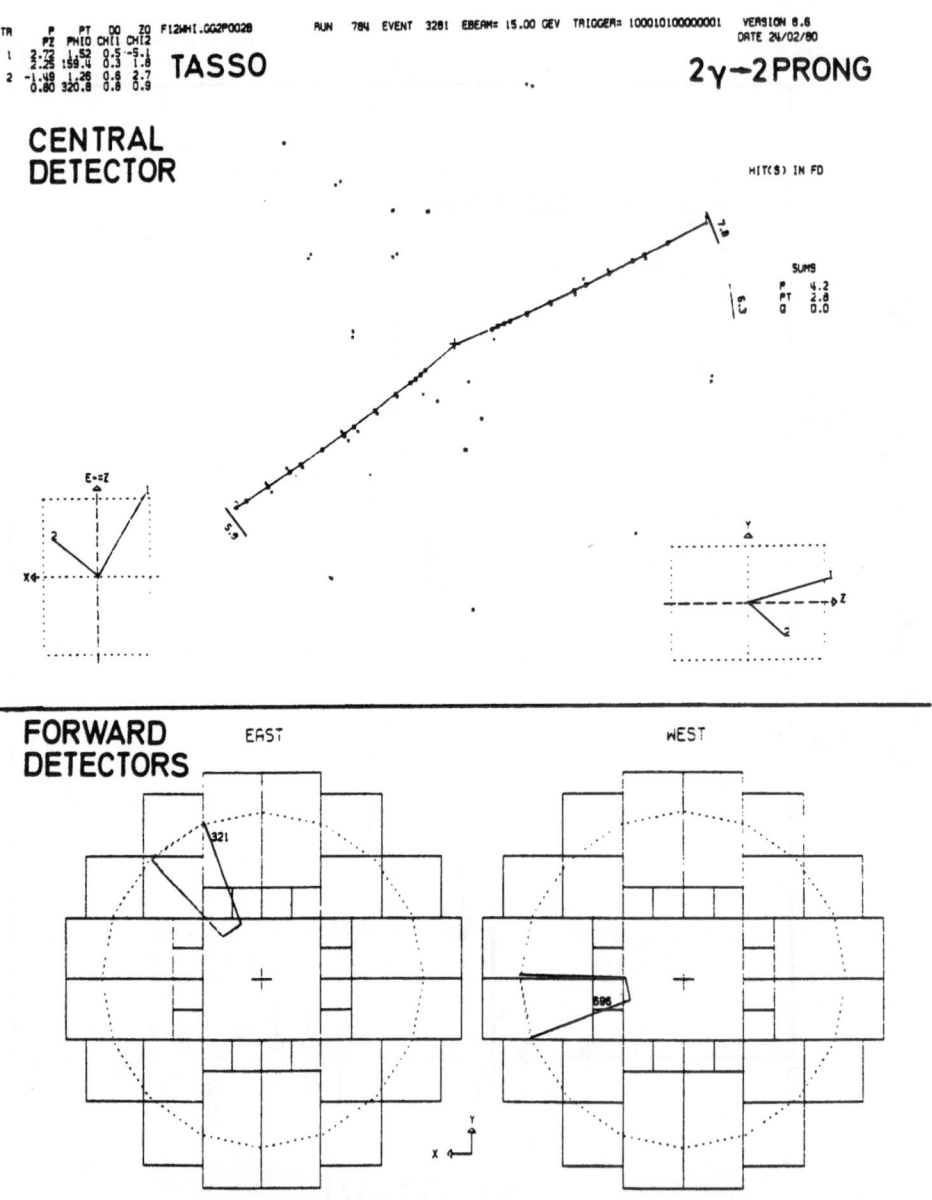

Fig.7

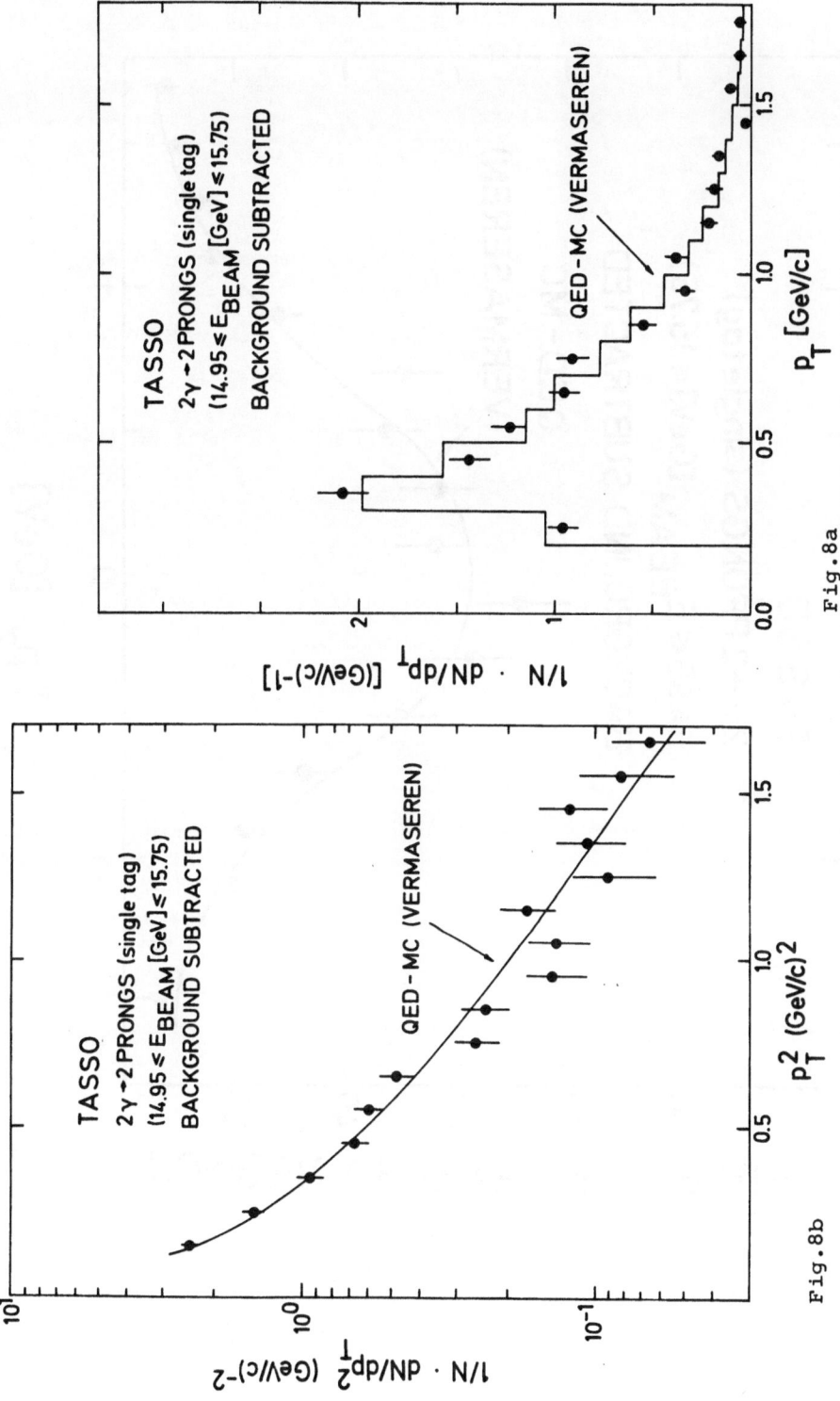

Fig.8a

Fig.8b

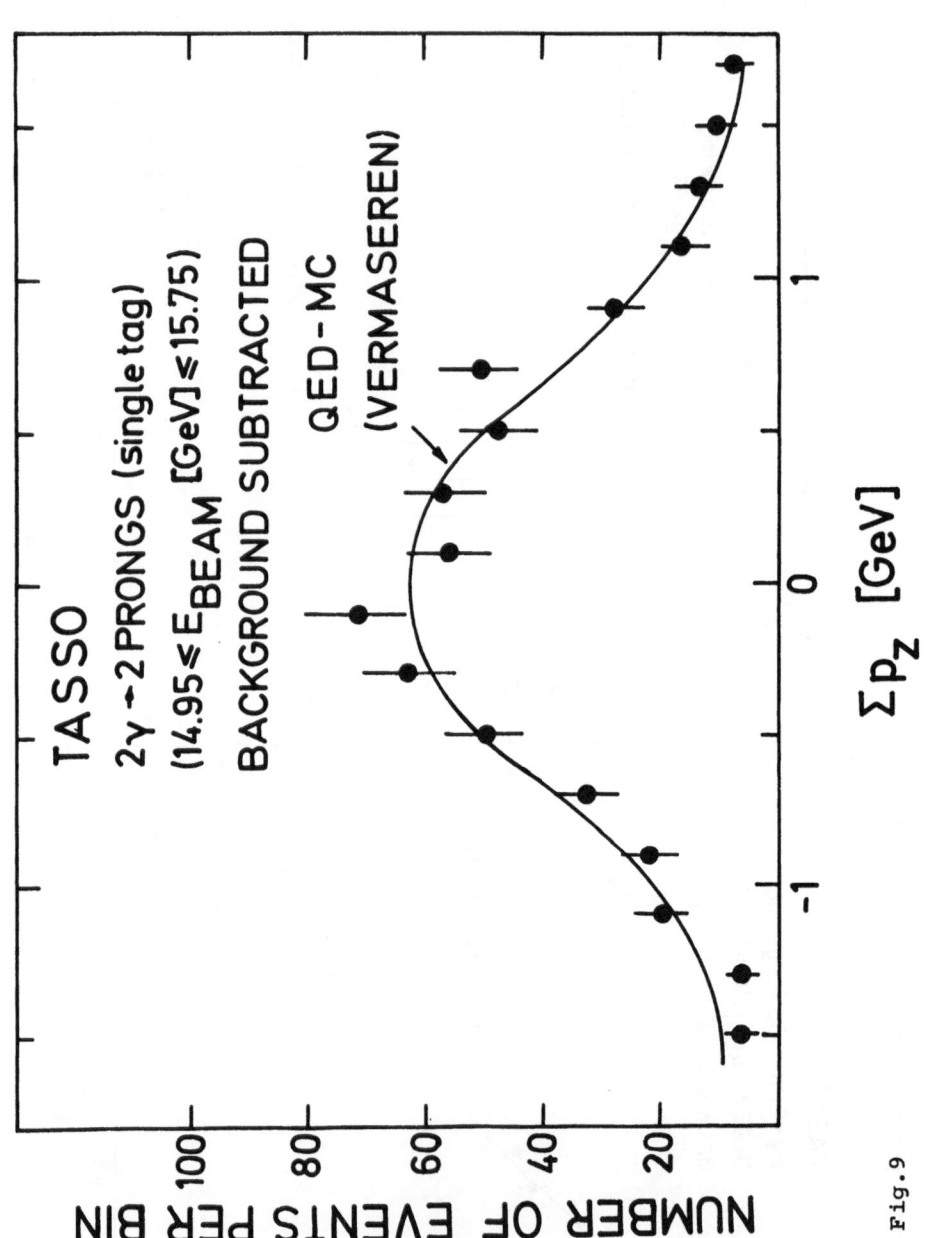

Fig.9

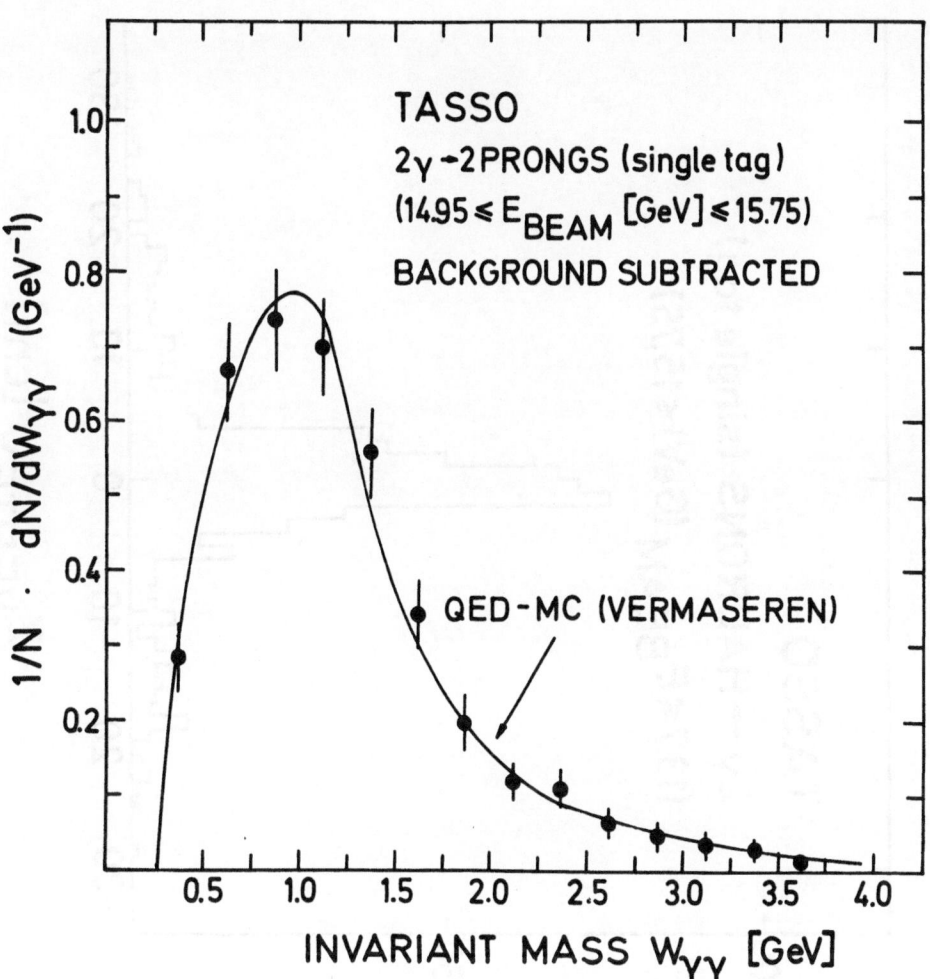

Fig.10

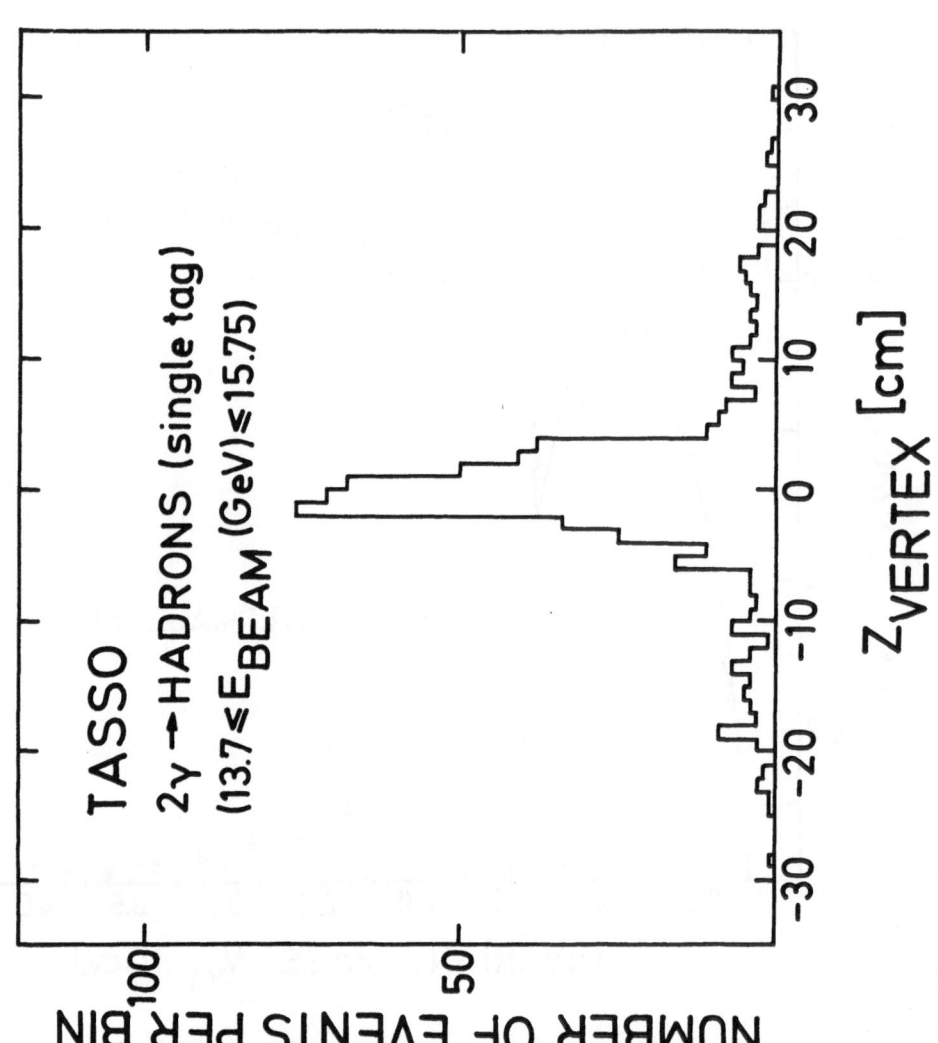

Fig.11

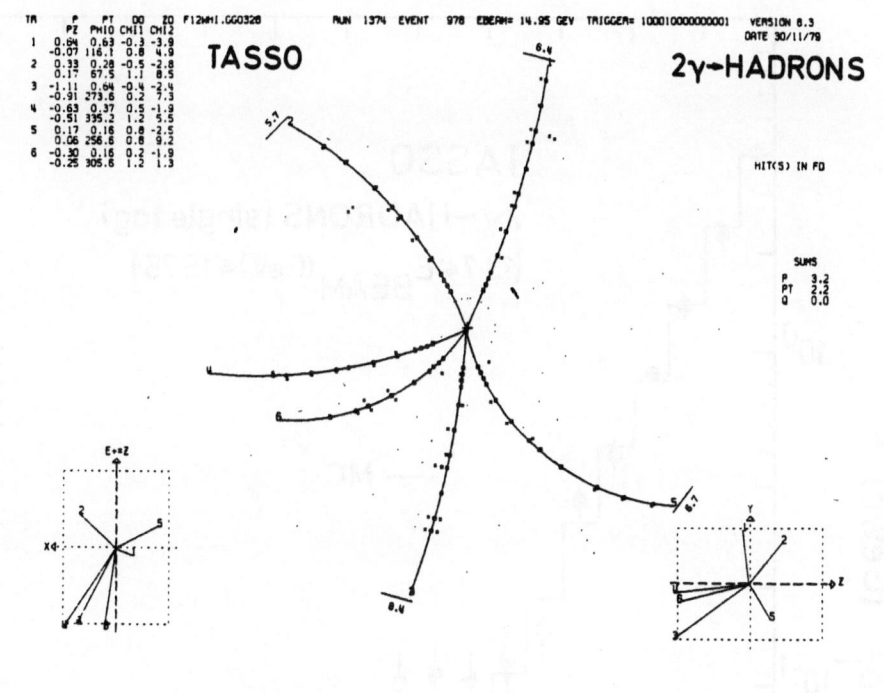

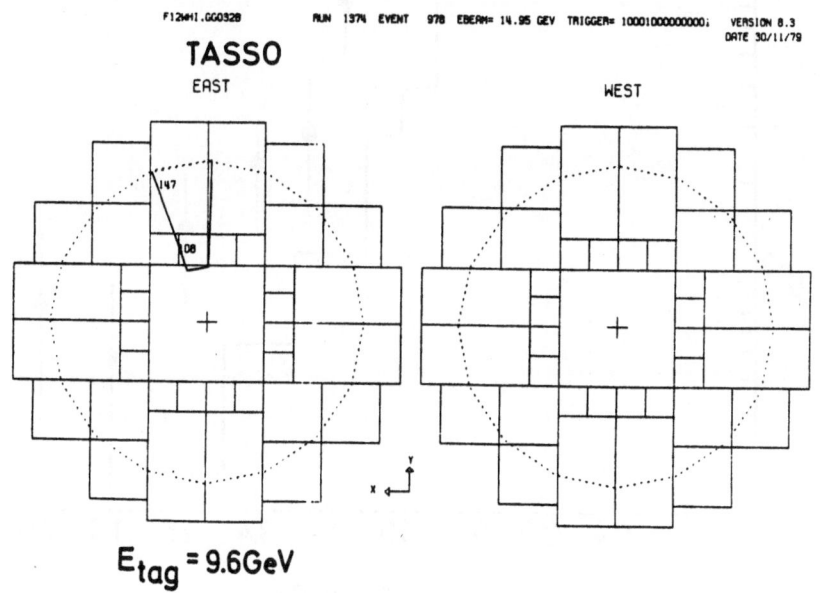

$E_{tag} = 9.6 GeV$

Fig.12

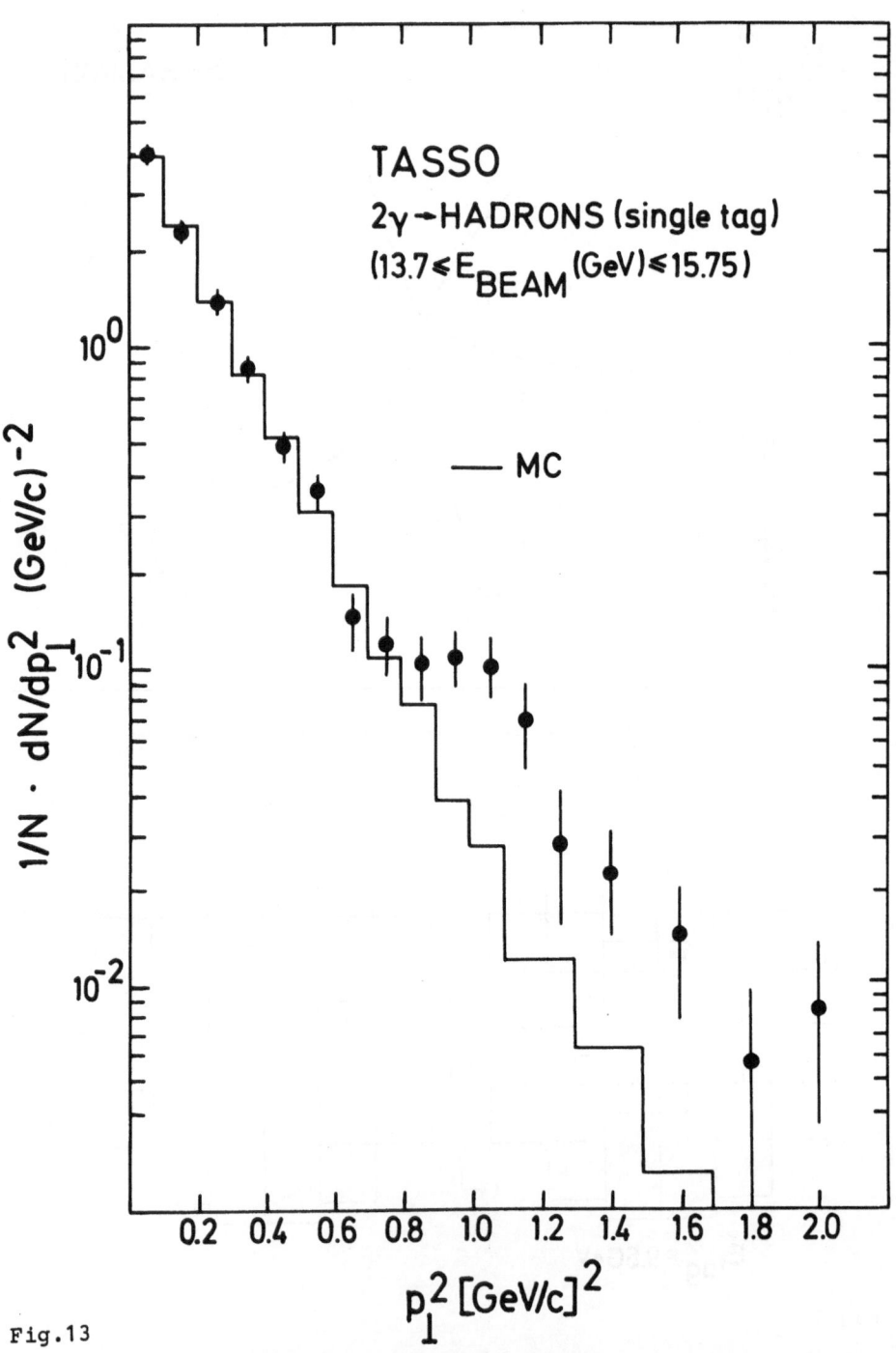

Fig.13

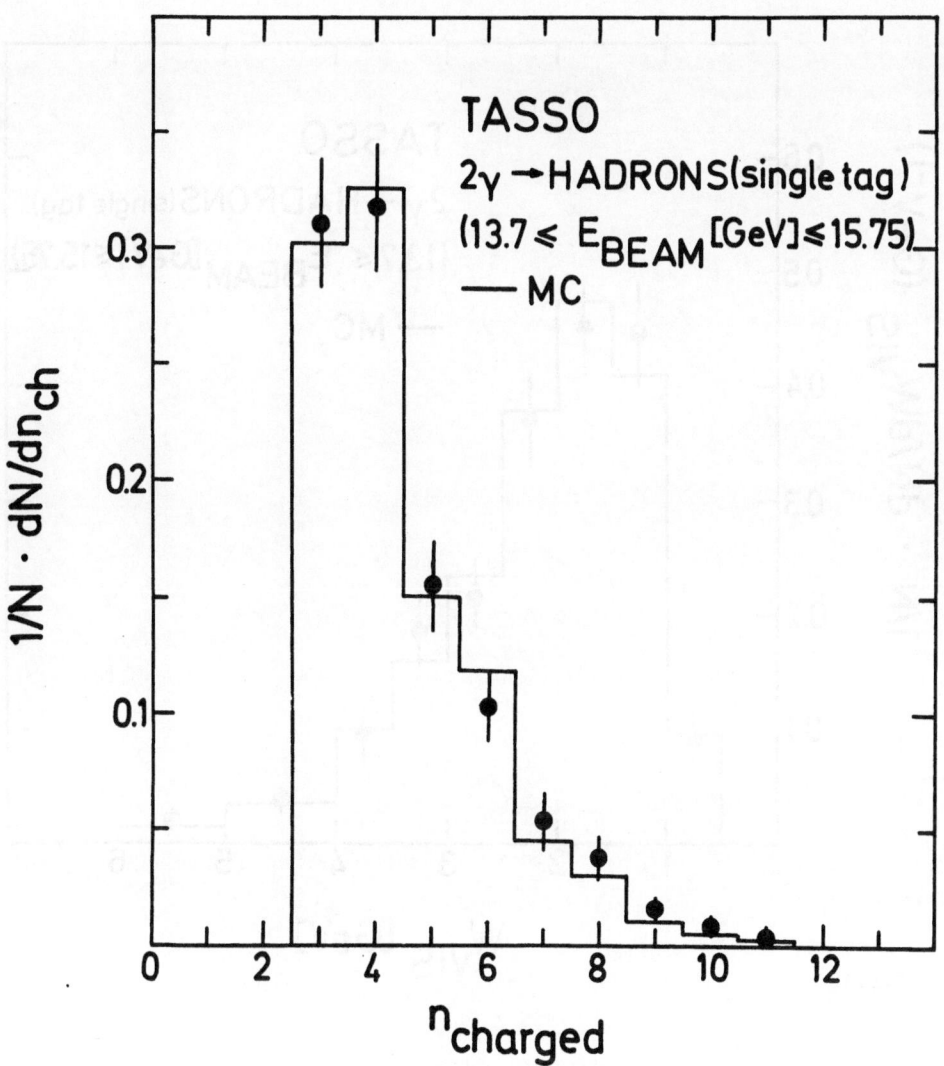

Fig.14

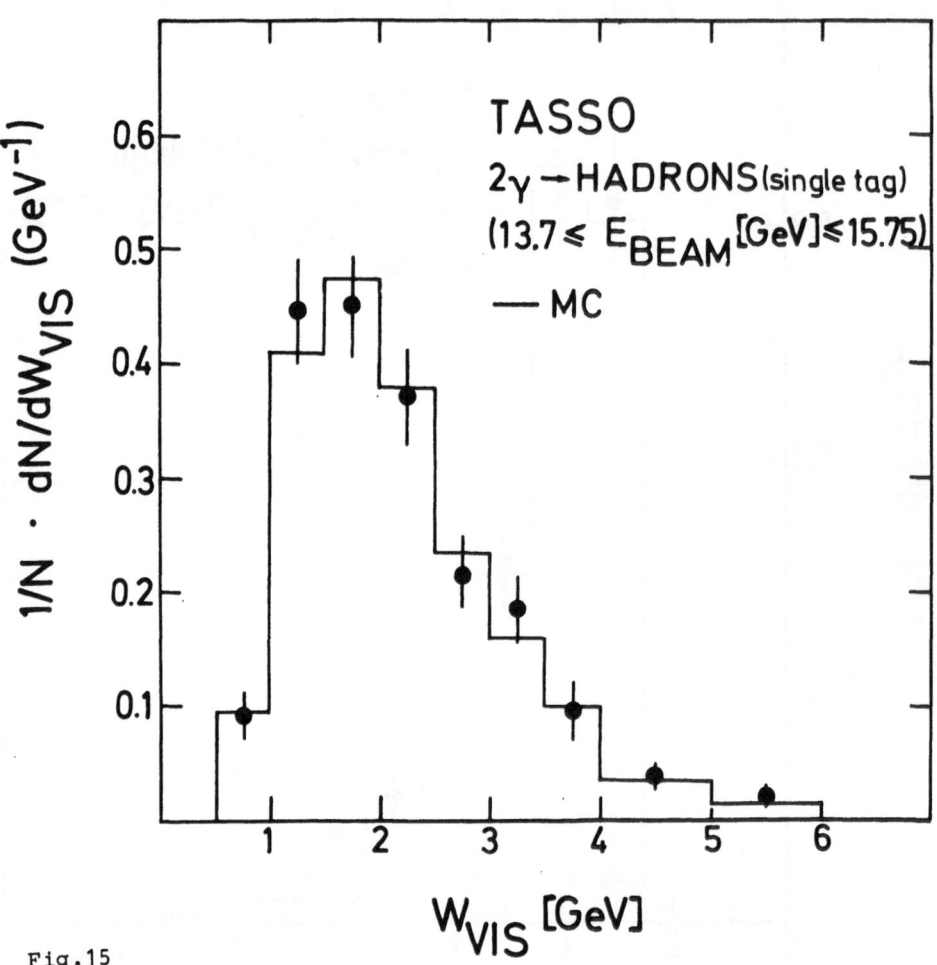

Fig.15

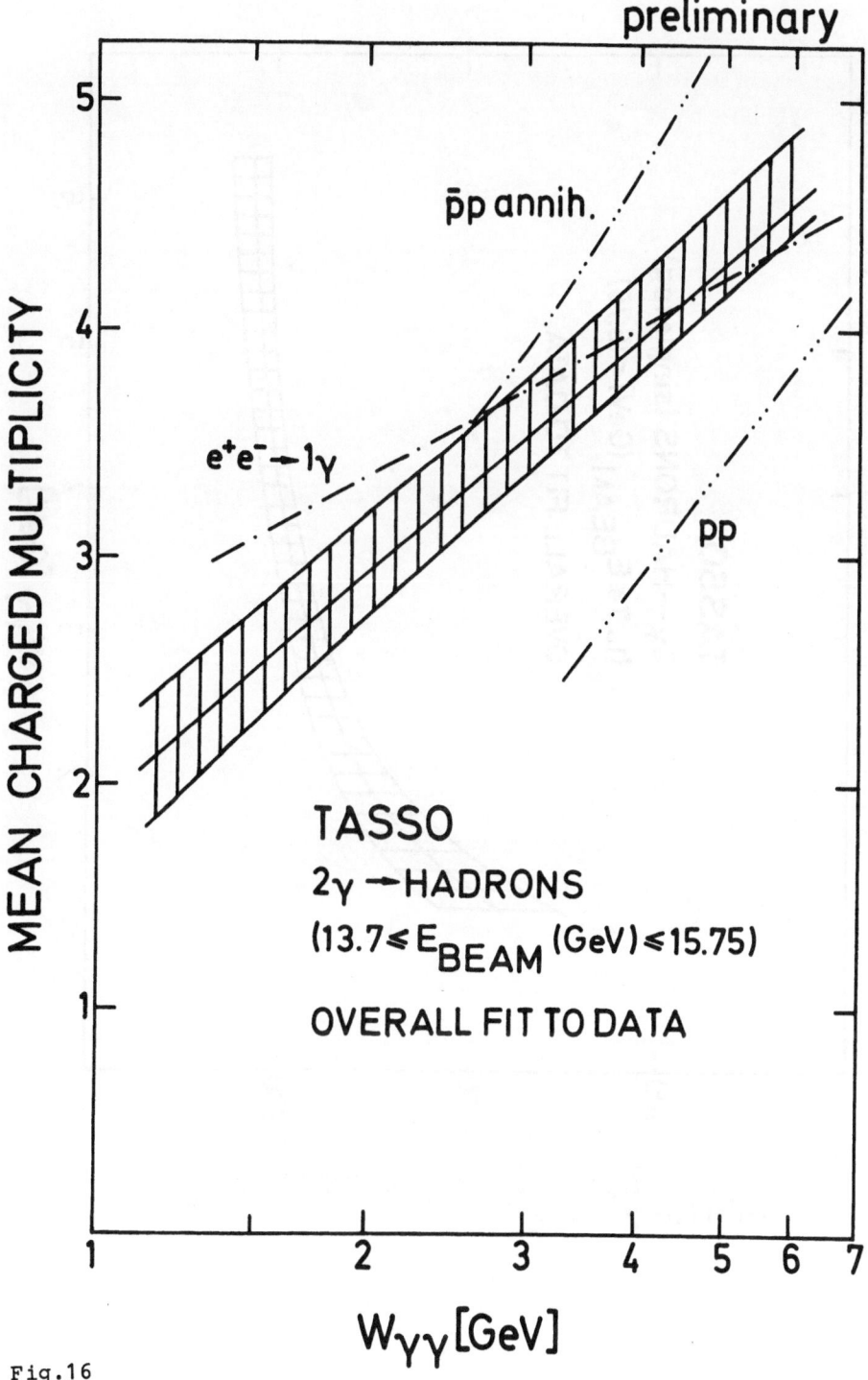

Fig.16

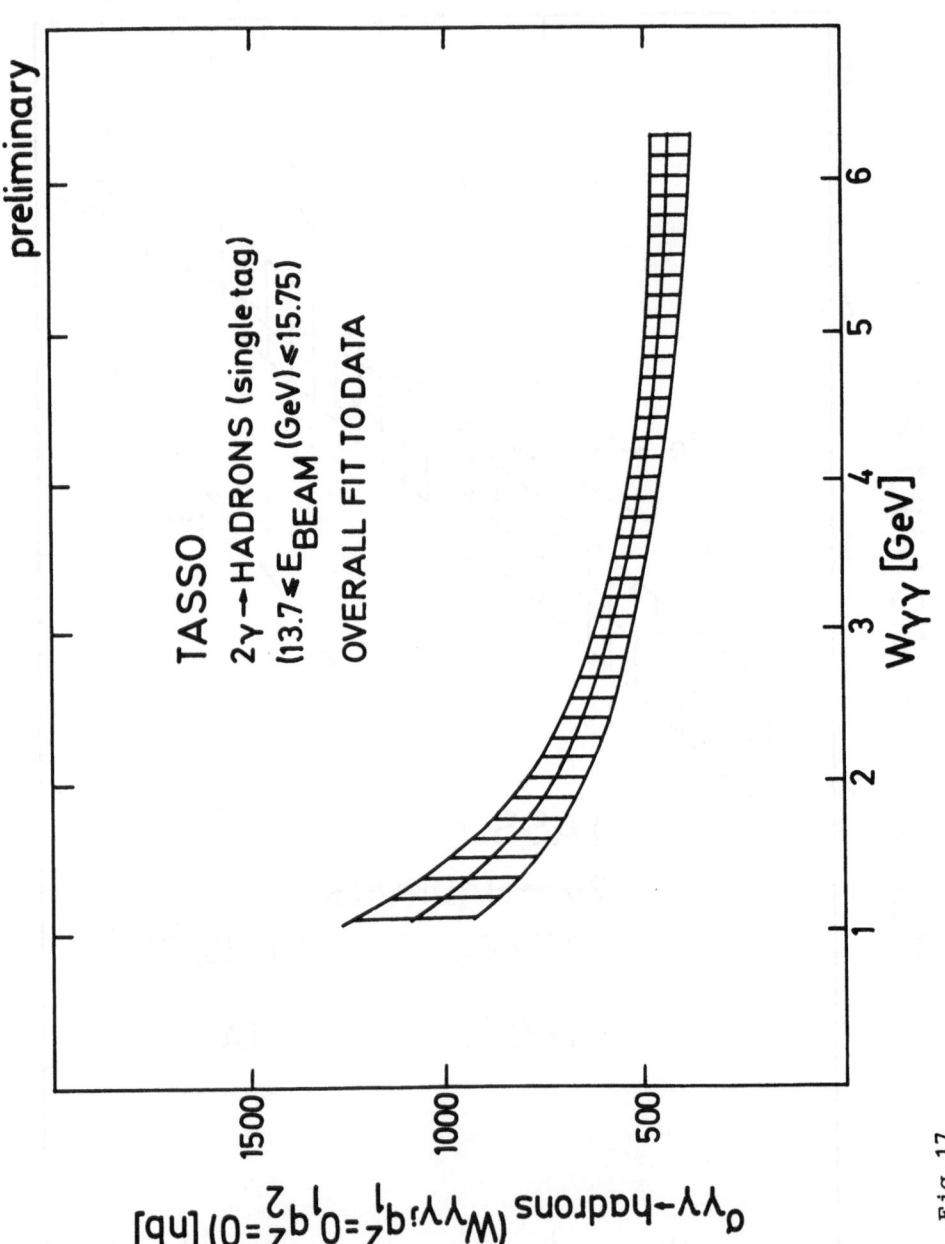

preliminary

TASSO
$2\gamma \rightarrow$ HADRONS (single tag)
$(13.7 \leqslant E_{BEAM}(GeV) \leqslant 15.75)$

OVERALL FIT TO DATA

$\sigma_{\gamma\gamma-hadrons}(W_{\gamma\gamma};q_1^2=0,q_2^2=0)$ [nb]

$W_{\gamma\gamma}$ [GeV]

Fig.17

2γ PROCESSES WITH THE MARK-J DETECTOR AT PETRA

J.D. Burger

Massachusetts Institute of Technology

The MARK-J experiment, built and operated by a group of fifty-six physicists from Aachen, DESY, MIT, NIKHEF, and Peking, occupies the southwest intersection at PETRA. The primary design goal of the experiment is to measure the forward-backward charge asymmetry in muon production which is expected from the interference of weak and electromagnetic interactions, but it is also well suited to investigate other physics processes and has carried out a diversified program during the first year and a half of operation of PETRA[1]. This has included a number of tests of quantum-electrodynamics in Bhabha scattering and in the production of all the known charged leptons (e, μ, τ) from one photon annihilation[2]. In addition we have measured events containing muons from the reaction

$$e^+e^- \rightarrow e^+e^- \mu^+\mu^-$$

from two photon processes and separated them from events containing muons from muon and tau pairs from single photon annihilation. The rates and angular distributions agree well with the prediction of quantum-electrodynamics calculated to the fourth order in α. These results are the subject of the present contribution. Our analysis of hadronic events from two photon processes will be reported later.

A side view of the MARK-J detector is shown in Figure 1. It is able to distinguish charged hadrons, electrons, muons, neutral hadrons and photons and to measure their directions and energies. It covers a solid angle of ϕ (the azimuthal angle) = 2π and θ (the polar angle) = $12°$ to $168°$. It consists of five magnetized iron toroids built around a non-magnetic inner detector and was designed to be insensitive to the effects of synchrotron radiation. The layered structure of the detector is shown schematically in Figure 2. It is described in more detail elsewhere[1],[2],[3].

Closest to the intersection are four layers of drift tubes (DT in Figures 1 and 2) which distinguish charged from neutral particles in the angular range $30° < \theta < 150°$ and reconstruct the position of the event vertex along the beam line to an accuracy of two millimeters. The inner shower counters (A,B and C in Figures 1 and 2) are equipped with a phototube at each end, so that in addition to measuring energy they can measure the azimuthal position of particles entering them by a weighted average of pulseheight and timing measurements. Figures 3a and 3b compare the position and angle of particles determined from these inner shower counters with the information obtained from tracks fitted in the drift tubes, and Figure 3c gives the observed energy spectrum, all for large angle Bhabha electrons at $\sqrt{s}$ = 30 GeV. The distribution of energy sampled in these counters and the K counters buried in the iron toroids, distinguishes electrons and hadrons.

The detector is well suited for muon identification and measurements. The following criteria are used to identify muons:

a) The particle track must be reconstructed in the inner drift chambers (S and T in Figures 1 and 2) which are located between the inner shower counters and the magnetized iron toroids, and the reconstructed track must point to the intersection region. Figure 4a shows the vertex distribution of hadron events, produced from tracks in the drift tubes, and Figure 4b shows the same distribution from μ pairs reconstructed in the S and T chambers. Both distributions are very clean, showing little background, and exhibit a width consistent with the crossing of the beam bunches.

b) The pulseheights in the calorimeter and trigger counters (A, B,C,D and K in Figures 1 and 2) must be consistent with a minimum ionizing particle and not with a shower.

c) There must be a signal in at least one inner shower counter (A or B in Figures 1 and 2) in time with the beam crossing signal, together with an in-time signal in a muon trigger counter buried halfway through the iron (D in Figures 1 and 2). These muon trigger counters are each 30 cm wide and 450 cm long with a phototube at each end. Their timing resolution is 400 ps. There is a layer of drift chambers (Q in Figures 1 and 2) adjacent to these counters which can measure a muon track in the bending coordinate halfway through the iron.

d) The particle must leave a track in the outer drift chambers (P in Figures 1 and 2) and thus must have penetrated 87 cm of iron. This restricts the polar angle acceptance for such muons to $45^{\circ} < \theta < 135^{\circ}$ and puts a cut on the transverse momentum of $P_{t_{\mu}} \gtrsim 2$ GeV/c. Since the iron is magnetized with a bending power of about 17 kG-m, the momentum is reconstructed to $\Delta p/p \simeq 20\%$. In the case of two photon processes, where the muon transverse momentum may be low, we relax this requirement for the second muon in the event in order to increase the acceptance. Then we merely require that the muon reach the D trigger counters, necessitating passage through only 42 cm of iron. This decreases the effective cut on transverse momentum to $P_t > 1$ GeV/c.and increases the angular acceptance to approximately $30° \lesssim \theta_\mu < 150°$ (ignoring bending). This also means we are not able to measure the momentum of this muon, only its direction.

Events containing muons from two photon processes (diagrams in Figure 5) must be separated from the two other major sources of events containing muons, single photon annihilation, and τ pair production (since the τ has a branching ratio of 16 percent into $\mu\nu\bar{\nu}$).

We separate our data into three samples: events in which we see only two muons, events in which we see one muon and a shower, and events in which we see two muons and a shower. The potential sources of·background are different in the three cases.

Muon pairs produced by single photon annihilation are expected to be back to back and to reconstruct the total energy of the beams except for radiative corrections. To identify these events we require that both muons leave a track in the outer drift chambers and have a reconstructible momentum greater than half the energy in one beam. We also require the acollinearity ($\xi = \cos^{-1}(-\hat{P}_{\mu 1} \cdot \hat{P}_{\mu 2})$) to be less than $20°$. We have compared the data passing these cuts to a Monte Carlo simulation including radiative corrections to order α^3 which were obtained from the programs of Berends and Kleiss[4]. Table Ia shows a comparison of the data taken with 27 GeV $\leq \sqrt{s} \leq 32$ GeV with the results of the Monte Carlo simulation as a function of acollinearity angle. Table Ib is a comparison of acoplanarity $\Delta\phi$, which is $180°$ minus the difference of the azimuthal angles of the two muons.

TABLE I a

Distribution in acollinearity (ξ) of muon pairs from
single photon annihilation ($\xi > 2°$)

ξ	$N_{observed}$	N_{QED}
$2 - 3°$	8	7.9
$3 - 4°$	6	4.6
$4 - 5°$	5	4.6
$5 - 10°$	10	9.6
$10 - 20°$	9	7.5
Total in $2 - 20°$	38	34.2

TABLE I b

Distribution in acoplanarity ($\Delta\phi$) of muon pairs from
single photon annihilation ($\Delta\phi > 2^{\circ}$)

$\Delta\phi$	$N_{observed}$	N_{QED}
$2 - 5^{\circ}$	9	5.9
$5 - 10^{\circ}$	4	2.1
$10 - 20^{\circ}$	3	1.2
Total in $2 - 20^{\circ}$	16	9.2

128 events were observed at all angles.

Figures 6 and 7 show the acollinearity and acoplanarity distributions for all muon pairs from one photon annihilation which were
collected up to December 1979, together with the prediction of the QED
Monte Carlo program. There is good agreement.

The events in which we see only the μ-pair from two photon
processes must be separated from the single photon annihilation events.
For the two photon μ-pairs we require that at least one of the two
muons must leave a track in the outer drift chambers and have its
momentum measured. The other muon is only required to reach the muon
trigger counters (D in Figures 1 and 2). Any measured muon-momentum
must be less than half the momentum in a single beam. There is no
acollinearity cut.

We have compared the data with the results of the Monte Carlo
program of Vermaseren[5] which is correct to order α^4 in the cross
section. At the time of this conference we have used an early version
of this program, which includes only the first two diagrams of Figure
5. The other diagrams are expected to be less important, although we
should see a few events from diagrams of the types c and d in Figure 5.
They should tend to contain μ-pairs with a narrow opening angle. We
do see a few such events. Up to now the Monte Carlo program does not
include the effects of radiative corrections.

Figure 8 shows a comparison of the Monte Carlo prediction for muon
acollinearity with the two photon data taken at all energies. Figure 9
is a plot of the acollinearity of muon pairs with both $P_{\mu's} \geq E_{beam}/2$
(one photon annihilation candidates, but no acollinearity cut) from the
same data sample as in Figure 8 and plotted on the same horizontal scale
to show the difference in the distribution.

Figure 10 shows the muon momentum distribution for the muons passing the single photon annihilation cuts, and for muon pairs outside these cuts, which are the preliminary candidates for pairs produced by two photon processes, before the momentum cuts are applied. In Figure 11 we plot the acollinearity of the pairs in Figure 10 versus the measured momentum. We use either the momentum of the single muon measured, or the average momentum for events in which both muons leave a track in the outer chambers. The vertical line at 20° is the acollinearity cut. The single photon annihilation pairs are concentrated in the upper lefthand corner.

Figure 12 is the acoplanarity distribution of the μ-pairs accepted as being produced by two photon processes together with the Monte Carlo prediction.

In Table II we compare our data with the prediction of Vermaseren's Monte Carlo program for two photon events in which we see only the two muons. The results are tabulated at five different center of mass energies. There are a small number of events from one photon annihilation μ-pairs and also from $\tau\bar{\tau}$ pairs in which both τ's decay into a muon and neutrinos, which pass our cuts for two photon pairs. We subtract Monte Carlo predictions for the numbers of events of these types before comparing the data with the results of the Monte Carlo.

TABLE II

$$e^+ e^- \rightarrow e^+ e^- \mu^+ \mu^-$$

Events in which only the μ^+ and μ^- are observed.

$\sqrt{s}$ GeV	12	27.4	29.9 to 31.6	35.0	35.8
$\int Ldt(nb^{-1})$	97.7	414	2691	763	128
$N_{obs}\mu\mu\,(2\gamma)$	0	19	80	22	5
$N_{background}(\mu\mu$ from $\tau\bar{\tau})$	∿ 0	0.2	1.4	0.4	0.0
$N_{background}\,(1\gamma)$	∿ 0	0.2	1.3	0.2	0.0
$N_{corrected}\,\mu\mu\,(2\gamma)$	∿ 0	18.6	77.3	21.4	4.9
$N_{predicted}\,\mu\mu\,(2\gamma)$	1.3	12.4	83.6	25.9	4.4

We also collect events in which we see only one muon and an electron or positron from $e^+ e^- \rightarrow e^+ e^- \mu^+ \mu^-$. The electron appears as a shower in the inner lead-scintillator shower counters (A,B and C in

Figures 1 and 2). We must distinguish these events from those of the type

$$e^+ + e^- \rightarrow \tau + \bar{\tau}$$

with the decay channels:
- $\rightarrow$ (h's or e) + ν's
- $\rightarrow \mu + \nu$'s

where we see the muon and a shower produced by the hadrons or electron. We have simulated these τ pair events using a Monte Carlo program containing the known branching ratios of $\tau \rightarrow \mu\nu\bar{\nu}$ (16%) and $\tau \rightarrow$ (e, lepton + hadron, or multi hadrons) + ν (84%). For events of either of these types we require that the muons leave a track in the outer chambers (and thus have $P_{t_\mu} \gtrsim 2$ GeV/c). The shower must have an energy $E_x > 4$ GeV in order to discriminate against showers from radiated photons in one photon annihilation events. Since τ's are light compared with the center-of-mass energy at PETRA, their decay products tend to be collimated back to back. We impose an acollinearity cut, requiring $\xi_{\mu\text{-shower}} < 30^\circ$ for the τ pair candidates and $\xi_{\mu\text{-shower}} \geq 30^\circ$ for the two photon candidates. In the two photon events we expect the detected electron to be produced preferentially at small angles with the beam direction, so we also impose a cut of $\theta_{\text{beam-shower}} < 30^\circ$ to identify two photon candidates. Figure 13 shows the acollinearity of μX pairs (X = hadron or e shower) for the data at all center-of-mass energies, together with the Monte Carlo predictions for $\tau^+\tau^-$ decays and two photon processes. Figure 14 shows the angle θ between the shower and the beam for the same data and also the Monte Carlo predictions. In Figure 15 both angles are shown for the data as a scatter plot along with the cuts. The points in the center bottom of the plot are accepted as $\tau^+\tau^-$ pairs and those in the upper left and right margins as two photon pairs. Table III compares our two photon data in the two energy regions with the greatest accumulated luminosities with the Monte Carlo predictions. Small corrections are made for one photon annihilation events and τ pair decays which fall within the cuts.

TABLE III

$$e^+e^- \rightarrow e^+e^- \mu^+\mu^-$$

Events in which only one $\mu^\pm$ and one $e^\pm$ are observed

$\sqrt{s}$ (GeV)	29.9-31.1	35 + 35.8
$\int$ Ldt(nb^{-1})	1832	892
$N_{obs}\mu e (2\gamma)$	28	7
Estimated Background $N_{\mu x}$ ($e^+e^- \rightarrow \mu^+\mu^- \gamma$)	1.1	0.4
Estimated Background $N_{\mu x}$ ($e^+e^- \rightarrow \tau^+\tau^-$)	0.5	0.2
$N_{corrected}$ μe (2γ)	26.4	6.4
$N_{predicted}$ μe (2γ)	23.8	8.9

There are also a few events in which we detect both muons and an electron shower. Here we apply the same muon cuts as for two photon events in which only the two muons are observed. We require the shower energy to be greater than half of the beam energy and that its direction does not lie within 15° of the plane in which the two muon momenta lie. This discriminates against showers from radiated photons in single photon annihilation events. The shower must also lie within 40° of the beam direction. Table IV shows a comparison of the Monte Carlo prediction of these events with the data.

TABLE IV

$$e^+e^- \rightarrow e^+e^- \mu^+\mu^-$$

Events in which both muons and one $e^\pm$ are observed

$\sqrt{s}$ (GeV)	12	27.4	30→31.6	35+35.8
$\int$ Ldt (nb^{-1})	97.7	414	2691	892
$N_{observed}\mu\mu e$	0	1	7	1
$N_{predicted}\mu\mu e$	0.2	1.9	11.3	2.7

There is good agreement between the data and the Monte Carlo in all three tables II, III and IV. Figure 16 is a graphical representation of the data from tables II, III and IV shown as cross sections measured within the acceptance of the detector.

In summary, there is good agreement in the acollinearity, acoplanarity, and azimuthal angle distributions between our data samples and the Monte Carlo predictions for mu-pairs from one photon annihilation, for mu-pairs or mu plus showers from two photon processes and for muons plus showers from $\tau\bar{\tau}$ pairs. This shows we are able to cleanly separate the muon events from two photon processes from the others by cuts on acollinearity, θ and muon momentum.

References

1)   D.P. Barber et al., The First Year of MARK-J at PETRA, MIT Laboratory for Nuclear Science Report 107,
     and
     D.P. Barber et al., Physics with High Energy Electron-Positron Colliding Beams, submitted to Physics Reports.

2)   D.P. Barber et al., Phys. Rev. Lett. 42, 1110 (1979).
     D.P. Barber et al., Phys. Rev. Lett. 43, 1915 (1979).

3)   D.P. Barber et al., Phys. Rev. Lett. 42, 1113 (1979).
     D.P. Barber et al., Phys. Rev. Lett. 43, 901 (1979).
     D.P. Barber et al., Phys. Lett. 85B, 463 (1979).
     D.P. Barber et al., Phys. Rev. Lett. 43, 830 (1979).
     D.P. Barber et al., Phys. Lett. 89B, 139 (1979).

4)   F.A. Berends et al., Nucl. Phys. B 68, 541 (1974).

5)   J.A.M. Vermaseren (CERN), contribution to this conference.

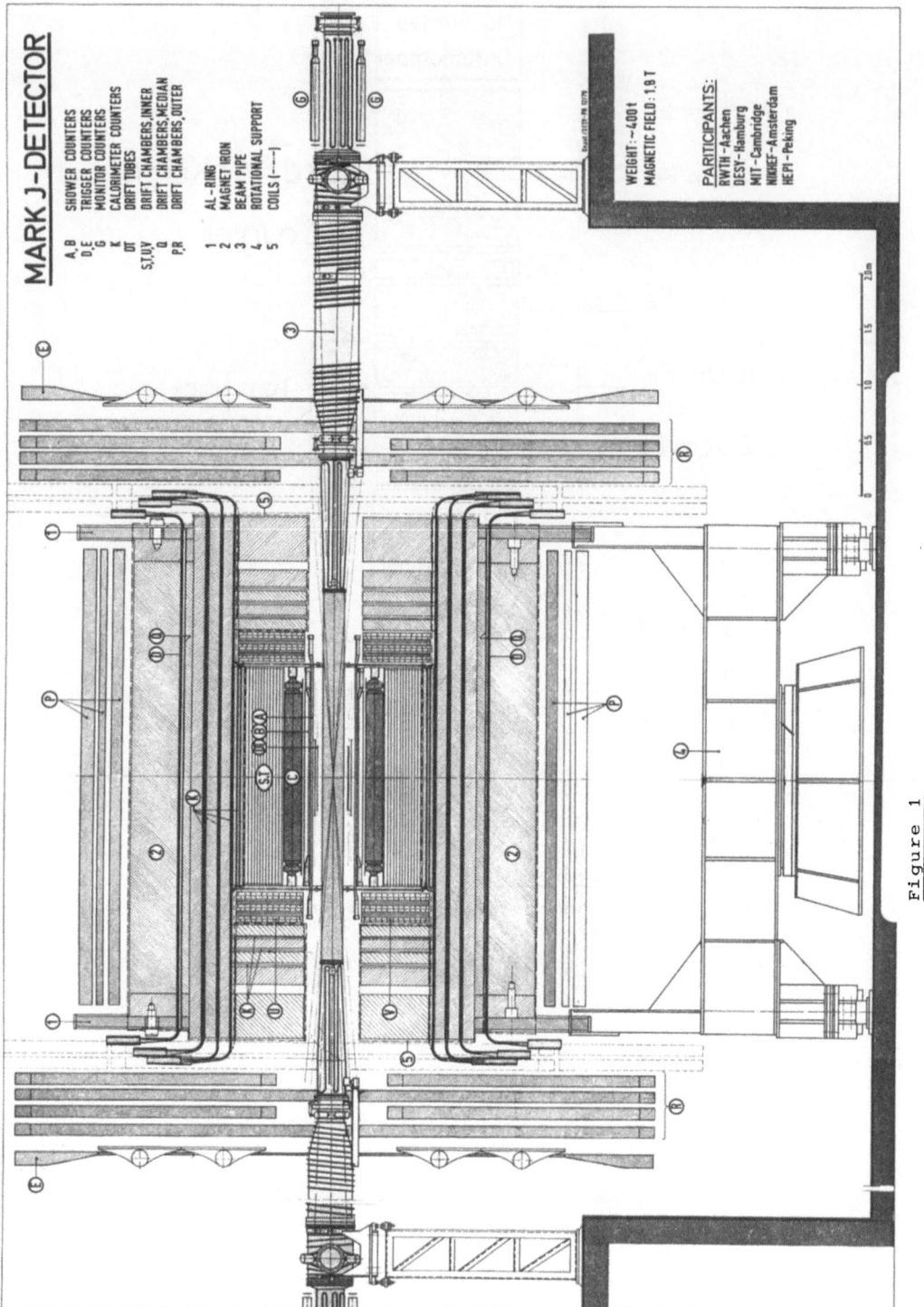

Figure 1
Side view of the MARK-J detector.

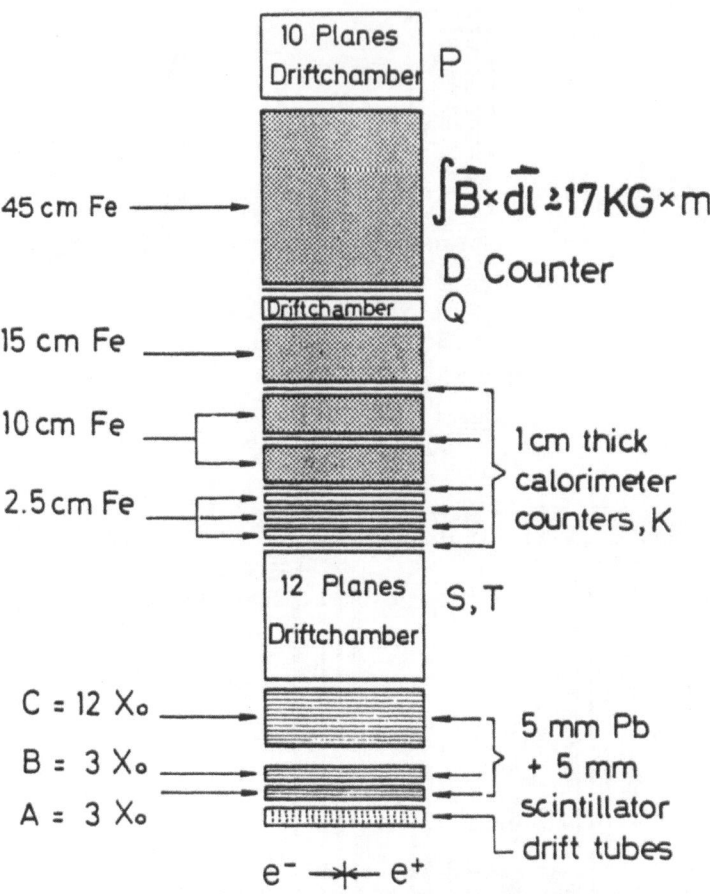

Figure 2

Schematic view of the layers of the detector.

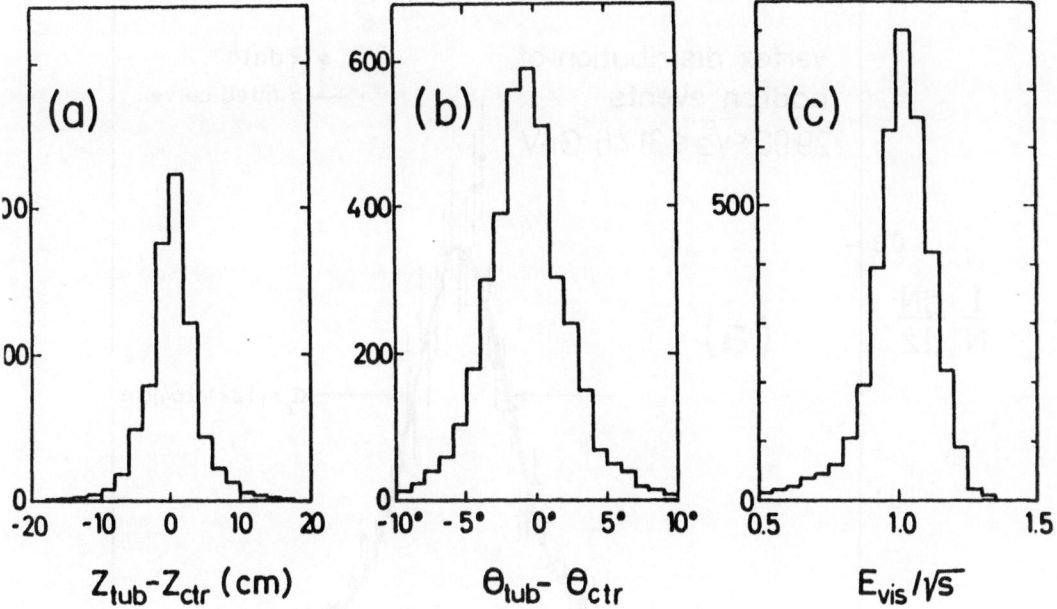

Figure 3

Counter resolution for large angle
Bhabha scattering at $\sqrt{s}$ = 30 GeV.

(a)   Position of hits deduced from TDC and ADC information
       (Z_{ctr}) in the A counters compared to extrapolated tracks
       (Z_{tub}) observed in drift tubes.

(b)   Polar angles of reconstructed counter tracks compared to
       tracks fitted with drift tube information.

(c)   Observed energy spectrum for large angle Bhabha scattering.

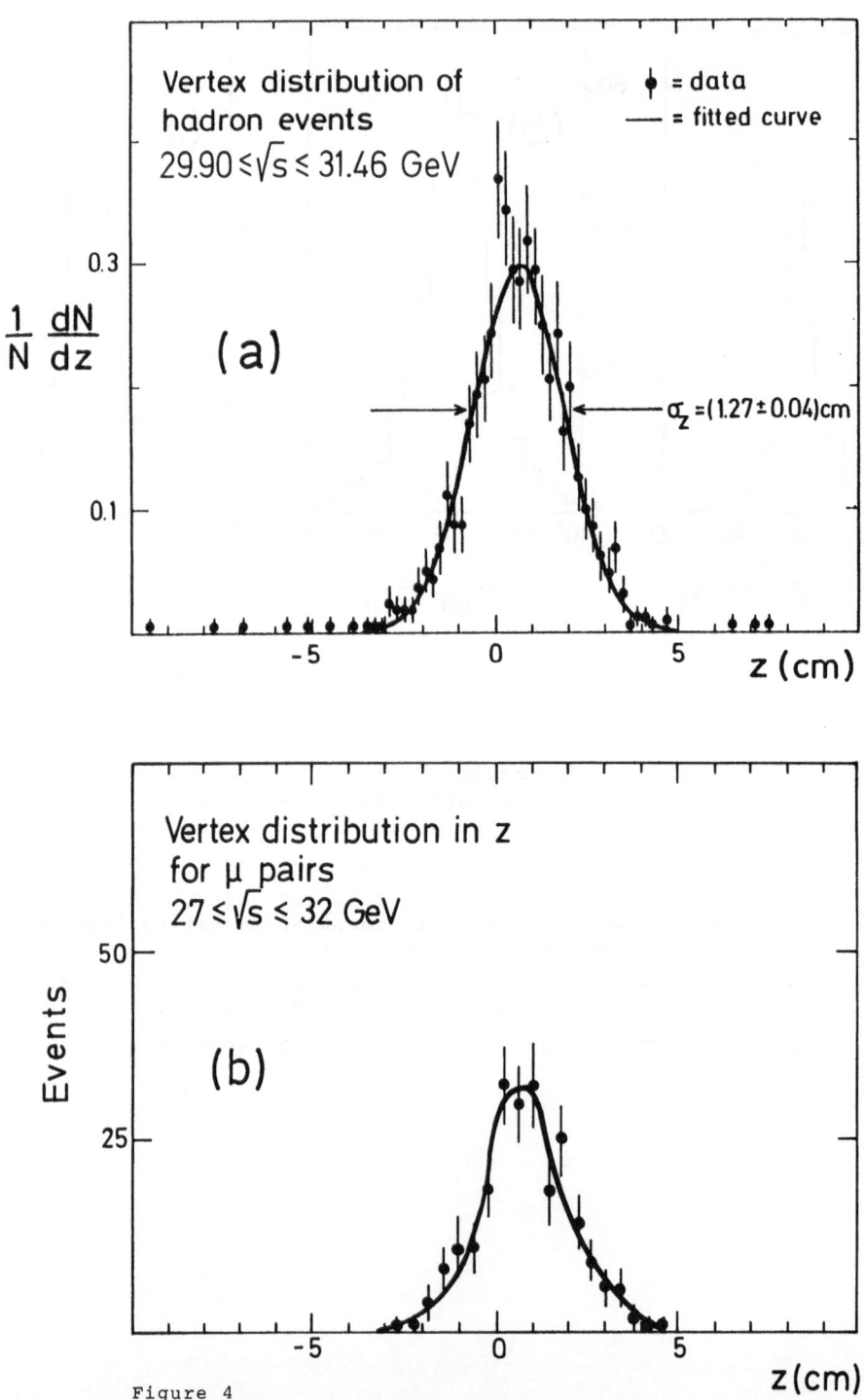

Figure 4

(a) Vertex distribution for hadron events
    from drift tubes.
(b) Vertex distribution for muon pairs
    from S and T chambers.

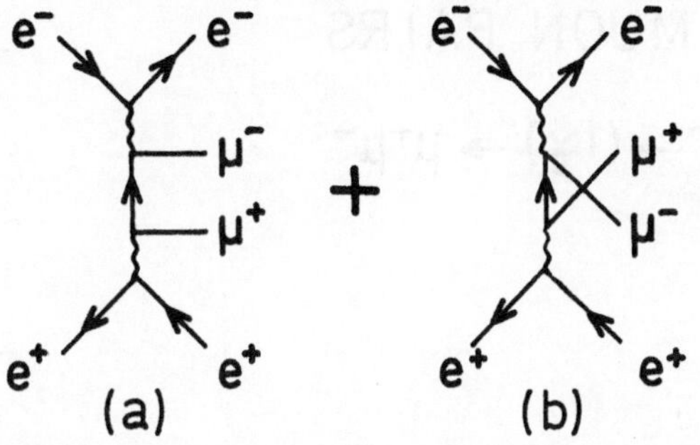

PLUS DIAGRAMS OF THE TYPES

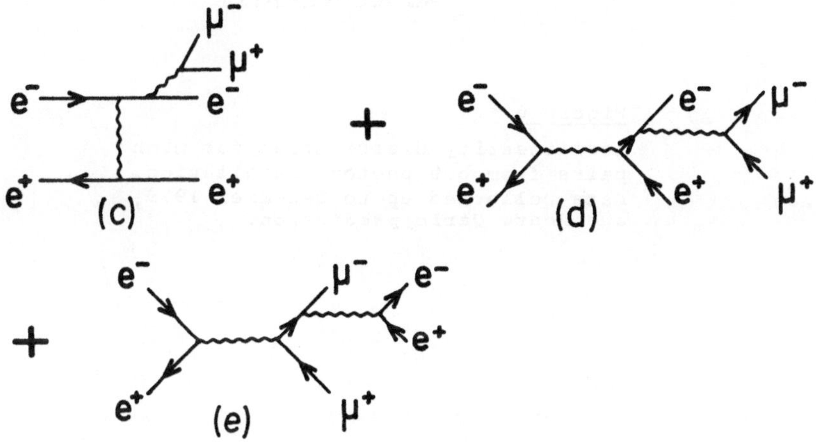

Figure 5

Representative Feynman diagrams for
muon pair production by two photon
processes. The various permutations
of diagrams c, d and e are not shown.

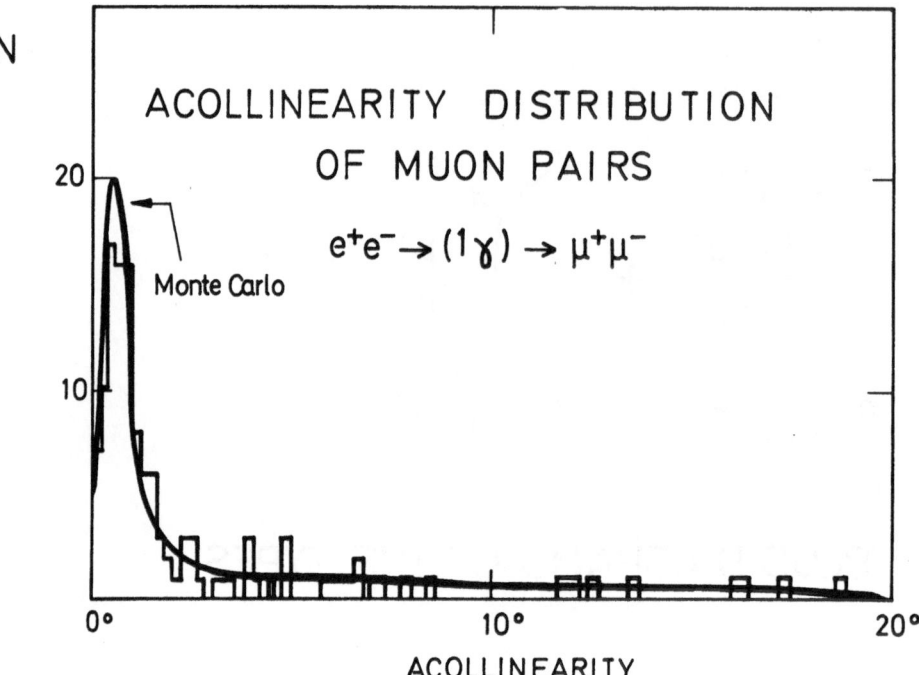

Figure 6

Acollinearity distribution for muon
pairs from one photon annihilation.
Data collected up to December 1979,
and Monte Carlo prediction.

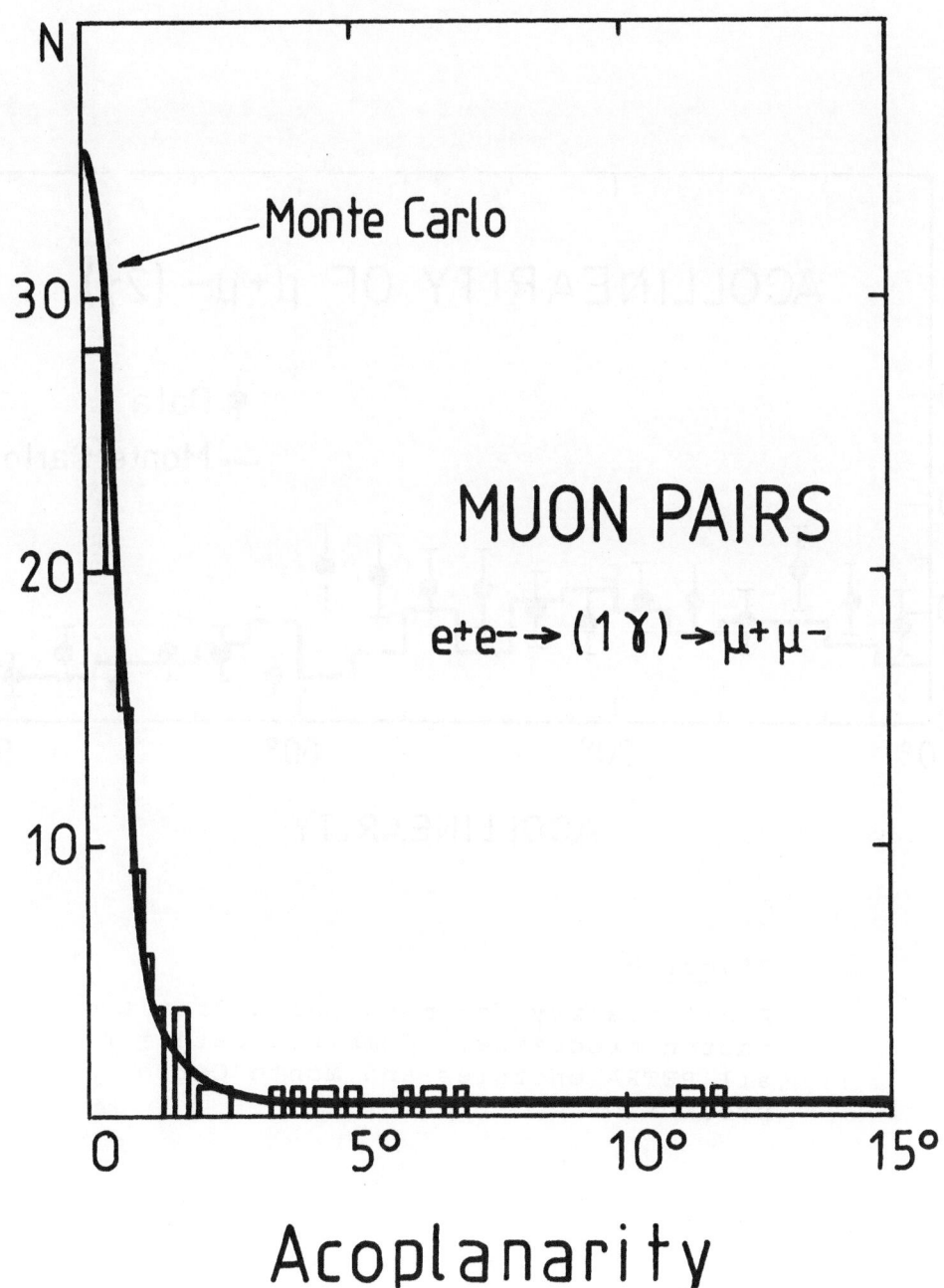

Figure 7

Acoplanarity distribution for muon
pairs from one photon annihilation.
Data collected up to December 1979,
and Monte Carlo prediction.

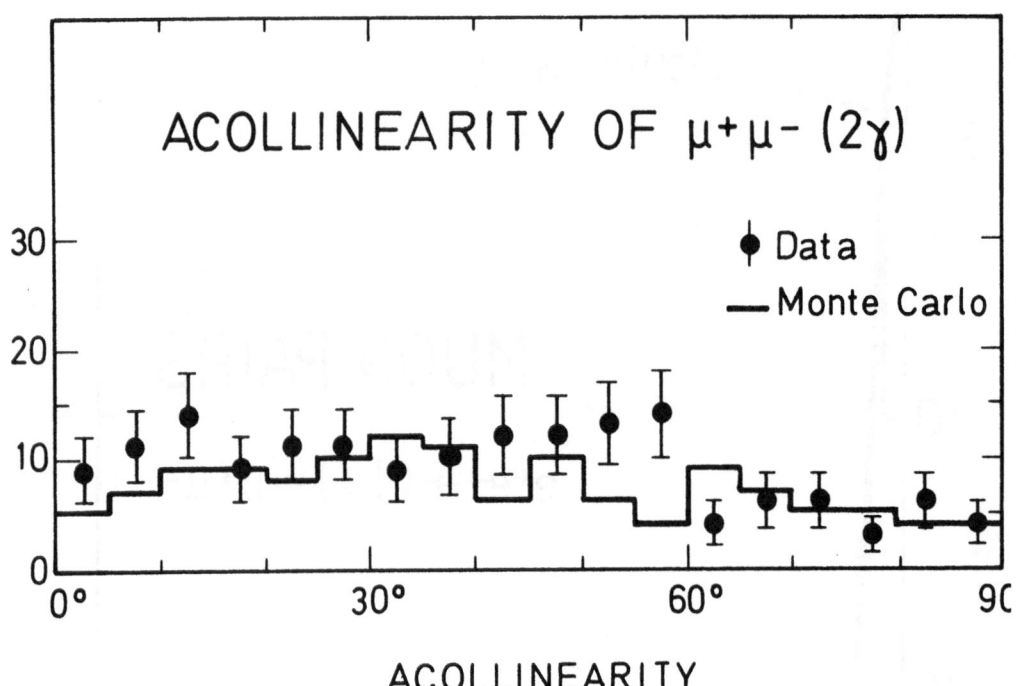

Figure 8

Acollinearity for muon pairs from two
photon processes. Combined data for
all PETRA energies and Monte Carlo
prediction.

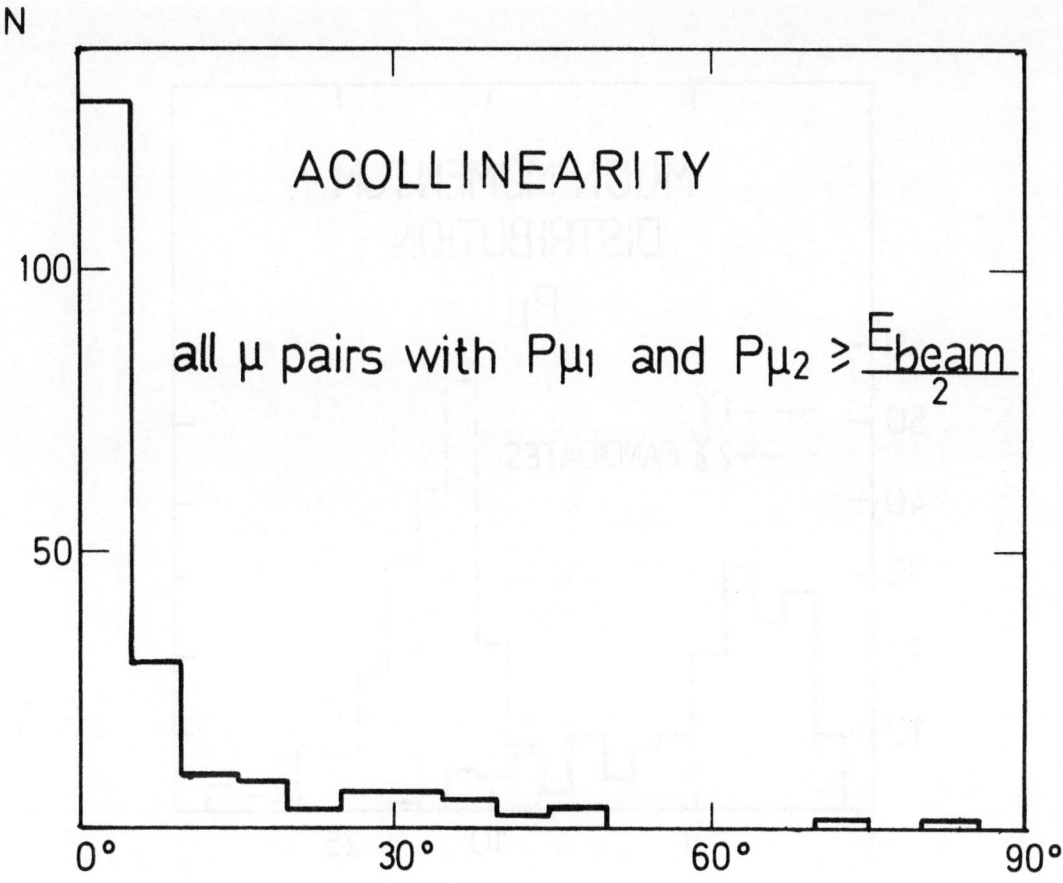

Figure 9

Acollinearity for muon pairs with both
P_μ's $\geq E_{beam}/2$. Combined data for all
PETRA energies.

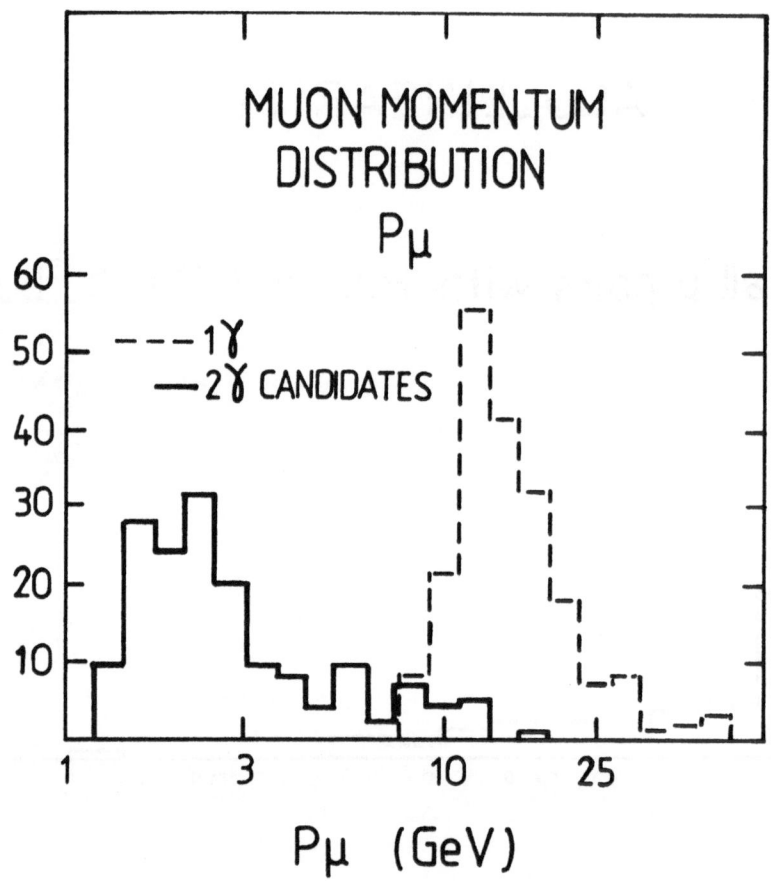

Figure 10

Momentum distribution for muons from
single photon annihilation and for
two photon process candidates. Combined
data for all PETRA energies.

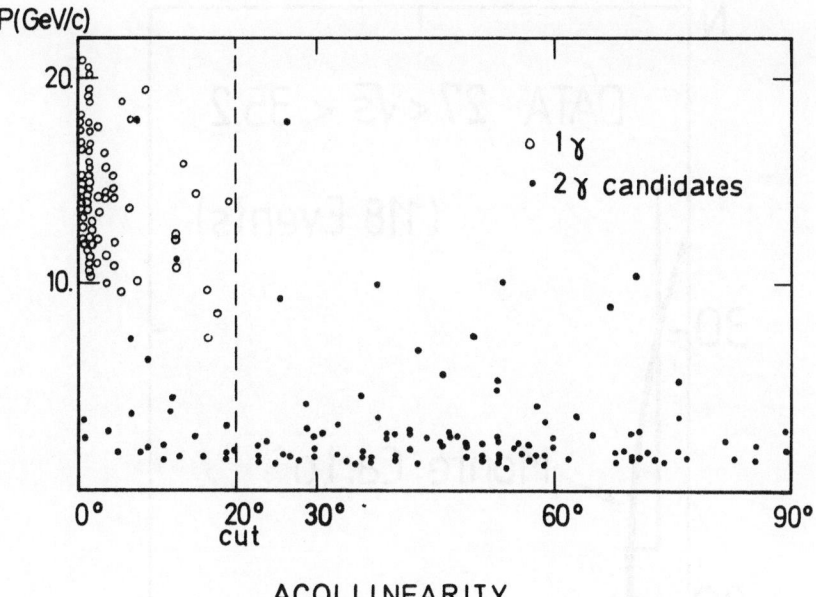

Figure 11

Acollinearity vs. muon momentum for
1γ and 2γ pairs.
Combined data for all PETRA energies.

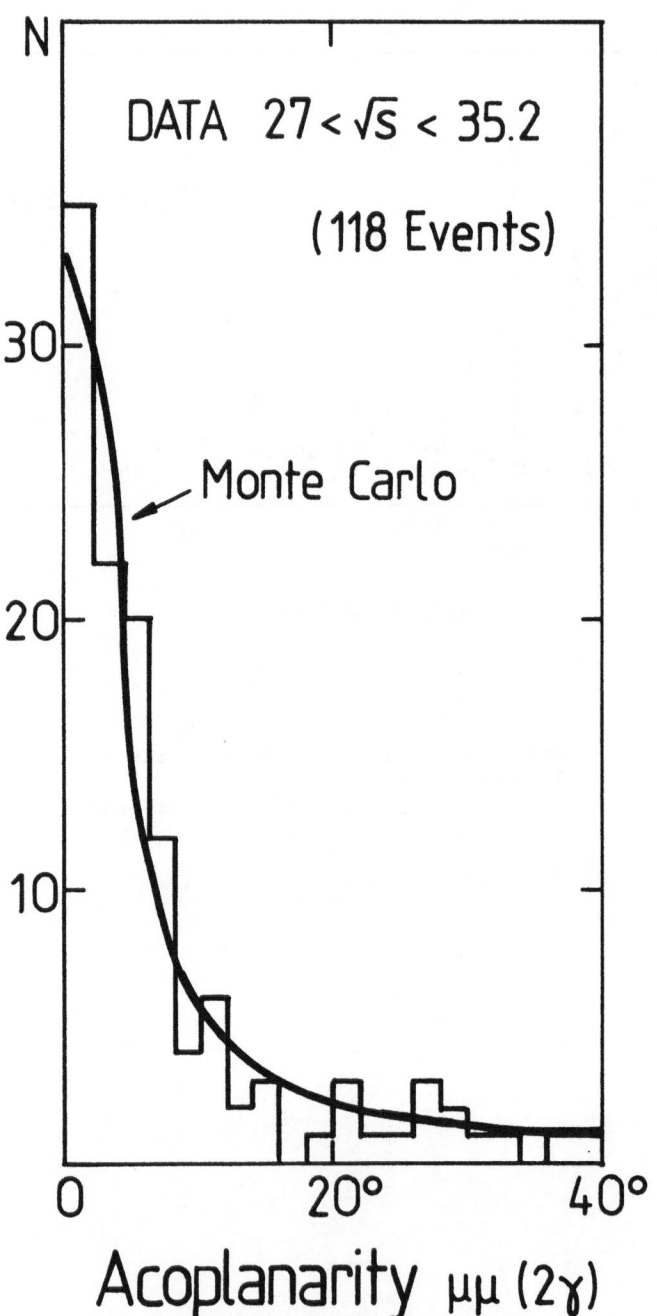

Figure 12

Acoplanarity of muon pairs from two photon processes.

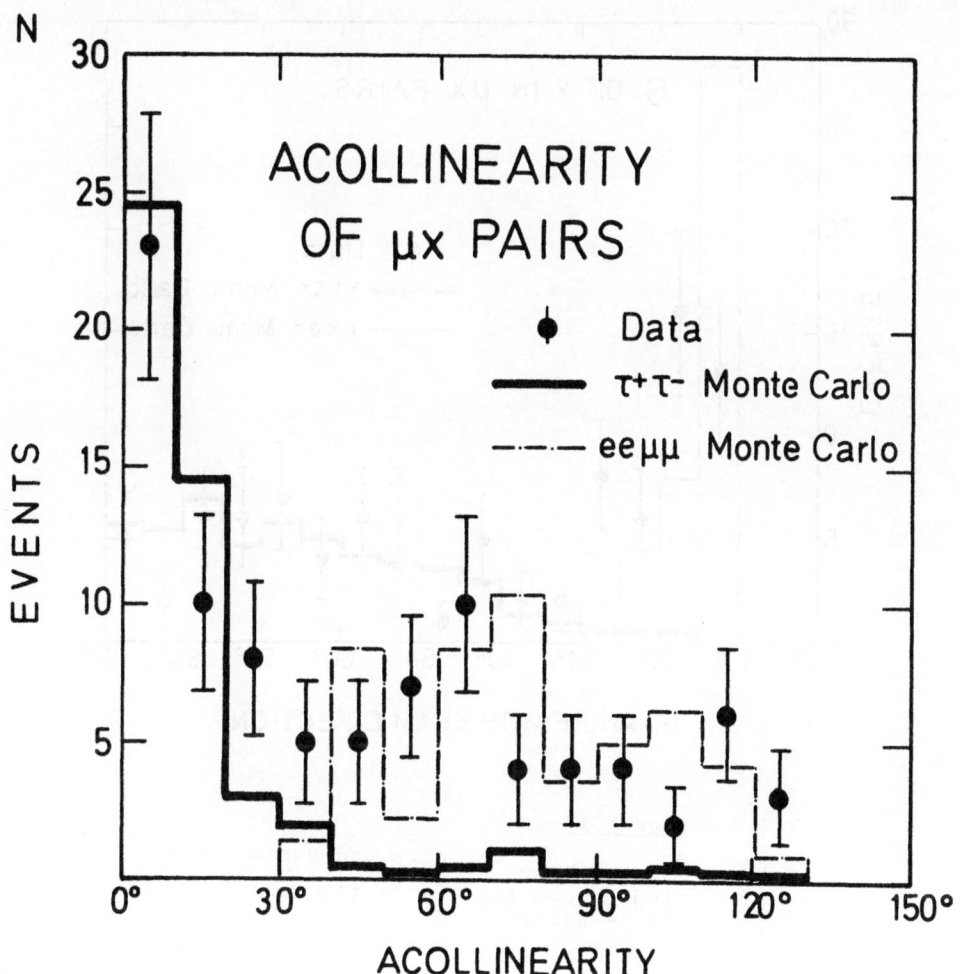

<u>Figure 13</u>

μ-X acollinearity for data and Monte Carlo predictions for $e^+e^- \to \tau^+\tau^-$ and $e^+e^- \to e(e\mu)\mu$.

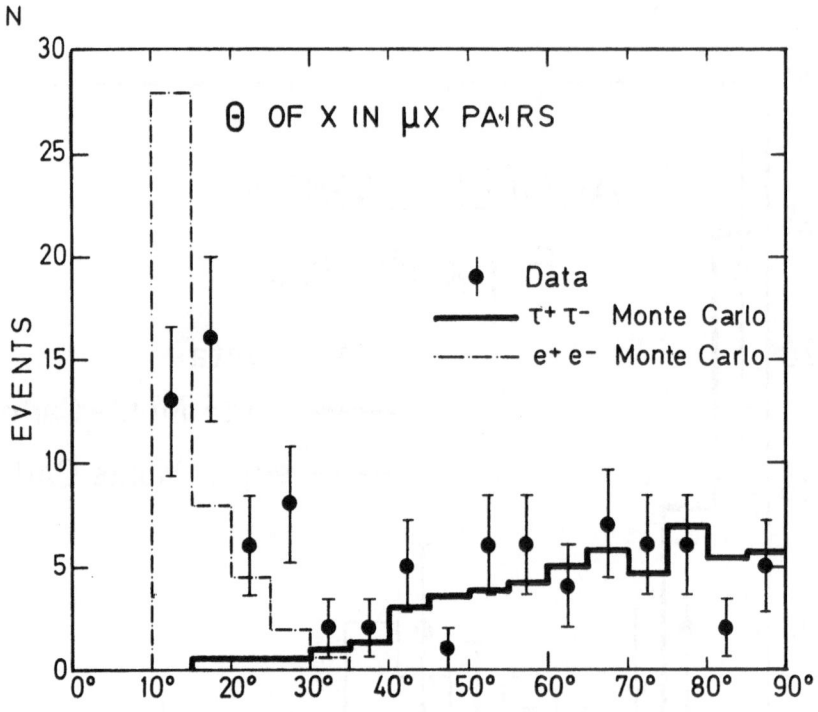

Figure 14

$\theta_{\text{shower-beam}}$ for data and Monte Carlo predictions for $e^+e^- \to \tau^+\tau^-$ and $e^+e^- \to e(e\mu)\mu$.

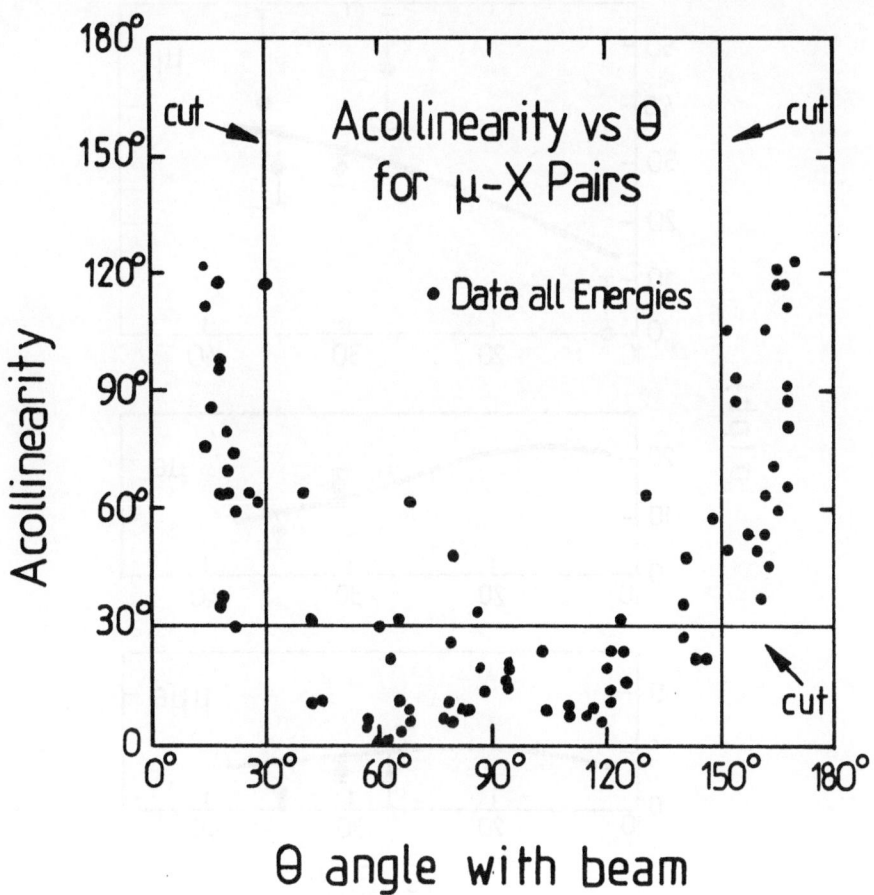

<u>Figure 15</u>

Scatterplot of μ-shower acollinearity
and θ$_{shower-beam}$ for the data at all
energies combined.

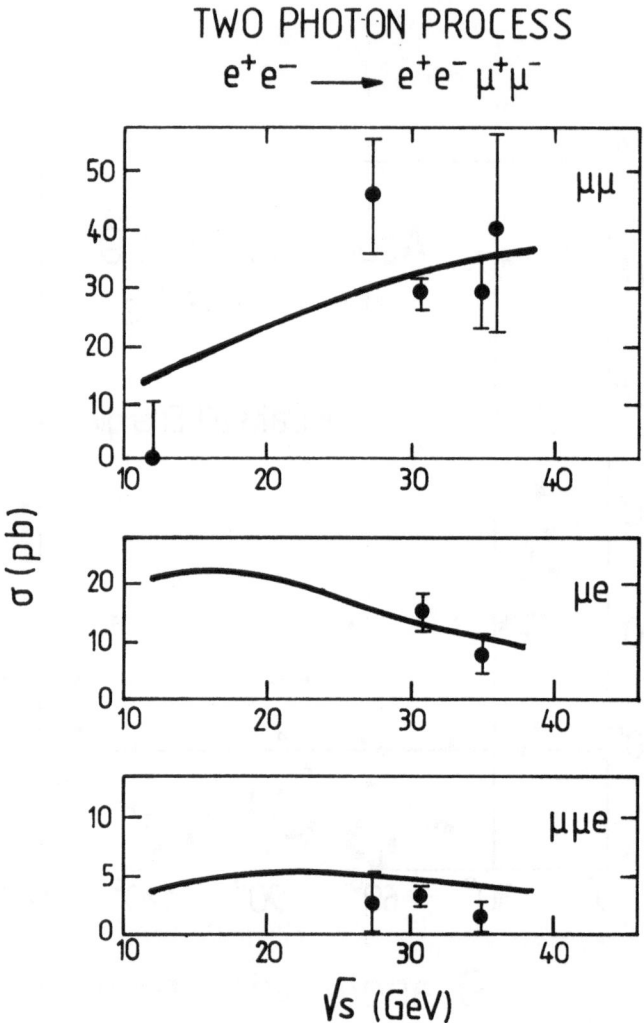

TWO PHOTON PROCESS

$$e^+e^- \longrightarrow e^+e^- \mu^+\mu^-$$

Figure 16

Cross sections within the acceptance of the detector for μμ, μe and μμe from two photon processes.

Present status of JADE's γγ-physics analysis

JADE Collaboration

Presented by H. Wriedt, University of Lancaster

Invited talk given at the International Workshop on γγ Collisions,
Amiens 1980

Abstract

The present status of the γγ-physics analysis performed by the JADE-Collaboration is
reported. The reaction ee → eeμμ is observed at beam-energies of 6 GeV and 15 GeV,
and it is compared with predictions from QED. The reaction ee → ee + hadrons is
observed at 15 GeV beam-energy. Several distributions of physical quantities are
shown.

The members of the JADE-Collaboration are:

W. Bartel, D. Cords, P. Dittmann, R. Eichler, R. Felst, D. Haidt, S. Kawabata,
H. Krehbiel, B. Naroska, L.H. O'Neill, J. Olsson, P. Steffen, W.L. Yen,
Deutsches Elektronen-Synchrotron DESY, Hamburg;
E. Elsen, M. Helm, K. Meier, A. Petersen, P. Warming, G. Weber, II. Institut für
Experimentalphysik der Universität Hamburg;
H. Drumm, J. Heintze, G. Heinzelmann, R.D. Heuer, J. von Krogh, P. Lennert,
H. Matsumura, T. Nozaki, H. Rieseberg, A. Wagner, Physikalisches Institut der
Universität Heidelberg;
D.C. Darvill, F. Foster, G. Hughes, H. Wriedt, University of Lancaster;
J. Allison, J. Armitage, A. Ball, I. Duerdoth, J. Hassard, F. Loebinger, H. McCann,
B. King, A. Macbeth, H. Mills, P.G. Murphy, H. Prosper, K. Stephens, University
of Manchester;
C. Clarke, M.C. Goddard, R. Marshall, G.F. Pearce, Rutherford Laboratory, Chilton;
M. Imori, T. Kobayashi, S. Komamiya, M. Koshiba, M. Minowa, S. Orito, A. Sato,
T. Suda, H. Takeda, Y. Totsuka, Y. Watanabe, S. Yamada, C. Yanagisawa, Laboratory
of International Collaboration on Elementary Particle Physics and Department of
Physics, University of Tokyo.

Introduction

The JADE experiment is being carried out at the e^+e^- storage ring PETRA. Data have been taken at total center-of-mass energies ($\sqrt{s}$) between 12 GeV and 35.8 GeV.

In section 1 the most important features of the JADE detector are summarized, in section 2 the specific $\gamma\gamma$-triggers are discussed and sections 3 and 4 deal with the present status of the analyses of the two-photon mediated $\mu\mu$ and multi-hadron final states, respectively.

1. Detector

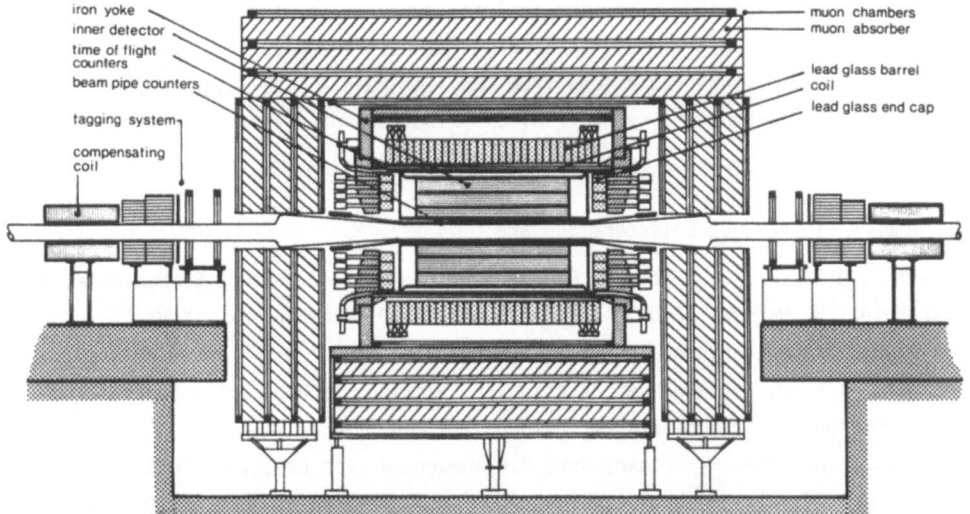

Fig. 1: Side view of the JADE detector at PETRA.

The JADE detector (Fig. 1) which is described in more detail elsewhere[1], consists of the following parts (see also Tab. 1):
- A normally conducting aluminium coil, which produces a solenoidal field parallel to the beam-axis with a maximal strength of 0.5 Tesla uniform within 0.7%.
- A set of beam-pipe and time-of-flight (TOF)scintillation counters, which are used for triggering on charged particles and for particle identification via TOF measurement.
- The inner detector[2]: a cylindrical drift chamber filled with a gas-mixture of argon, methane, and isobutane, operated at a pressure of 4 atm.
  Charged particles coming from the beam-pipe and traversing the whole inner detector give a signal on 48 wires which leads to a momentum resolution of 3.5% x p for particles above 1 GeV/c momentum. The inner detector covers 97% of 4 π if one demands that a particle must at least pass the innermost 8 drift

spaces.

Another important feature of the inner detector is its capability of identifying particles by a dE/dx-measurement.

dE/dx (keV/cm)

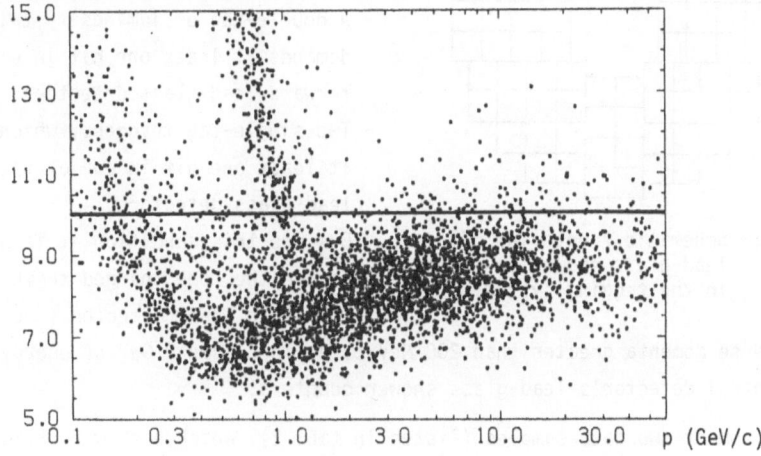

Fig. 2: dE/dx versus momentum for particles from beam-gas reactions. Solid time: theoretical curve for electrons.

Fig. 2 shows for a beam-gas event sample dE/dx versus momentum of each particle. Contributions from electrons, pions and protons are visible. With this method it is possible to separate electrons from pions even in the low-momentum region, where separation by shower counters is not possible.

- A system of lead-glass shower counters which allow the identification of electrons and gamma-rays and the rejection of non-showering particles. The lead-glass, which is divided into a barrel- and an end-cup part, covers 90% of the solid angle.

- A muon detection system which consists of a muon filter to absorb hadrons and drift-chambers to record muons. The filter is made up of three layers of iron boxes loaded with concrete. In front of, between, and behind the filter are 5 layers of drift-chambers installed (only 4 layers at the front ends of the detector). If one demands a particle to pass at least through 3 layers, the muon system covers 92% of 4 π.

- A small angle tagging system which consists of drift-chambers, scintillation counters and an array of 92 lead-glass blocks (Fig. 3) at each end of the central detector, 4.85 m away from the interaction point and covering a polar angle between 34 and 75 mrad.

The tagging system is used for the on-line luminosity-measurement via small angle Bhabha-scattering and for triggering on γγ-processes.

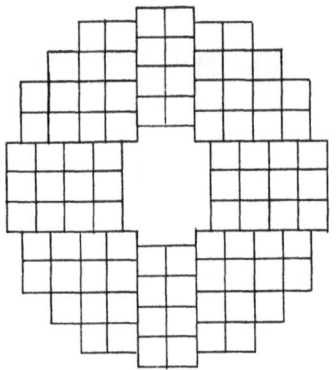

Fig. 3: Schematic front view of lead-glass shower counters in the tagging system.

2. Two-photon triggers and event samples

Besides several other hardware trigger modes the JADE-experiment has the following specific triggers for two-photon reactions in use:
- A double-tag or luminosity trigger, which demands at least one hit in each of the forward lead-glass detectors.
- Two single-tag triggers, which both demand at least one hit in one of the forward lead-glass detectors.

Besides this condition it is necessary to have either one charged track traversing the whole inner detector (such tracks have transverse momenta greater than 200 MeV/c) or more than 2 GeV of energy deposited in the central detector's lead-glass shower counters.

At the moment two data samples (listed in tab. 2), which both were taken in autumn 1979, are being analysed.

3. ee → ee + μμ

Fig. 4 shows an example of a tagged μμ-event.

The analysis of the QED-reaction

$$ee \rightarrow ee\mu\mu$$

has been restricted to tagged events. To belong to this category, an event has to fulfil the following conditions:
   (i)   more than 1 GeV in at least one of the tagging lead-glass arrays;
   (ii)   two opposite charged tracks in the central detector with minimum momentum of 130 MeV/c;
   (iii) the opening angle of the two tracks must be bigger than 27° (to remove converted gamma-rays) and smaller than 175° (to exclude muons from cosmic rays);
   (iV)   if one of the two particles is identified (by dE/dx-measurement, by time-of-flight-measurement, or by the shower-counters) as a proton or both as electrons, the event is rejected;
   (v)   no energy from neutral particles must be deposited in the central detector's shower counters (up to 100 MeV are allowed to account for electronic noise).

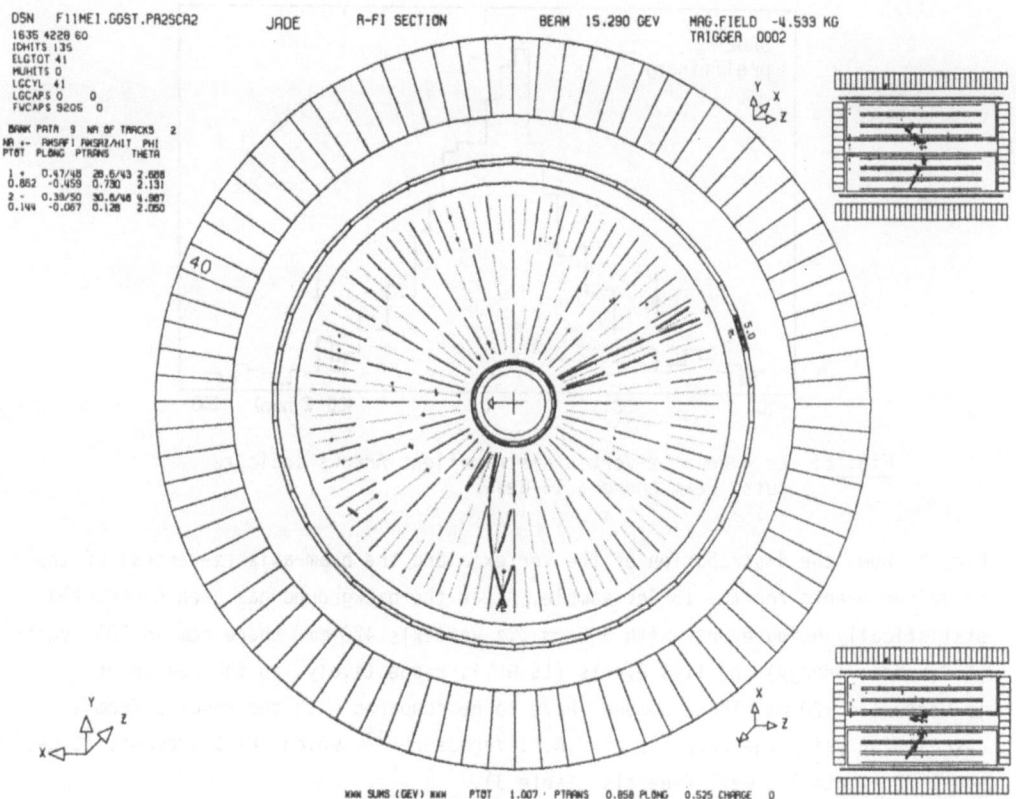

Fig. 4: Event of the type ee → eeμμ (view along the beam). Trajectories of charged
particles are represented by full lines. Hit beam-pipe- and time-of-flight
counters are marked in dark. In the inner detector the mirror hits are also
plotted. The outer circle represents the shower counters, each wedge representing a row of 30 lead-glass counters. The energy deposited in each of
these rows appears in units of MeV. On the right a side view and a top
view of the event is shown. On the left the momenta of the charged particles
are printed.

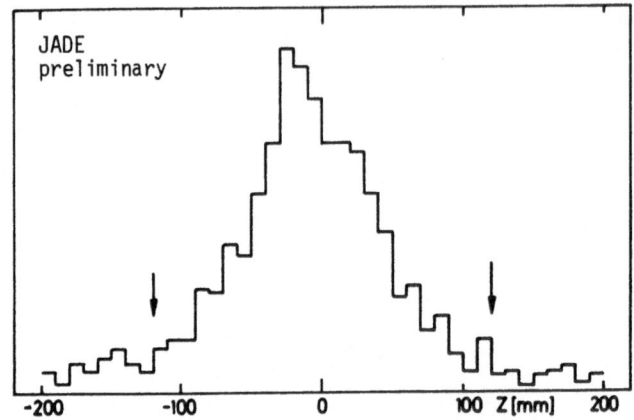

Fig. 5: ee → eeμμ: z-vertex distribution. Arrows indicate
cuts. Beam energy: 15 GeV.

Fig. 5 shows the distribution of the vertex along the beam-axis (z-vertex) of the
so gained events for the 15 GeV sample. After the background has been subtracted
statistically using events with 300 mm ≤ |z-vertex| ≤ 420 mm, there remain 104 events
(6 GeV beam-energy) and 1149 events (15 GeV), respectively, in the region of
|z-vertex| < 120 mm. These numbers have to be compared with the results from a
calculation using the programs of J.A.M. Vermaseren[3], which are 111 events (6 GeV)
and 1177 events (15 GeV) (see also table 3).

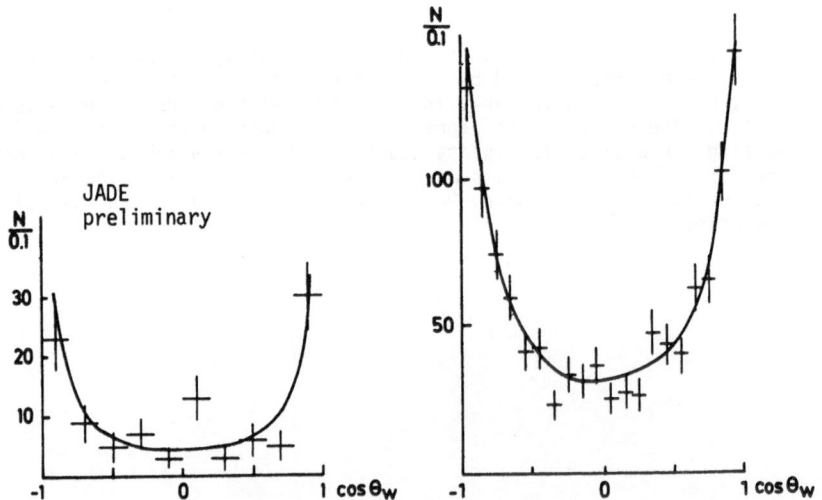

Fig. 6a,b: ee → eeμμ: Polar angle (cos θ) distributions of produced two-
muon system after background subtraction. Error bars indicate
statistical errors only. Solid lines: prediction by simulation.
Beam energies: a) 6 GeV, b) 15 GeV.

Figs. 6a and 6b show the distributions of the polar angle of the produced two-muon
system for the 6 GeV- and the 15 GeV-sample. They peak at small angles, which is
predicted by the simulation (indicated by the full curves).

It has to be pointed out that the predictions are done on the basis of an absolute
normalization, i.e. the numbers of eeμμ-events in the experiment have not been used
as input-parameters to the simulation.

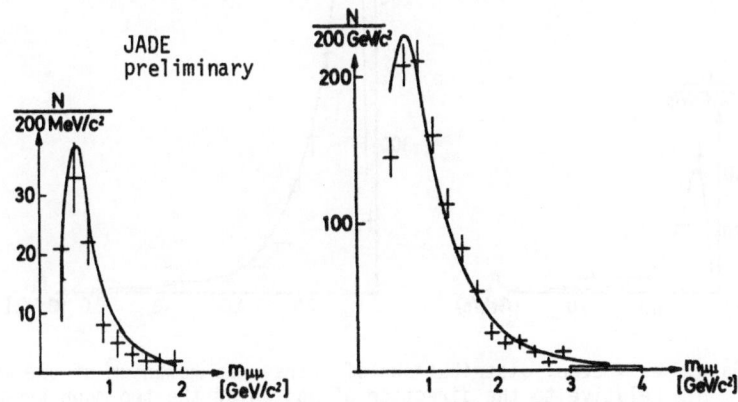

Figs. 7a,b: ee → eeμμ: Invariant mass distributions of produced two-muon
system after background subtraction. Error bars indicate statistical errors only. Solid lines: predictions by simulation.
Beam energies: a) 6 GeV, b) 15 GeV

In figs. 7a and 7b the distributions of the invariant mass of the two-muon system
are plotted. The range of the observed masses reaches up to 2 GeV/c^2 (6 GeV) and
4 GeV/c^2 (15 GeV), respectively.

Figs. 8a and 8b show the distributions of the transverse momentum of one muon
relative to the direction of motion of the two-muon system.

For all distributions the experimental data and the predictions are in good
agreement.

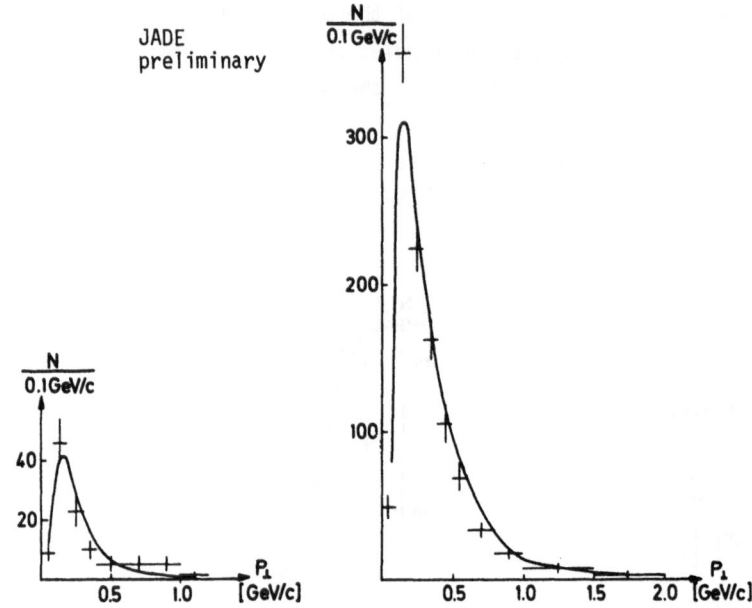

Figs. 8a,b: ee → eeμμ: distributions of transverse momentum of one muon
relative to the direction of motion of the two-muon system.
Background subtracted. Error bars indicate statistical errors only.
Solid lines: predictions by simulation. Beam energies: a) 6 GeV,
b) 15 GeV.

4. ee → ee + hadrons

Fig. 9 shows an example of a tagged multi-hadron event, which has got 6 charged
and 11 neutral tracks in the central detector.

As in the QED-case the analysis has been restricted to tagged events, which must
fulfil the following software selection criteria in order to be accepted:

   (i) more than 1 GeV in at least one of the tagging lead-glass arrays;
   (ii) at least three charged tracks in the central detector originating from the
        interaction region.

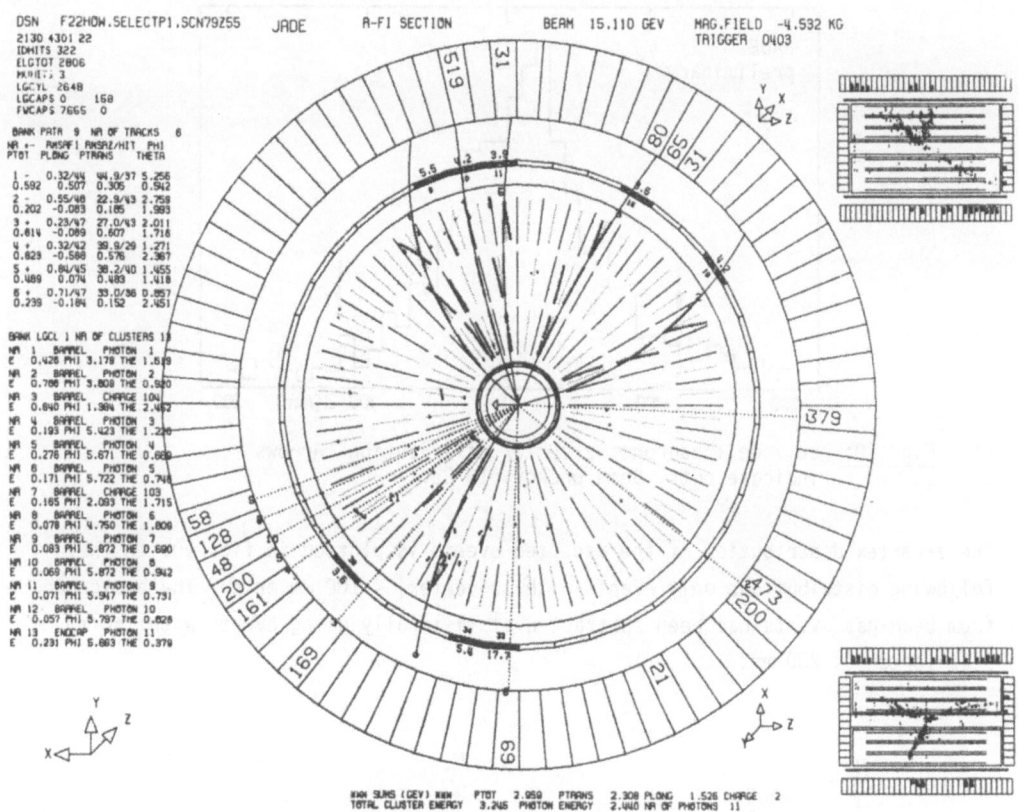

Fig. 9: Event of the type ee → ee + hadrons (view along the beam). Trajectories of charged and neutral particles are represented by full and dotted lines respectively. (For further description see fig. 4).

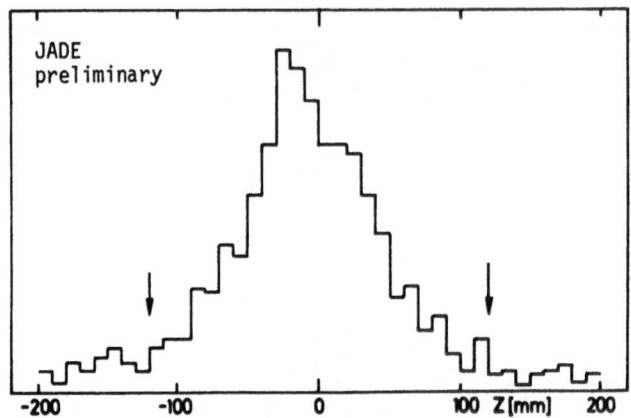

Fig. 10:   ee → ee + hadrons: z-vertex distribution. Arrows
indicate cuts. Beam energy: 15 GeV

The z-vertex distribution of the accepted events is plotted in fig. 10. In the
following distributions only events with $|z\text{-vertex}| < 100$ mm enter. The background
from beam-gas events has been subtracted statistically using events with 150 mm
$\leq |z\text{-vertex}| < 200$ mm.

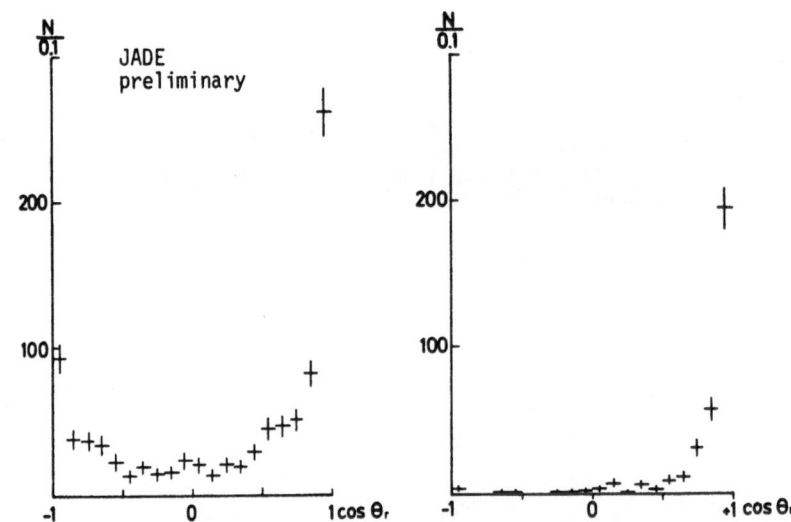

Fig. 11a,b: ee → ee + hadrons: distributions of polar angle between
direction of motion of multi-hadron system and direction
of z-axis marked by the tagged electron. a) distribution
after background subtraction, b) distribution of sample
used for background subtraction (beam-gas events).
Error bars indicate statistical errors only.
Beam energy: 15 GeV.

Fig. 11a shows the distribution of cos θ_r, θ_r being the polar angle between the direction of motion of the produced multi-hadron system and that direction of the z-axis which is marked by the tagged electron. For comparison shows fig. 11b the same distribution for those events, which were used for subtracting the background. The latter show a forward peak only as expected for beam-gas events whereas the former also show a backward peak.

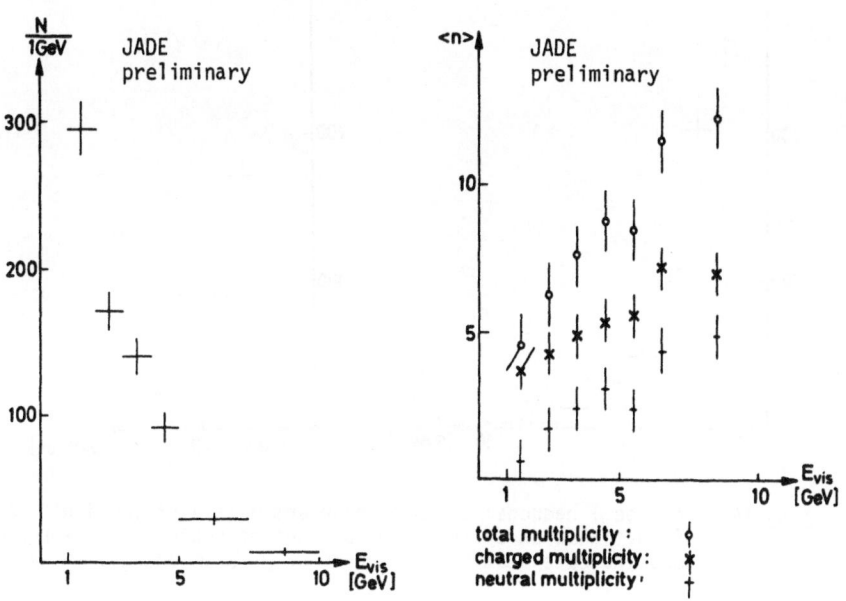

Fig. 12: ee → ee + hadrons: Total visible energy distribution of hadrons after background subtraction. Error bars indicate statistical errors only. Beam energy: 15 GeV.

Fig. 13: ee → ee + hadrons: Distributions of observed charged (x), neutral (+), and total (o) multiplicities of hadrons as function of observed energy. Error bars indicate statistical and estimated systematic errors. Beam energy: 15 GeV.

In fig. 12 is plotted the distribution of the total visible energy, which peaks for small energies and reaches up to 10 GeV. Fig. 13 shows the distributions of the observed charged, neutral and total multiplicities of the hadronic final state as a function of the observed energy.

The distribution of the visible invariant mass of the hadrons is plotted in fig. 14. The bulk of the events has got small invariant masses, but some events with masses up to 10 GeV/c^2 are observed.

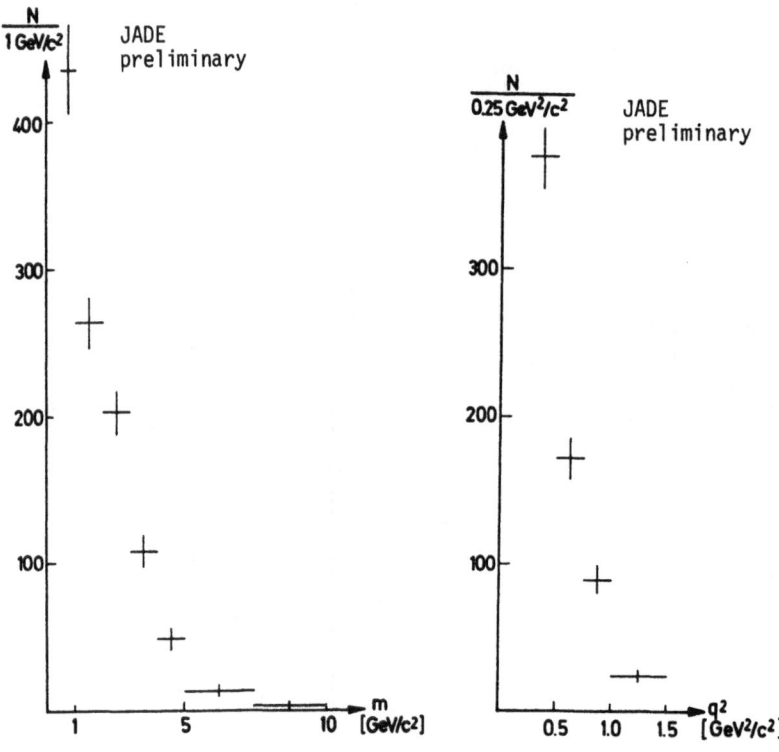

Fig. 14: ee → ee + hadrons: Visible invariant mass distribution of multi-hadron system after background subtraction. Error bars indicate statistical errors only. Beam energy: 15 GeV.

Fig. 15: ee → ee + hadrons: Distributions of squared four-momentum transfer q^2 of that virtual photon which belongs to the tagged electron. Background subtracted. Error bars indicate statistical errors only. Beam energy: 15 GeV.

Fig. 15 shows the distribution of the squared four-momentum transfer

$$q^2 = 2\, E_b E' \,(1-\cos\theta)$$

$(E_B$: beam energy,

E': energy of the tagged electrons,

θ : polar angle of the tagged electron)

of that virtual photon, which belongs to the tagged electron. With the above described tagging system a range in q^2 up to 1.5 GeV/c^2 at beam energies of 15 GeV is covered.

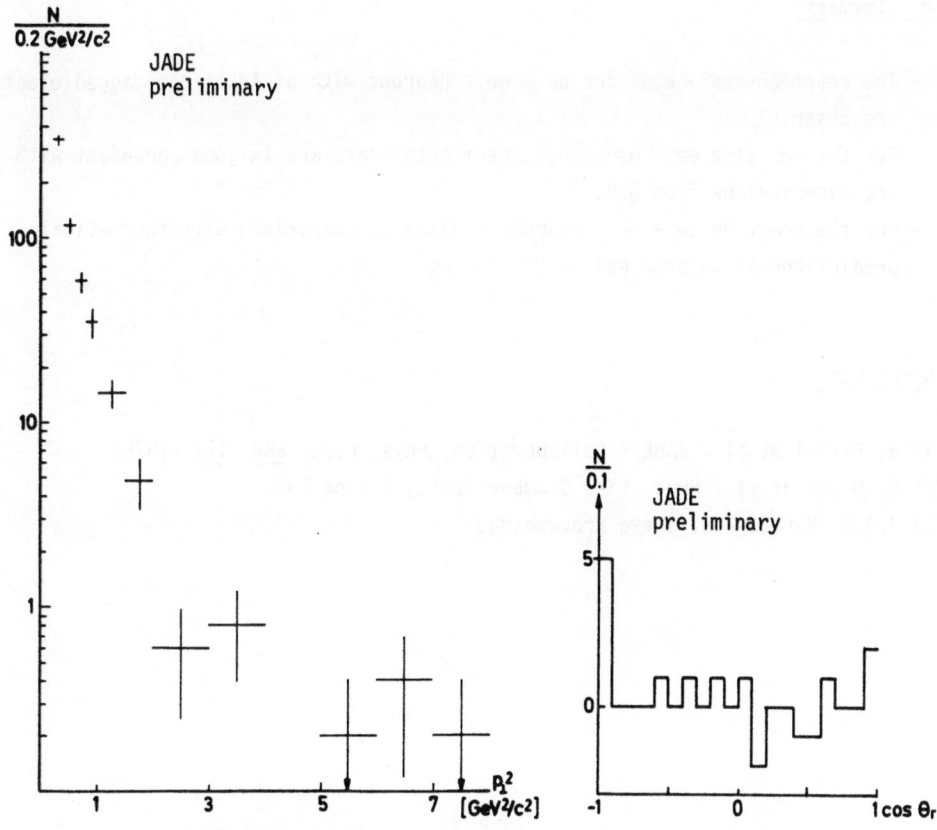

Fig. 16: ee → ee + hadrons: Distribution of squared transverse momentum, $p_\perp^2$, of each hadron relative to the direction of motion of the visible hadronic system. Background subtracted. Error bars indicate statistical errors only. Beam energy: 15 GeV.

Fig. 17: ee → ee + hadrons: Same distribution as fig. 11a for events with at least one hadron with $p_\perp^2 > 2$ GeV^2/c^2.

In fig. 16 is plotted the distribution of the square of the transverse momentum, $p_\perp^2$, of each hadron relative to the direction of motion of the visible hadronic system. It shows a steep slope for small values of $p_\perp^2$, whereas for higher values of $p_\perp^2$ the slope becomes less steep. This effect is not due to beam-gas background as can be seen from fig. 17, the $\cos \theta_r$-distribution of events with at least one particle with $p_\perp^2 > 2$ GeV^2/c^2: it shows a backward peak, which is not present in background events (cf. figs. 11a,b).

5. Summary

- The reactions ee → eeμμ and ee → ee + hadrons with at least one tagged electron are observed.
- For the reaction ee → eeμμ the experimental data are in good agreement with the expectations from QED.
- For the reaction ee → ee + hadrons a detailed comparison with theoretical predictions is in progress.

References

1) W. Bartel et al., JADE - Collaboration, Phys. Lett. 88B, 171 (1979)
2) H. Drumm et al., Proc. Wire Chamber Conf., Vienna 1980
3) J.A.M. Vermaseren, these proceedings

Table 1

Components of the JADE - detector

Component	geometrical dimensions	characteristics

solenoid magnet	diameter: 2 m
	length: 3.5 m
	thickness of coil: 7 cm
		maximal field strength:
		  0.5 Tesla
		uniformity of field:
		  99.3%

beam-pipe counters		no. of counters: 24
time-of-flight counters		no. of counters: 42

inner detector	diameter: 1.6 m
	length: 2.4 m
		no. of signal wires: 1536
		gas: argon/methane/iso-
		    butane (0.887:0.085:
		    0.028) at 4 atm
		    pressure
		covered solid angle: 97%

lead-glass shower counters	inner surface of one block: 85 x 102 mm^2 depth of one block: 300 mm	no. of blocks: 2715 covered solid angle: 90%
muon filter		no. of drift chambers: 634 gas: argon/ethane (0.9:0.1) at atmospheric pressure covered solid angle: 92%
tagging system (lead-glass shower counters, scintillation counters, drift chambers)	inner surface of one lead- glass block: 80 x 80 mm^2 depth of one block: 400 mm distance of lead-glass from interaction region: 4850 mm	no. of drift chambers: 2 x 12, no. of scintil- lation counters: 2 x 8, no. of lead-glass blocks: 2 x 92, range of polar angle covered by the lead-glass: 34-75 mrad

Table 2

Data samples used in the analyses

Beam energy E_b [GeV]	Luminosity $\int L_{ee}$ [nb^{-1}]	analysed reaction
"15"	1830	ee → eeμμ
(14.95 - 15.73)		ee → eeh
6	83	ee → eeμμ

Table 3

Numbers of tagged ee → eeμμ-events

Beam energy E_b [GeV]	number of events	
	data	simulation
6	104	111
15	1149	1177

THE SMALL ANGLE TAGGING SYSTEM OF CELLO
presented by M.Goldberg-LPNHE-Paris

A small angle detector (25 to 50 mR) has been built, with the CELLO spectrometer at Petra, in order to provide:

- Luminosity Monitoring
- Hints for separation between 1 photon and 2 photon processes.
- A tool for 2 photon interaction study.

This is a realisation of the Paris Group within the CELLO collaboration which includes the laboratories of DESY,KARLSRUHE,MUNICH,ORSAY, PARIS and SACLAY.

In the following we provide before a brief description of this tagging-system a rapid overview of the CELLO Spectrometer.

THE CELLO SPECTROMETER

The CELLO Spectrometer aims to identify and to measure the energy of leptons, photons and hadrons. It consists of (see fig.1) a Central Detector, end caps and the small angle tagging system. The overall acceptance is .96 of 4π str.

The central detector includes a supraconducting solenoïd 1.5m in diameter and 3.5m Long. Its maximum field is 1.3 T. The material thickness is 0.47 X_o.

The tracking Device is made with cylindrical Proportional and Drift Chambers. There are 5 Proportional chambers with anode and Cathode Read Out and 7 Drift Chambers for fine resolution of the order of .2mm.

Around the magnet there is an Electromagnetic Shower Detector of the Lead Liquid Argon type. There are 16 modules of Lead stacks with strips at 0°,45° and 90°. The depth is 20 Xo with a fine longitudinal and transversal sampling.

An iron hadron filter of 5 to 8 Absorption length surrounded by a set of 32 wire chamber allows muon identification.

The End Caps consist of proportional chambers and lead liquid argon calorimeters.

THE SMALL ANGLE TAGGING DETECTOR

Its acceptance is limited from 25 to 50 mrad due to the beam pipe diameter and to the compensating magnets. The schematic of the Detector is shown in fig.2. Its position is between 5 and 6.7 m from the interaction point.

The tracking of the electrons is performed by a set of 6 planes of Drift chambers. The drift space is 22 mm. A spatial resolution of .3mm is obtained with wire oriented at 45°(u) and 135°(v). Three sets of u.v. planes are positionned respectively about 510,550,580 cm from the interaction point. In the middle set there are two arrays of scintillators (u and v orientation) to disentangle left Right ambiguities, to define the acceptance and for precise timing.

Electron identification and energy measurements are performed by a lead glass spectrometer with three layers (vertical 3 Xo, Horizontal 3 Xo,Longitudinal 14 Xo). This sampling gives a e/hadron discrimination around 1%.

Due to the current luminosity($\mathcal{L} \sim 10^{30}$ cm^{-2} sec^{-1}) at Petra, the main use of this detector for 2 photon physics will be in the single tagging mode, the number of double tagged events being marginal.

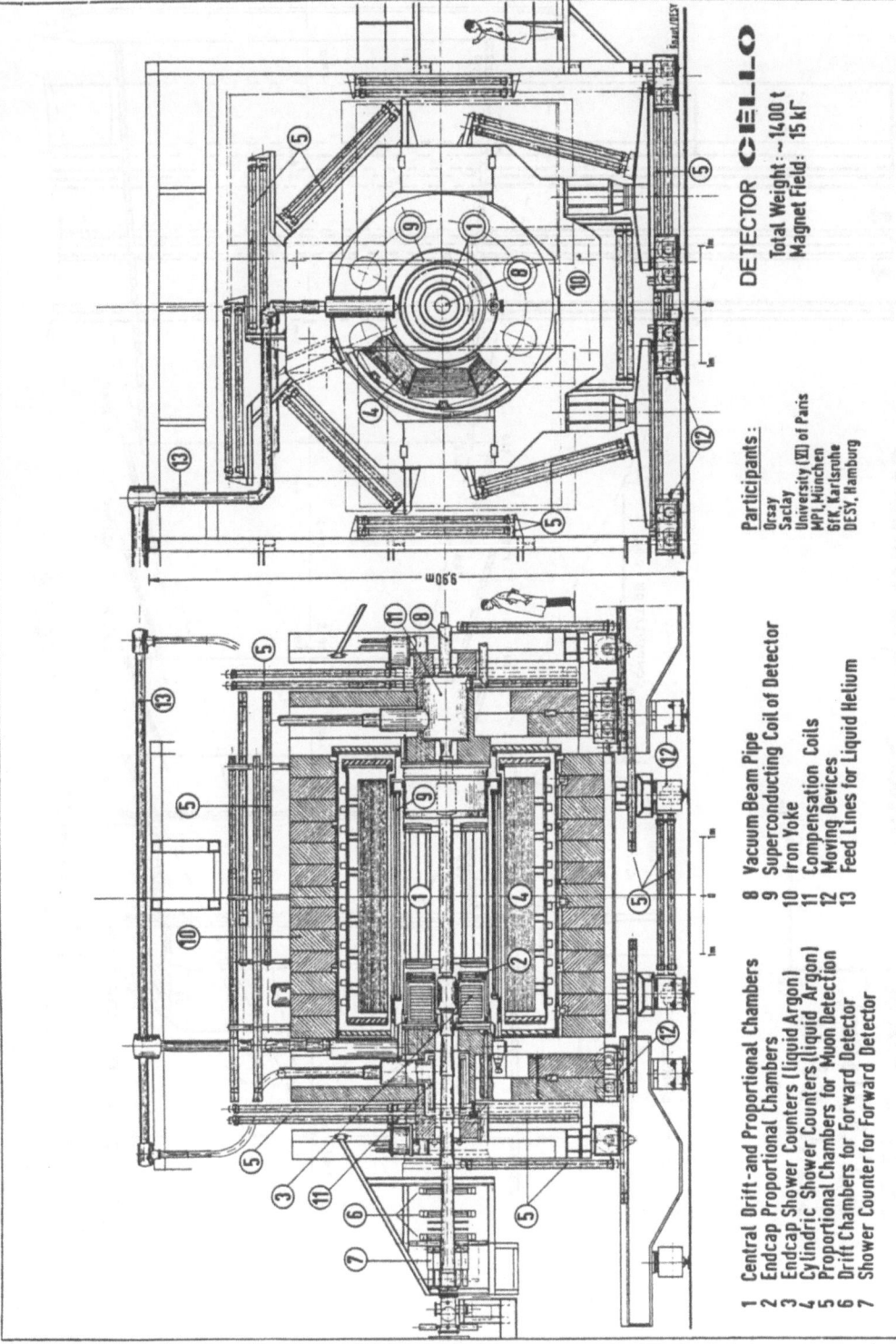

DETECTOR GELLO

Total Weight: ~ 1400 t
Magnet Field: 15 kr

Participants:
Orsay
Saclay
University (VII) of Paris
MPI, München
GfK, Karlsruhe
DESY, Hamburg

1 Central Drift- and Proportional Chambers
2 Endcap Proportional Chambers
3 Endcap Shower Counters (liquid Argon)
4 Cylindric Shower Counters (liquid Argon)
5 Proportional Chambers for Muon Detection
6 Drift Chambers for Forward Detector
7 Shower Counter for Forward Detector
8 Vacuum Beam Pipe
9 Superconducting Coil of Detector
10 Iron Yoke
11 Compensation Coils
12 Moving Devices
13 Feed Lines for Liquid Helium

FIG. 1

180

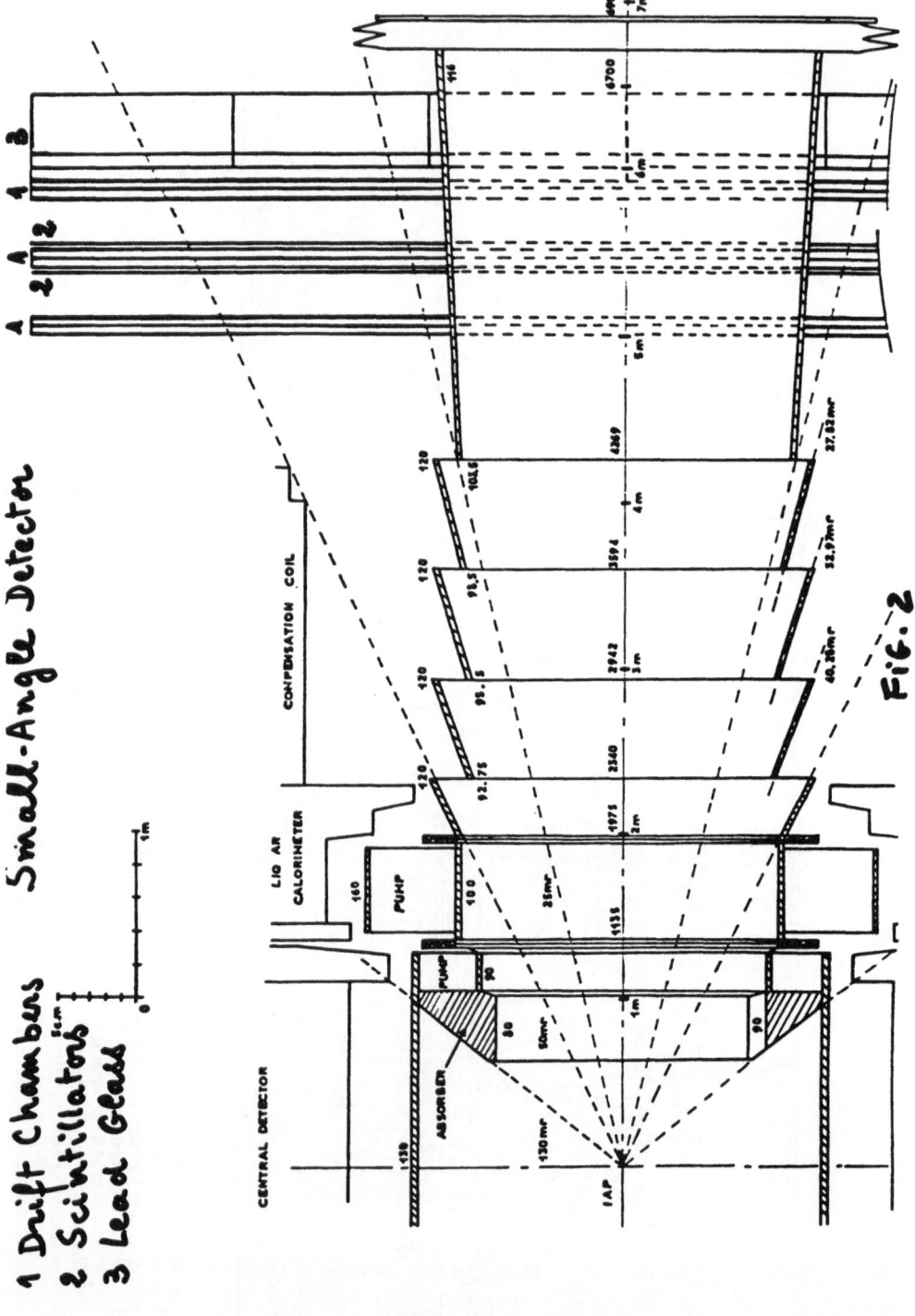

1 Drift Chambers   Small-Angle Detector
2 Scintillators
3 Lead Glass

Fig. 2

BACK-FACTORIZATION PROCEDURES

C. Carimalo, P. Kessler and J. Parisi [+)]

Laboratoire de Physique Corpusculaire, Collège de France, Paris, France

Abstract

We here discuss the problem of back-factorization in $\gamma\gamma$ collision processes; i. e.,
how to extract, from the experimental measurement of a reaction e e $\longrightarrow$ e e X, the
information on the physically interesting process $\gamma\gamma \rightarrow$ X in the form of only one
term or very few terms (cross section, polarization terms, structure functions).
Back-factorization procedures are basically associated with the equivalent-photon
approximation.

In part I, we define the principles of back-factorization for the reactions considered, and we discuss its application to experiments with finite-angle tagging of both
outgoing electrons. Using a helicity formalism, we show that, under certain conditions - which appear to be quite stringent, as is proven by our numerical checks -,
the completely differential cross section of e e $\longrightarrow$ e e X may be expressed as a
five-term formula; after integration over azimuthal angles, it becomes a one-term
formula, i. e. the double equivalent-photon approximation. In addition to double-
tagging, measurements based on single-tagging and on double anti-tagging are also
discussed from the point of view of back-factorization.

In part II, we treat the special configuration to be considered for a determination
of the structure functions of the photon. Using the naive quark model without higher-
order corrections, we are studying analytically and checking numerically the conditions to be satisfied in order to obtain a one-term formula for extraction of the
information on the photon's quark content from the data measured.

+) Presented by C. Carimalo (Part I) and J. Parisi (Part II).

I. Principles, and application to finite-angle tagging measurements

Back-factorization procedures are basically associated with the equivalent-photon approximation (E. P. A.), formerly called Williams-Weizsäcker approximation [1]. This approximation method allows one to connect electron-induced processes with photon-induced ones (fig. 1) in the following way.

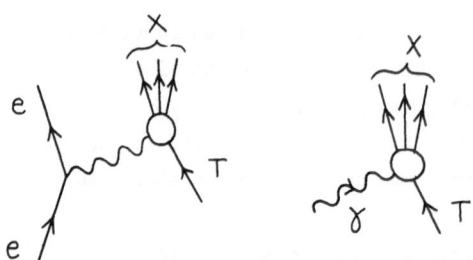

Fig. 1. Electron induced and photon-induced processes of target excitation

$$\sigma_{eT \to eX} \simeq \int N(\omega) \, \sigma_{\gamma T \to X}(\omega) \, d\omega$$

with:
$$N(\omega) = \frac{2 \alpha}{\pi} \frac{1}{\omega} \left[(1 - \frac{\omega}{E_0} + \frac{\omega^2}{E_0^2}) \ln \frac{E_0}{m_e} - \frac{1}{2} (1 - \frac{\omega}{E_0}) \right]$$

where E_0 is the electron's energy, and ω is its energy loss (or the equivalent photon's energy).

This very simple formula (or some slightly different versions of it) was extensively used for many years, allowing for easy computation of cross sections for many processes, and thus for an economy of human effort and computing time, and also for greater physical transparency.

In particular, F. Low in 1960 [2], using a double equivalent-photon approximation, was able to show how the life-time of the π^0 may be connected with the cross section of the process e e $\longrightarrow$ e e π^0:

$$\sigma_{ee \to ee\pi^0} \simeq \int N(\omega) \, d\omega \, N(\omega') \, d\omega' \, \sigma_{\gamma\gamma \to \pi^0}$$

where:
$$\sigma_{\gamma\gamma \to \pi^0} \sim (1/\tau_{\pi^0}) \, \delta(m_\pi^2 - 4 \omega \omega')$$

When 10 years later photon-photon collisions became fashionable as a new area in high-energy physics, this approach was used more or less by all authors involved [3].

We are going to show that, for $\gamma\gamma$ collision processes, the E. P. A. is actually much more than a simplifying procedure for computations: It becomes an irreplaceable tool for the analysis of experimental measurements.

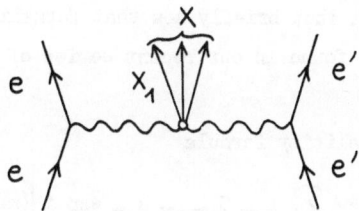

Fig. 2. General Feynman diagram for $\gamma\gamma$ collision processes

Any exact theoretical calculation of the type of process represented in fig. 2 shows the following structure:

$$d\sigma_{ee \to eeX} \sim L_{\mu\nu}\, C^{\mu\nu\rho\sigma}\, R_{\rho\sigma}\, dP_{L.I.}$$

where the second-rank tensors $L_{\mu\nu}$ and $R_{\rho\sigma}$ pertain to the left-hand and right-hand vertex respectively, and the fourth-rank tensor $C^{\mu\nu\rho\sigma}$ to the central vertex; $dP_{L.I.}$ is the Lorentz-invariant phase-space.

Now, every time we wish to compare an experimental measurement with theory, we may choose between two procedures: model-fitting, or back-factorization.

(i) <u>Model-fitting</u>. Using a model to compute $C^{\mu\nu\rho\sigma}$, we factorize the expression thus obtained for that fourth-rank tensor with $L_{\mu\nu}$ and $R_{\rho\sigma}$ (known from QED), we possibly integrate over some variables, and we compare the result of this calculation with the measured cross section for $e\,e \to e\,e\,X$. If $\gamma\gamma \to X$ is a QED process, there is no problem. If it is a very simple hadronic process such as resonance production, that procedure might still be applicable without major difficulties. But what about multi-hadron production? No Regge-pole model would probably allow one to write down an expression for $C^{\mu\nu\rho\sigma}$. As for QCD, as soon as we depart from the most naive models, the procedure here defined would be an utterly heavy and hazardous one. Any disagreement with the measurement would be difficult to analyze, and one might have to start again and again building up new models to fit the experimental data.

(ii) <u>Back-factorization</u>. We call "back-factorization" <u>the (in principle model-independent) extraction of the information on $\gamma\gamma \to X$ from the experimental measurement of $ee \to eeX$</u>, in order to compare that information with theoretical predictions. Any theorist, at least, would agree that this should be the better solution. Since, as we said, $L_{\mu\nu}$ and $R_{\rho\sigma}$ are known from QED, this is formally possible. But only formally! As a matter of fact, the fourth-rank tensor $C^{\mu\nu\rho\sigma}$ has $4^4 = 256$ elements, and no computer in the world would be able to disentangle them properly. Therefore, our problem is the following: How can we reduce the above factorization formula from 256 to a small number of terms?

There is, as far as we know, only one way to achieve that goal: namely, by using the helicity formalism. We shall show briefly how that formalism works (more details, and all formulas needed, can be found in our recent series of papers in the "Physical Review" [4]).

We start with the 81-term helicity formula

$$d\sigma_{ee \to eeX}/dP_{L.I.} \sim \sum_{\substack{m,\overline{m} \\ n,\overline{n}}} L_{m\overline{m}} \; C_{m\overline{m},n\overline{n}} \; R_{n\overline{n}} \; \exp i\left[(m-\overline{m})\varphi_1 + (n-\overline{n})(\varphi-\varphi_1)\right]$$

where $m(\overline{m})$, $n(\overline{n})$ are the helicities of both virtual photons, defined in the $\gamma\gamma$ c. m. frame; φ and φ_1 are, in that frame, the relative azimuthal angles between e and e', and between e and X_1 respectively (see fig. 2). To go down from 256 to 81 terms, we implicitly made use of gauge invariance. Each helicity variable takes the values ± 1 (transverse polarization) or 0 (longitudinal polarization).

Hermiticity further reduces the formula to 36 terms. If in addition we assume that the $\gamma\gamma$ process is a 2-body reaction (or a quasi-2-body reaction such as $\gamma\gamma \to X_1$ plus anything), parity and rotational invariance (or angular-momentum conservation) lead us down to a 20-term formula that was first given by Carlson & Tung [5].

To go farther, we must choose specific kinematic conditions. We shall assume Q, Q' $\simeq 0$, defining $Q = \sqrt{-q^2}$, $Q' = \sqrt{-q'^2}$ where q and q' are the virtual photons' four-momenta; we notice that, because of the photons' propagators, an overwhelming contribution to the cross section proceeds, anyway, from those small-transfer values.

Now it is easily shown that the ratio of longitudinal to transverse amplitudes (for the left-hand photon, for instance) is given by

$$\left|\frac{j^0_\nu}{j^{\pm 1}_\nu}\right| = \frac{Q}{q_0} \left|\frac{j_{3\nu}}{j_{\pm,\nu}}\right| \quad \text{with } j_{\pm,\nu} = -\frac{1}{\sqrt{2}}(j_{1\nu} \pm i\, j_{2\nu})$$

where the 3-axis is the $\gamma\gamma$ collision axis, and q_0 is the left-hand photon's energy in the $\gamma\gamma$ c. m. frame.

Assuming the electromagnetic current $j_{\mu\nu}$ to be not too anisotropic in 3-space, we conclude that longitudinal contributions of both photons may be neglected, provided we have: $Q \ll q_0$, $Q' \ll q'_0$. This small-transfer condition may be simply expressed by:

$$Q, \; Q' \ll M_X/2$$

Assuming that condition to be satisfied, one gets the following five-term formula for processes involving the collision of two transverse (quasi-real) polarized photons (here using the helicity subscripts $\pm$ for ± 1):

$$d\sigma_{ee \to eeX}/dP_{L.I.} \sim L_{++}\,(C_{++,++} + C_{++,--})R_{++} + 2L_{++}\,(\mathrm{Re}\,C_{++,+-})\,R_{+-}\cos 2(\varphi - \varphi_1)$$
$$+ 2L_{+-}\,(\mathrm{Re}\,C_{+-,++})\,R_{++}\cos 2\varphi_1 + L_{+-}\,C_{+-,+-}\,R_{+-}\cos 2\varphi$$

$$+ L_{+-} \ C_{+-,-+} \ R_{+-} \ \cos 2 \ (2\varphi_1 - \varphi)$$

where the first term is the unpolarized $\gamma\gamma$ cross section, the others being polarization terms.

We checked that formula for the case of production of a pair of massless spin-1/2 fermions (leptons or quarks) in lowest-order perturbation theory. Polar and azimuthal angles were defined as in fig. 3 below.

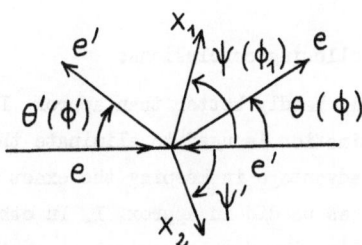

Fig. 3. Specification of polar and azimuthal angles (the latter are given in parentheses next to the corresponding polar ones) for $\gamma\gamma$ collisions involving pair production ($X = X_1 + X_2$).

In addition, we defined:

$$x = (E_o - E)/E_o = \omega/E_o \qquad\qquad x' = (E_o - E')/E_o = \omega'/E_o$$

where E_o is the beam energy; E, E' are the energies of the outgoing electrons, and ω, ω' those of the virtual photons. We also introduced following definition:

$$\mathcal{E} = \frac{2 \ Q}{(M_X)_{Q,Q' \to 0}} = \theta \sqrt{\frac{1 - x}{x \ x'}} \qquad\qquad \mathcal{E}' = \frac{2 \ Q'}{(M_X)_{Q,Q' \to 0}} = \theta' \sqrt{\frac{1 - x'}{x \ x'}}$$

so that our a priori condition of validity of the 5-term formula practically became: $\mathcal{E}, \mathcal{E}' \ll 1$.

What exactly means "much smaller than 1"? That is what we were going to check.

We considered the differential cross section

$$\bar\sigma = \frac{d\sigma_{ee \to eeX}}{dE \ d\Omega \ dE' \ d\Omega' \ d\Omega_1} = K \ d\sigma_{ee \to eeX}/dP_{L.I.}$$

where K is a kinematic factor, and Ω, Ω', Ω_1 are the solid angles of e, e' and X_1 respectively (see fig. 3).

We computed the relative error

$$\Delta = (\bar\sigma_{approx} - \bar\sigma_{exact})/\bar\sigma_{exact}$$

for fixed values of x, x'; $\mathcal{E}, \mathcal{E}'$ (or θ, θ'), letting ψ and the azimuthal angles go through a wide variety of values (with the restriction: $30° < \psi < 150°$). As is easily seen, our results are independent of beam energy.

Two different approximations were used: Approx. I, where the 5-term formula was computed exactly; approx. II where - for coherence, and for simplification - we made $\mathcal{E}$, $\mathcal{E}' = 0$ wherever permitted inside the 5-term formula.

As an example, we show the cases: x = 0.1, x' = 0.1 (Table 1); x = 0.1, x' = 0.4 (Table 2).

From Tables 1 and 2, we draw following conclusions:

(i) In the average, approx. I is hardly better than approx. II. We thus conclude that, once the small-transfer approximation is used to eliminate the longitudinal contributions, there is no particular advantage in keeping the exact expression of the remaining (purely transverse) terms, as we did in approx. I. In other words, since here approx. II is equivalent to applying the double equivalent-photon approximation (including, however, polarization terms) in its most simplified form as given by the small-transfer approximation, we may make following statement: <u>Generally speaking, not much can be gained by "refining" the equivalent-photon approximation with respect to the simplest formula (the standard formula) obtained by making use of the small-transfer approximation; looking for such refinements is a useless complication.</u>

(ii) The error range, in both approximations, grows rapidly with increasing θ, θ'. If we wish to keep Δ systematically smaller than, let us say, a factor of 2, we should stick to values of θ and θ' smaller than about 20 mrad. In other words, tagging systems like those existing at PETRA and PEP would be unfit for such a determination of the five independent transverse terms ($\gamma\gamma$ cross section, plus four polarization terms) by back-factorization.

We shall now show how one arrives at the one-term approximation which is, properly speaking, the double E. P. A. Noticing that, for $\mathcal{E}, \mathcal{E}'$ going to zero, azimuthal angles in the $\gamma\gamma$ c. m. frame on one hand and in the lab frame on the other hand tend to become the same, i. e.

$$\varphi \simeq \phi' - \phi \qquad \varphi_1 \simeq \phi_1 - \phi$$

and noticing in addition that, in the 5-term formula above, all quantities involved except for the cosine factors then become independent of ϕ, ϕ', integration over ϕ and ϕ' leads to

$$\tilde{\sigma} = \int \bar{\sigma} \, d\phi \, d\phi' = \frac{d\sigma_{ee \to eeX}}{dE \, d(\cos\theta) \, dE' \, d(\cos\theta') \, d\Omega_1} \simeq \int K \, L_{++} \, (c_{++,++} + c_{++,--}) \, R_{++} \, d\phi \, d\phi'$$

Since now the polarization terms have all vanished, that formula describes the scattering of two unpolarized quasi-real photons.

That one-term formula was again checked numerically, choosing - as before - fixed va-
lues for x, x'; $\varepsilon, \varepsilon'$ (or θ, θ'), and letting ψ pass through a set of values lying
between 30° and 150°. For each configuration considered, we again computed the range
of variation of the relative error

$$\Delta = (\widetilde{\sigma}_{approx} - \widetilde{\sigma}_{exact})/\widetilde{\sigma}_{exact}$$

Again we used both types of approximation: Approx. I where the remaining term was computed exactly, and approx. II where we made $\varepsilon, \varepsilon' = 0$ wherever permitted inside that
term. It should be noticed that approx. I does not allow for back-factorization, since the exact kinematic correlations used there involve the azimuthal angles. Approx.
II, on the other hand, may be written in the form:

$$\frac{d\sigma_{ee \to eeX}}{dx\, d\theta\, dx'\, d\theta'\, d\Omega_1} \simeq N(x,\theta)\ N(x',\theta')\ \frac{d\sigma_{\gamma\gamma \to X}}{d\Omega_1}$$

with the standard equivalent-photon spectrum

$$N(x,\theta) = \frac{2\alpha}{\pi}\frac{1}{\theta}\frac{1}{x}\left(1 - x + \frac{x^2}{2}\right)$$

and the analogous expression of $N(x',\theta')$, and with

$$\frac{d\sigma_{\gamma\gamma \to X}}{d\Omega_1} = \frac{d\sigma_{\gamma\gamma \to X}}{d\Omega_1^*}\frac{d\Omega_1^*}{d\Omega_1}$$

where Ω_1^* is the solid angle of X_1 in the $\gamma\gamma$ c. m. frame. One has:

$$\frac{d\Omega_1^*}{d\Omega_1} \simeq \left[\frac{M_X}{2\,E_0 - E\,(1 - \cos\psi) - E'\,(1 + \cos\psi')}\right]^2$$

Again we considered the cases: x = 0.1, x' = 0.1 (Table 3); x = 0.1, x' = 0.4 (Table
4).

From Tables 3 and 4, we draw following conclusions:

(i) While the error range has been considerably reduced in both approximations by integrating over ϕ, ϕ', approx. I is now distinctly better than approx. II. A close investigation shows that the still large errors in approx. II proceed from the fact that,
in some configurations, the Lorentz transformation between the lab frame and the $\gamma\gamma$
c. m. frame is badly approximated by neglecting terms depending on the azimuthal angles. (These large errors can manifestly not be reduced by modifying the equivalent-
photon spectrum).

(ii) Approx. I, as already said, can only be used for model-fitting. At least, it
would be model-fitting involving only one term, i. e. the $\gamma\gamma$ cross-section, instead
of the fourth-rank tensor $c^{\mu\nu\rho\sigma}$.

(iii) If one wishes to back-factorize, using approx. II - i. e. the double equivalent-photon formula -, and if one wishes to keep the error smaller than a factor of

about 2, one should keep both $\mathcal{E}$ and $\mathcal{E}'$ smaller than $\approx 1/3$, i. e. one should not exceed θ, $\theta' \approx$ 2-4°.

It should be noticed that, by integrating over ψ, one may expect the error range to be considerably reduced again, due to averaging and to possible cancellations between positive and negative ψ values. However, it might be interesting to study angular distributions.

Our general conclusion is the following:

In finite-angle tagging measurements, such as are being performed or foreseen at PETRA and PEP, not only one looses most of the events, i. e. those where the electrons are emitted at small angles (less than $\approx 1°$); but part of the remaining ones, where the electrons are coming out at large angles (more than a few degrees) are also "lost" in some sense, since they don't allow for any simple analysis by back-factorization.

The above discussion was for double-tagging measurements. We shall now briefly discuss the other possibilities to be considered with finite-angle tagging counters: single-tagging and double anti-tagging measurements.

In single-tagging, one of the outgoing electrons remains unseen. Assuming the tagging counters to be almost 100% efficient in the range between the minimal tagging angle θ_{min} and some large angle θ_{max} (where the electron's scattering probability becomes exceedingly small), one may consider, with almost 100% certainty, that the unseen electron was emitted between 0° and θ_{min}. That means that, basically, the scattering angle of that electron may be treated as extremely small. Indeed, since the electron's angular spectrum is $\propto \theta^3 \, d\theta/(\theta^2 + \theta_o^2)^2$ with $\theta_o \approx m_e/E_o$, one gets the average value of θ as

$$\langle \theta \rangle \simeq \int_0^{\theta_{min}} \frac{\theta^4 \, d\theta}{(\theta^2 + \theta_o^2)^2} \bigg/ \int_0^{\theta_{min}} \frac{\theta^3 \, d\theta}{(\theta^2 + \theta_o^2)^2} \simeq \frac{\theta_{min}}{\ln (\theta_{min}/\theta_o)}$$

which is of the order of $\theta_{min}/10$ at high energy, i. e. $\langle \theta \rangle \simeq 2$ mrad at PETRA or PEP.

That means that the corresponding virtual photons are essentially "good" ones, i. e. quasi-real ones. In other words, single-tagging measurements under such conditions correspond to electroproduction on a free-photon target. Using the helicity formalism, they may then be analyzed through a 6-term formula in the general case, or through a 4-term formula in the case of a 2-body (or quasi-2-body) $\gamma\gamma$ reaction.

The latter formula has a structure familiar from, for instance, electroproduction of a pion from a nuclear (or nucleon) target:

$$\frac{d\sigma_{ee \to eeX}}{dx \, dx' \, d\Omega' \, d\Omega_1} \simeq N(x) \; N'(x', \theta', \phi') \; \frac{d\Sigma}{d\Omega_1^*} \frac{d\Omega_1^*}{d\Omega_1}$$

where N(x) is the equivalent-photon spectrum (integrated over angles) for the quasi-real photon, $N'(x',\theta',\phi')$ is a slightly modified equivalent-photon spectrum for the other (far off-shell) photon, and one defines

$$\frac{d\Sigma}{d\Omega_1^*} = \frac{d\sigma_U^{\gamma\gamma}}{d\Omega_1^*} + \eta\frac{d\sigma_L^{\gamma\gamma}}{d\Omega_1^*} + \eta\frac{d\sigma_P^{\gamma\gamma}}{d\Omega_1^*}\cos 2\varphi_1' + 2\sqrt{\eta(1+\eta)}\frac{d\sigma_I^{\gamma\gamma}}{d\Omega_1^*}\cos\varphi_1'$$

where η is the off-shell photon's polarization parameter; φ_1' is, in the $\gamma\gamma$ c. m. frame, the relative azimuthal angle between the tagged electron and the particle X_1 produced (assuming again: $X = X_1 + X_2$, where X_2 is another particle or "anything"). The four terms considered above are: transverse contribution, longitudinal contribution, transverse-transverse interference and transverse-longitudinal interference, respectively.

The above formula for the differential cross section is quite accurate, as we checked numerically. For a beam energy of about 10 GeV and for invariant masses produced that are larger than 1 GeV, assuming a minimal tagging angle of about 20 mrad, it was verified that the error with respect to an exact computation stays lower than 4% in all kinematic configurations considered.

A 6- or 4-term formula is obviously much simpler than the 36- or 20- term formula one has in the general double-tagging case (as long as one cannot make use of the small-transfer approximation). Thus, it may be said - generally speaking - that single-tagging measurements are much simpler to back-factorize than double-tagged ones. However, disentangling 6 or 4 terms may still be a complicated task. The ideal situation, here again, would be to reduce the number of terms to one, i. e. to be able to use the double E. P. A. Again, that situation occurs only where the tagged electron's scattering angle remains small (less than a few degrees).

Finally, the most satisfactory type of measurement - from the point of view of back-factorization - to be achieved with finite-angle tagging counters, is double anti-tagging, i. e. measurement of events where neither of both electrons is seen in the tagging counters; in other words, the latter are used as veto counters on both sides. As just shown, the mean scattering angle of either unseen electron is $\langle\theta\rangle \simeq 2$ mrad under the conditions of PETRA or PEP. The double E. P. A. may then be expected to work with an accuracy better than 1%.

Of course, single-tagging and - even more - double anti-tagging measurements have their shortcomings as far as other problems (background rejection, event reconstitution) are concerned [6]. The ideal kind of experiment, providing an adequate answer to all problems, remains - as far as it is technically possible - double-tagging at 0°.

190

Table 1. Range of $\Delta = (\bar{\sigma}_{approx} - \bar{\sigma}_{exact})/\bar{\sigma}_{exact}$ in the process $e\,e \longrightarrow e\,e\,X_1\,X_2$ (where $X_1\,X_2$ is a pair of massless spin-1/2 particles), according to approximation I or II, for various kinematic configurations with fixed values of x, x'; ε,ε' (or θ, θ''), letting γ and all azimuthal angles go through a wide variety of values ($30° < \gamma < 150°$). Here: x = 0.1, x' = 0.1.

		$\varepsilon = 1/100$ ($\theta \simeq 1$ mrad)	$\varepsilon = 1/30$ ($\theta \simeq 3$ mrad)	$\varepsilon = 1/10$ ($\theta \simeq 10$ mrad)	$\varepsilon = 1/3$ ($\theta \simeq 30$ mrad)
$\varepsilon' = 1/100$ ($\theta' \simeq 1$ mrad)	approx. I	$-7\% < \Delta < +7\%$	$-13\% < \Delta < +16\%$	$-31\% < \Delta < +43\%$	$-75\% < \Delta < +227\%$
	approx. II	$-20\% < \Delta < +5\%$	$-23\% < \Delta < +14\%$	$-36\% < \Delta < +40\%$	$-83\% < \Delta < +224\%$
$\varepsilon' = 1/30$ ($\theta' \simeq 3$ mrad)	approx. I	$-13\% < \Delta < +16\%$	$-19\% < \Delta < +24\%$	$-36\% < \Delta < +54\%$	$-77\% < \Delta < +254\%$
	approx. II	$-23\% < \Delta < +14\%$	$-23\% < \Delta < +17\%$	$-38\% < \Delta < +42\%$	$-83\% < \Delta < +228\%$
$\varepsilon' = 1/10$ ($\theta' \simeq 10$ mrad)	approx. I	$-31\% < \Delta < +43\%$	$-36\% < \Delta < +54\%$	$-48\% < \Delta < +87\%$	$-81\% < \Delta < +342\%$
	approx. II	$-36\% < \Delta < +40\%$	$-38\% < \Delta < +42\%$	$-45\% < \Delta < +53\%$	$-85\% < \Delta < +241\%$
$\varepsilon' = 1/3$ ($\theta' \simeq 30$ mrad)	approx. I	$-75\% < \Delta < +227\%$	$-77\% < \Delta < +227\%$	$-81\% < \Delta < +342\%$	$-91\% < \Delta < +770\%$
	approx. II	$-83\% < \Delta < +224\%$	$-83\% < \Delta < +228\%$	$-85\% < \Delta < +241\%$	$-87\% < \Delta < +288\%$

Table 2. Same as Table 1, except that here: x = 0.1, x' = 0.4

		$\varepsilon = 1/100$ ($\theta \simeq 2$ mrad)	$\varepsilon = 1/30$ ($\theta \simeq 7$ mrad)	$\varepsilon = 1/10$ ($\theta \simeq 20$ mrad)	$\varepsilon = 1/3$ ($\theta \simeq 70$ mrad)
$\varepsilon' = 1/100$ ($\theta' \simeq 3$ mrad)	approx. I	$-7\% < \Delta < +6\%$	$-13\% < \Delta < +12\%$	$-30\% < \Delta < +29\%$	$-68\% < \Delta < +101\%$
	approx. II	$-8\% < \Delta < +8\%$	$-11\% < \Delta < +9\%$	$-26\% < \Delta < +18\%$	$-63\% < \Delta < +66\%$
$\varepsilon' = 1/30$ ($\theta' \simeq 9$ mrad)	approx. I	$-13\% < \Delta < +11\%$	$-19\% < \Delta < +18\%$	$-35\% < \Delta < +37\%$	$-69\% < \Delta < +116\%$
	approx. II	$-23\% < \Delta < +25\%$	$-24\% < \Delta < +24\%$	$-27\% < \Delta < +29\%$	$-64\% < \Delta < +74\%$
$\varepsilon' = 1/10$ ($\theta' \simeq 26$ mrad)	approx. I	$-29\% < \Delta < +32\%$	$-34\% < \Delta < +40\%$	$-46\% < \Delta < +62\%$	$-72\% < \Delta < +164\%$
	approx. II	$-58\% < \Delta < +83\%$	$-59\% < \Delta < +81\%$	$-60\% < \Delta < +84\%$	$-57\% < \Delta < +98\%$
$\varepsilon' = 1/3$ ($\theta' \simeq 86$ mrad)	approx. I	$-62\% < \Delta < +137\%$	$-64\% < \Delta < +152\%$	$-68\% < \Delta < +199\%$	$-80\% < \Delta < +440\%$
	approx. II	$-95\% < \Delta < +373\%$	$-96\% < \Delta < +369\%$	$-95\% < \Delta < +368\%$	$-87\% < \Delta < +391\%$

Table 3. Range of $\Delta = (\delta_{approx} - \delta_{exact})/\delta_{exact}$ in the process e e $\longrightarrow$ e e X_1 X_2 (where X_1 X_2 is a pair of massless spin-1/2 particles), according to approximation I or II, for various kinematic configurations with fixed values of x, x'; ε, ε' (or θ, θ'), letting ψ go through a wide variety of values (between 30° and 150°). Here: x = 0.1, x' = 0.1.

		ε = 1/100 ($\theta \simeq 1$ mrad)	ε = 1/30 ($\theta \simeq 3$ mrad)	ε = 1/10 ($\theta \simeq 10$ mrad)	ε = 1/3 ($\theta \simeq 30$ mrad)
ε' = 1/100 ($\theta' \simeq 1$ mrad)	approx. I	$-1\% < \Delta < +1\%$	$-1\% < \Delta < +1\%$	$-1\% < \Delta < +1\%$	$-7\% < \Delta < 0$
	approx. II	$-1\% < \Delta < +1\%$	$-1\% < \Delta < +1\%$	$-4\% < \Delta < +1\%$	$-33\% < \Delta < +5\%$
ε' = 1/30 ($\theta' \simeq 3$ mrad)	approx. I	$-1\% < \Delta < +1\%$	$-1\% < \Delta < +1\%$	$-1\% < \Delta < +1\%$	$-7\% < \Delta < 0$
	approx. II	$-1\% < \Delta < +1\%$	$-1\% < \Delta < 0$	$-3\% < \Delta < +1\%$	$-33\% < \Delta < +4\%$
ε' = 1/10 ($\theta' \simeq 10$ mrad)	approx. I	$-1\% < \Delta < +1\%$	$-1\% < \Delta < +1\%$	$-2\% < \Delta < 0$	$-7\% < \Delta < 0$
	approx. II	$-4\% < \Delta < +1\%$	$-3\% < \Delta < +1\%$	$-3\% < \Delta < 0$	$-33\% < \Delta < +2\%$
ε' = 1/3 ($\theta' \simeq 30$ mrad)	approx. I	$-7\% < \Delta < 0$	$-7\% < \Delta < 0$	$-7\% < \Delta < 0$	$-7\% < \Delta < +1\%$
	approx. II	$-33\% < \Delta < +5\%$	$-33\% < \Delta < +4\%$	$-33\% < \Delta < +2\%$	$-26\% < \Delta < -6\%$

Table 4. Same as Table 3, except that here: x = 0.1, x' = 0.4

		ε = 1/100 ($\theta \simeq 2$ mrad)	ε = 1/30 ($\theta \simeq 7$ mrad)	ε = 1/10 ($\theta \simeq 20$ mrad)	ε = 1/3 ($\theta \simeq 70$ mrad)
ε' = 1/100 ($\theta' \simeq 3$ mrad)	approx. I	$-1\% < \triangle < +1\%$	$-1\% < \triangle < +1\%$	$-1\% < \triangle < +1\%$	$-7\% < \triangle < 0$
	approx. II	$-1\% < \triangle < +1\%$	$-1\% < \triangle < +1\%$	$-1\% < \triangle < +1\%$	$-4\% < \triangle < +7\%$
ε' = 1/30 ($\theta' \simeq 9$ mrad)	approx. I	$-1\% < \triangle < +1\%$	$-1\% < \triangle < +1\%$	$-1\% < \triangle < +1\%$	$-7\% < \triangle < 0$
	approx. II	$-2\% < \triangle < +1\%$	$-2\% < \triangle < +1\%$	$-1\% < \triangle < +1\%$	$-4\% < \triangle < +6\%$
ε' = 1/10 ($\theta' \simeq 26$ mrad)	approx. I	$-1\% < \triangle < +1\%$	$-1\% < \triangle < +1\%$	$-1\% < \triangle < +1\%$	$-7\% < \triangle < +2\%$
	approx. II	$-12\% < \triangle < +1\%$	$-12\% < \triangle < +1\%$	$-12\% < \triangle < 0$	$-5\% < \triangle < +3\%$
ε' = 1/3 ($\theta' \simeq 86$ mrad)	approx. I	$-6\% < \triangle < +1\%$	$-6\% < \triangle < +1\%$	$-6\% < \triangle < +1\%$	$-23\% < \triangle < +1\%$
	approx. II	$-54\% < \triangle < +1\%$	$-54\% < \triangle < +1\%$	$-51\% < \triangle < +1\%$	$-28\% < \triangle < -3\%$

II. Deep-inelastic configuration

The configuration we are going to consider corresponds to the simplest type of measurement that would allow one in principle to determine the structure functions, i. e. the quark content, of the photon (or of the electron) [7]. In that configuration (see fig. 4), one electron would be emitted at small angle, the other one at large angle; as well, one quark-jet would come out at small angle ("beam-pipe jet"), the other one at large angle (high-p_T jet).

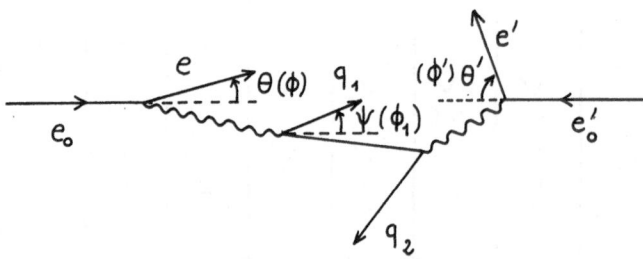

Fig. 4. Deep-inelastic configuration for a measurement of the photon's structure functions (azimuthal angles in parentheses, next to the corresponding polar ones)

The question we are going to examine is whether a simple factorization formula may be used with sufficient accuracy in order to allow one to extract from the experimental data a function measuring the photon's quark content. To perform this check, we shall use the most naive quark model - simply treating the quark as a spin-1/2 fermion of given mass - without any higher-order correction.

Our notations will be the same as in part I (as is seen by comparing fig. 4 with fig. 3, q_1, q_2 now take the place of X_1, X_2).

Using again the helicity formalism, we start with the 20-term formula of Carlson and Tung [5]. Since θ is small, q^2 is small, and therefore all longitudinal contributions for the left-hand photon tend to vanish (how small θ or q^2 should be to make them negligible will be later shown by our numerical checks).

We thus get the following 10-term formula (where $\bar{K}$ is basically a kinematic factor):

$$
\frac{d\sigma_{ee \rightarrow eeq_1 q_2}}{dE\, d\Omega\, dE'\, d\Omega'\, dq_1} = \bar{K} \Big[L_{++} \, (C_{++,++} + C_{++,--}) \, R_{++} + 2 \, L_{++} \, (\text{Re } C_{++,+-}) \, R_{+-} \cos 2\,(\varphi - \varphi_1)
$$
$$
+ \, 2 \, L_{+-} \, (\text{Re } C_{+-,++}) \, R_{++} \cos 2\,\varphi_1 + L_{+-} \, C_{+-,+-} \, R_{+-} \cos 2\,\varphi
$$
$$
+ \, L_{+-} \, C_{+-,-+} \, R_{+-} \cos 2\,(2\,\varphi_1 - \varphi) + 2 \, L_{++} \, (\text{Re } C_{++,+0} - \text{Re } C_{++,0-})
$$
$$
R_{+0} \cos\,(\varphi - \varphi_1) + 2 \, L_{+-} \, (\text{Re } C_{+-,+0}) \, R_{+0} \cos\,(\varphi + \varphi_1)
$$
$$
+ \, 2 \, L_{+-} \, (\text{Re } C_{+-,0+}) \, R_{+0} \cos\,(3\,\varphi_1 - \varphi) + L_{++} \, C_{++,00} \, R_{00}
$$

$$+ 2 L_{+-} C_{+-,00} R_{00} \cos 2 \varphi_1 \Big]$$

The above formula may be further simplified as follows. It can be shown that the condition $\theta \simeq 0$ is sufficient to obtain $\varphi \simeq \phi' - \phi$; i. e. the azimuthal angle between the outgoing electrons becomes the same in the lab frame and in the $\gamma\gamma$ c. m. frame. On the other hand, it is easily shown that all factors in the above formula, except for the cosine functions, become independent of ϕ' in the limit $\theta, \psi \simeq 0$. Therefore, integration over ϕ' leads to vanishing of all terms containing a dependence on φ. We stay with a 4-term formula, which may be written in the following way:

$$\sigma^* = \frac{d\sigma_{ee \to eeq_1 q_2}}{dx \, dx_B \, d(\cos\theta) \, d(\cos\psi) \, d(\cos\theta')} \simeq \int d\phi \, d\phi_1 \, d\phi' \, \widetilde{K} \Big[L_{++} (C_{++,++} + C_{++,--}) R_{++}$$

(formula I)

$$+ 2 L_{+-} (\operatorname{Re} C_{+-,++}) R_{++} \cos 2\varphi_1$$

$$+ L_{++} C_{++,00} R_{00} + L_{+-} C_{+-,00} R_{00} \cos 2\varphi_1 \Big]$$

where we replaced the variable x' by the scaling parameter

$$x_B = \left[\frac{Q'^2}{2 \, q \cdot q'}\right]_{Q^2=0} = \left[\frac{Q'^2}{M_X^2 + Q'^2}\right]_{Q^2=0} = \frac{(1 - x') \sin^2 (\theta'/2)}{x \left[1 - (1 - x') \cos^2 (\theta'/2)\right]}$$

and we defined: $\widetilde{K} = E_0^2 \, (dx'/dx_B) \, \overline{K}$.

Further reduction of the number of terms may be achieved as follows. Using the dynamics of $\gamma\gamma \to q \, \bar{q}$ for spin-1/2 quarks (at lowest order in perturbation theory), one checks that $C_{++,00}$ and $C_{+-,00}$ are negligible with respect to the remaining, purely transverse, terms; thus we stay with the two-term formula

$$\sigma^* \simeq \int d\phi \, d\phi_1 \, d\phi' \, \widetilde{K} \Big[L_{++} (C_{++,++} + C_{++,--}) R_{++} + 2 L_{+-} (\operatorname{Re} C_{+-,++}) R_{++} \cos 2 \varphi_1 \Big]$$

At this point, we shall make use of following expressions:

$$C_{++,++} + C_{++,--} \simeq \frac{16}{(M_X^2 + Q'^2)^2 \, (\chi^2 + \mu^2 + \rho^2)} \Big[(M_X^2 + Q'^2)^2 \, (\chi^2 + \mu^2) - 2\chi^2 M_X^2 \, Q'^2 \Big]$$

$$C_{+-,++} \simeq \frac{16 \, \chi^2 \, M_X^2 \, Q'^2}{(M_X^2 + Q'^2)^2 \, (\chi^2 + \mu^2 + \rho^2)^2}$$

(as computed at lowest order in perturbation theory), where we define:

$$\chi^2 \simeq \frac{2 \, M_X \, E_0}{M_X^2 + Q'^2} \Big[x^2 \psi^2 + 2 \, x \, (1 - x) \, \theta \psi \, \cos (\phi_1 - \phi) + (1 - x)^2 \theta^2 \Big]$$

(χ being the emission angle of X_1 in the $\gamma\gamma$ c. m. frame), and

$$\mu^2 = 4 \, m_q^2/M_X^2 \qquad \rho^2 = 4 \, E_0^2 \, \theta^2 \, Q'^2/(M_X^2 + Q'^2)^2$$

In addition, we use the kinematic relation

$$\cos 2\, \varphi_1 \simeq \frac{x^2 \gamma^2 \cos 2\,(\phi_1 - \phi) + 2\,x\,(1-x)\,\theta\,\gamma \cos\,(\phi_1 - \phi) + (1-x)^2 \theta^2}{x^2 \gamma^2 + 2\,x\,(1-x)\,\theta\,\gamma\,\cos\,(\phi_1 - \phi) + (1-x)^2 \theta^2}$$

Now, using all those formulas, we may proceed to further simplification. If we assume θ to be "smaller than anything else" (we shall later define precisely what that means), all factors in the above 2-term formula become independent of the azimuthal angles, except for $\cos 2\,\varphi_1$ which becomes $\simeq \cos 2\,(\phi_1 - \phi)$. Thus the term containing $\cos 2\,\varphi_1$ vanishes by integration over ϕ or ϕ_1, and finally we get the one-term formula

$$\sigma^* \simeq 8\,\varkappa^3\,\tilde{K}\,L_{++}\,(C_{++,++} + C_{++,--})\,R_{++}$$

which may be expressed as well in the following form (formula II):

$$\sigma^* \simeq N_{\gamma/e}(x,\cos\theta)\,N_{q/\gamma}(x_B,\cos\gamma)\,\frac{d\sigma^{eq}(xx_B,\theta')}{d(\cos\theta')} \qquad\text{, with}$$

$$N_{\gamma/e}(x,\cos\theta) = \frac{2\,\alpha}{\varkappa\,x}\left[\left(1 - x + \frac{x^2}{2}\right)\frac{1}{\theta^2 + \theta_o^2} - (1-x)\frac{\theta_o^2}{(\theta^2 + \theta_o^2)^2}\right]$$

$$N_{q/\gamma}(x_B,\cos\gamma) = \frac{\bar{\alpha}}{2\,\varkappa}\left[\frac{(1-x_B)^2 + x_B^2}{\gamma^2 + \bar{\mu}^2} + \frac{2\,x_B\,(1-x_B)\,\bar{\mu}^2}{(\gamma^2 + \bar{\mu}^2)^2}\right]$$

where one defines $\theta_o = \frac{m_e}{E_o}\frac{x}{1-x}$, $\bar{\mu} = \frac{m_q}{E_o\,x\,(1-x_B)}$, and $\bar{\alpha}$ is derived from α by taking account of fractional charge and colour of the quarks. $d\sigma^{eq}(xx_B,\theta')/d(\cos\theta')$ is the differential cross section for scattering of a quark of energy $x\,x_B\,E_o$ with an electron of energy E_o.

Integrating formula II over θ and γ, one is led to the double leading log approximation, known from the literature.

The quark-distribution function in the electron is given by convolution over x, defining $z = x\,x_B$:

$$N_{q/e}(z) \simeq \int N_{\gamma/e}(x,\cos\theta)\,N_{q/\gamma}(z/x,\cos\gamma)\,dx\,d(\cos\theta)\,d(\cos\gamma)$$

As far as our approximation works, that distribution function is very simply extracted from the measured data, since one has

$$\frac{d\sigma_{ee\to eeq_1 q_2}}{dz\,d(\cos\theta')}\Bigg/\frac{d\sigma^{eq}(z,\theta')}{d(\cos\theta')} \simeq N_{q/e}(z)$$

(Notice that the parameter z is easily derived from the angle and momentum of the outgoing electron e', or as well of the high-p_T jet).

In addition to formula II, we shall try as well a modified version of that formula, obtained by keeping correction terms in θ and θ^2 in the expression of $C_{++,++} + C_{++,--}$ given above. This modified version is (formula II'):

$$\sigma^* \simeq N_{\gamma/e}(x, \cos\theta) \; N'_{q/\gamma}(x_B, \cos\psi) \; \frac{d\sigma^{eq}(xx_B, \theta')}{d(\cos\theta')}$$

where the first and the third factor at r. h. side stay unmodified with respect to formula II, and the modified factor $N'_{q/\gamma}(x_B, \cos\psi)$ is given by

$$N'_{q/\gamma}(x_B, \cos\psi) = \frac{\alpha}{2\pi} \frac{1}{(a^2 - b^2)^{1/2}} \left\{ \left[(1 - x_B)^2 + x_B^2 \right] \left(1 - \frac{1-x}{x^2} \frac{x_B}{1 - x_B} \frac{a}{a^2 - b^2} \theta^2 \right) \right.$$
$$\left. + \frac{2 x_B (1 - x_B) \bar{\mu}^2 a}{a^2 - b^2} \right\}$$

where we define:

$$a = \gamma^2 + \bar{\mu}^2 + \frac{(1-x)\theta^2}{x^2}\left(1 - x + \frac{x_B}{1 - x_B}\right) \qquad b = 2 \frac{1-x}{x}\theta\gamma$$

Obviously, when θ goes to zero, the expression of $N'_{q/\gamma}$ is reduced to that of $N_{q/\gamma}$, and formula II' becomes identical with formula II. More precisely, the condition for identifying both formulas is (considering the fact that the main contribution comes from x values much smaller than 1, whereas x_B should not be too small in the average):

$$\theta^2 \ll x^2 (\gamma^2 + \bar{\mu}^2)$$

That condition thus defines the meaning of our assumption "θ smaller than anything else", as made above. Obviously, as long as one has $\theta \ll m_q/E_o$, the condition is satisfied, whatever the value of γ may be; otherwise, it involves the ratio between θ and $x\gamma$.

In Tables 5 and 6, we compared the three approximations I, II and II' with an exact computation for fixed values of x, x_B, θ and γ, letting θ' vary in wide limits (160 mrad $< \theta' <$ 400 mrad). The beam energy was fixed at 15 GeV (Table 5) or 70 GeV (Table 6). The quark mass was set at 300 MeV. The Tables show the range of variation of the relative error

$$\Delta = (\sigma^*_{approx} - \sigma^*_{exact})/\sigma^*_{exact}$$

From those Tables, we draw following conclusions:

(i) Approx. II' differs in general very little from approx. I; thus the procedure applied in deriving approx. II' was indeed correct.

(ii) Approx. II' is all right, up to $\approx 20\%$, at E_o = 15 GeV; it is worse, although not dramatically bad, at E_o = 70 GeV; that energy-dependent effect is obviously due to the quark mass introducing a mass (or energy) scale.

(iii) As for approx. II, it is all right - as we predicted - wherever one has $\theta \ll m_q/E_o$ (i. e. $\theta \ll 0.02$ at 15 GeV, and $\theta \ll 0.004$ at 70 GeV). Actually, we notice that it is good at θ = 2 mrad for both energies, and still not too bad at 12 mrad when E_o

= 15 GeV. Wherever the condition $\theta \ll m_q/E_o$ is not satisfied, it is the ratio $\theta/(x\psi)$ that becomes critical; therefore we are not surprised to find the most extreme (and indeed quite catastrophic) discrepancies in the right-hand upper region of the Tables (where θ is relatively large, and ψ relatively small).

Our _general conclusion_ is the following: It appears possible to extract the quark-distribution function in the electron, at least roughly, from the type of measurement considered, by using back-factorization. (Limiting ourselves to lowest order, we have a one-term formula; otherwise, i. e. not neglecting $C_{\pm,00}$, it would be a 2-term formula). If one wants to go farther, i. e. to determine the photon's quark content, the value or range of θ plays a critical role: If θ is not limited to extremely small values, what one gets is the quark content of an off-shell photon $(N'_{q/\gamma})$ instead of that of a quasi-real one $(N_{q/\gamma})$. In order to avoid that complication, it appears advisable - assuming that finite-angle tagging counters are available - to use "anti-tagging" of the corresponding electron (remember that, in that case, one has $\langle\theta\rangle \simeq 2$ mrad at PETRA or PEP).

Table 5. Range of $\Delta = (\sigma^{*}_{approx} - \sigma^{*}_{exact})/\sigma^{*}_{exact}$ in the process $e\,e \longrightarrow e\,e\,q\,\bar{q}$, according to approximation I, II or II', for various kinematic configurations with fixed values of x, x_B, θ, γ, letting θ' go through a variety of values (160 mrad $< \theta' <$ 400 mrad). Here we choose: $x = 0.1$, $x_B = 0.4$. Beam energy $E_o = 15$ GeV.

		$\theta = 2$ mrad	$\theta = 12$ mrad	$\theta = 32$ mrad
$\gamma = 16$ mrad	approx. I	$-2\% <\Delta< 0$	$-7\% <\Delta< -5\%$	$-23\% <\Delta< -18\%$
	approx. II	$+8\% <\Delta< +9\%$	$+35\% <\Delta< +37\%$	$+217\% <\Delta< +222\%$
	approx. II'	$+7\% <\Delta< +9\%$	$+2\% <\Delta< +3\%$	$-17\% <\Delta< -15\%$
$\gamma = 80$ mrad	approx. I	$-2\% <\Delta< 0$	$-7\% <\Delta< -5\%$	$-24\% <\Delta< -19\%$
	approx. II	$+8\% <\Delta< +9\%$	$+30\% <\Delta< +32\%$	$+192\% <\Delta< +197\%$
	approx. II'	$+7\% <\Delta< +8\%$	$+1\% <\Delta< +3\%$	$-18\% <\Delta< -16\%$
$\gamma = 160$ mrad	approx. I	$-1\% <\Delta< 0$	$-7\% <\Delta< -6\%$	$-26\% <\Delta< -21\%$
	approx. II	$+7\% <\Delta< +8\%$	$+20\% <\Delta< +22\%$	$+132\% <\Delta< +137\%$
	approx. II'	$+6\% <\Delta< +7\%$	$0 <\Delta< +1\%$	$-21\% <\Delta< -19\%$

Table 6. Same as Table 5, except that here: E_o = 70 GeV

		θ = 2 mrad	θ = 12 mrad	θ = 32 mrad
γ = 16 mrad	approx. I	$-4\% < \Delta < -3\%$	$-38\% < \Delta < -37\%$	$-48\% < \Delta < -41\%$
	approx. II	$+12\% < \Delta < +14\%$	$+558\% < \Delta < +561\%$	$+4200\% < \Delta < +4300\%$
	approx. II'	$-4\% < \Delta < -3\%$	$-38\% < \Delta < -37\%$	$-50\% < \Delta < -48\%$
γ = 80 mrad	approx. I	$-4\% < \Delta < -3\%$	$-48\% < \Delta < -46\%$	$-52\% < \Delta < -45\%$
	approx. II	$0 < \Delta < +1\%$	$+107\% < \Delta < +108\%$	$+1320\% < \Delta < +1350\%$
	approx. II'	$-3\% < \Delta < -2\%$	$-48\% < \Delta < -47\%$	$-54\% < \Delta < -52\%$
γ = 160 mrad	approx. I	$-2\% < \Delta < -1\%$	$-43\% < \Delta < -41\%$	$-61\% < \Delta < -55\%$
	approx. II	$-2\% < \Delta < 0$	$-7\% < \Delta < -6\%$	$+284\% < \Delta < +290\%$
	approx. II'	$-2\% < \Delta < -1\%$	$-44\% < \Delta < -43\%$	$-63\% < \Delta < -60\%$

References

(1) E. J. Williams, Proc. Roy. Soc. London A 139, 163 (1933); Phys. Rev. 45, 729 (1934); Mat. Fys. Medd. 13, 4 (1935). C. F. von Weizsäcker, Zs. Phys. 88, 612 (1934). Still earlier approaches can be found in: N. Bohr, Philos. Mag. 25, 10 (1913) and 30, 581 (1915). E. Fermi, Zs. Phys. 29, 315 (1924). For a general review of the Williams-Weizsäcker method and its applications, see: P. Kessler, Acta Physica Austriaca 41, 141 (1975).

(2) F. E. Low, Phys. Rev. 120, 582 (1960).

(3) See e. g. the review paper by V. M. Budnev, I. F. Ginzburg, G. V. Medelin and V. G. Serbo, Phys. Rep. 15 C, 181 (1975).

(4) C. Carimalo, P. Kessler and J. Parisi, Phys. Rev. D 20, 1057 (1979); Phys. Rev. D 20, 2170 (1979); Phys. Rev. D 21, 669 (1980).

(5) C. E. Carlson and W. K. Tung, Phys. Rev. D 4, 2873 (1971).

(6) See: N. Arteaga-Romero et al., Report LPC 80-06 (1980).

(7) The deep-inelastic configuration was studied in particular by: S. J. Brodsky, T. Kinoshita and H. Terazawa, Phys. Rev. Lett. 27, 280 (1971). T. F. Walsh and P. Zerwas, Phys. Rev. Lett. 44 B, 195 (1973). R. L. Kingsley, Nucl. Phys. B 60, 45 (1973). M. A. Ahmed and G. G. Ross, Phys. Lett. 59 B, 369 (1975). E. Witten, Nucl. Phys. B 120, 189 (1977). C. Llewellyn Smith, Phys. Lett. 79 B, 83 (1978). K. Kajantie, Acta Physica Austriaca, Suppl. XXI, 663 (1979); Phys. Lett. 83 B, 413 (1979).

RADIATIVE CORRECTIONS TO $\gamma\gamma$ PROCESSES

IN e^+e^-, e^-e^-, e^+e^+ COLLISION RINGS

G. Cochard and S. Ong
Laboratoire de Physique Théorique des
Particules, Université de Picardie,
Amiens (France)
&
Laboratoire de Physique Corpusculaire,
Collège de France, Paris (France)

We consider here the radiative corrections relative to $\gamma\gamma$ processes in e^+e^-, e^-e^-, e^+e^+, collision rings, that is to the processes described to α^4 order by the Feynman graph (P) of figure 1.

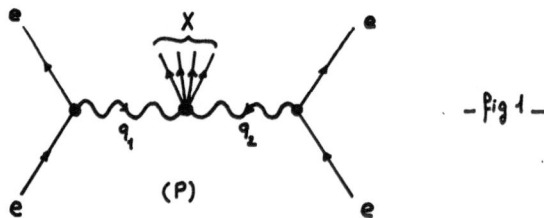

— Fig 1 —

We shall consider the radiative corrections to α^5 order arising only from electron (or positron) lines. Computation of this kind of corrections is made for two different experimental situations:

- $\gamma\gamma$ processes with double tagging of the scattered electrons[1].
- $\gamma\gamma$ processes without tagging of the scattered electrons[2].

1-BASIC ASSUMPTIONS

The radiative corrections belong to two categories:

a) Virtual corrections arising from the interference between the graph (P) and the graphs of the figure 2.

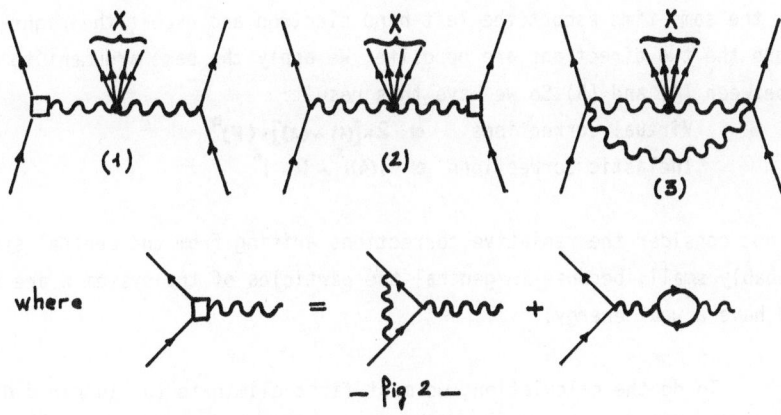

— Fig 2 —

This kind of corrections is due to the exchange of a third photon and to the vacuum polarization.

    b)Inelastic corrections which correspond to the emission of a real photon not observed by the detection system. This kind of corrections is described by figure 3.

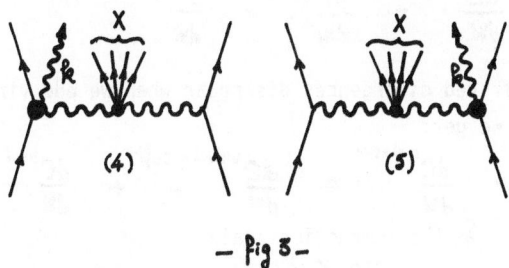

— Fig 3 —

In order to compute these corrections, we will make the following approximations:

APP1:The electrons are considered as extreme relativistic: $\frac{m}{E}, \frac{m}{E_1}, \frac{m}{E_2'} \ll 1$

APP2:The exchanged photons are considered as quasi-real. More explicitly we assume that the condition

$$\frac{t_1}{W^2}, \frac{t_2}{W^2} \ll 1$$

(where W is the invariant mass of the X system and $t_1 = -q_1^2$, $t_2 = -q_2^2$) is satisfied. Under this assumption, we can use the equivalent photon approximation.

APP3:We assume negligible the contributions coming from the graph (3) of figure 2 on one hand, and the interference between the graphs (4) and (5) of figure 3 on the other hand. This assumption is based on the following qualitative argument. When a photon is emitted or absorbed by an electron at high energy, photon and electron are moving in the same direction. Thus, in (3), the photon, exchanged between the ee vertices, cannot at the same time escort the left-hand electron and escort the right-hand electron because the two directions are opposite. We apply the same argument to the inteference between (4) and (5). So we have this result:

$$\text{Virtual corrections} \simeq 2 \times [(4) + (2)] \times (P)^*$$
$$\text{Inelastic corrections} \simeq |(4)|^2 + |(5)|^2$$

APP4: We do not consider the radiative corrections arising from the central system X, probably small, because in general the particles of the system X are massive and have a weak energy.

To do the calculation, we must first eliminate the infrared divergences contained in the virtual and in the inelastic corrections. In order to do that, we divide the inelastic corrections into 2 categories:

- contribution dC^{soft}/dW of the soft photons which have an energy comprised between λ ($\lambda \rightarrow 0$) and a cut-off Λ , chosen arbitrary small.
- contribution dC^{hard}/dW of the hard photons which have an energy higher than Λ.

$$\frac{dC^{inelastic}}{dW} = \frac{dC^{soft}}{dW} + \frac{dC^{hard}}{dW}$$

The infrared divergences disappear when we add virtual and soft photon contributions and we get:

$$\frac{dC}{dW} = \frac{dC^{virtual}}{dW} + \frac{dC^{inelastic}}{dW} = \frac{dC^{virtual+soft}}{dW} + \frac{dC^{hard}}{dW}$$

We call δ the correction rate

$$\delta = \frac{dC/dW}{d\sigma_0/dW}$$

and we get for the corrected cross-section:

$$\frac{d\sigma^{corr.}}{dW} = (1+\delta) \cdot \frac{d\sigma_0}{dW}$$

2- $\gamma\gamma$ PROCESSES WITH DOUBLE TAGGING AT SMALL ANGLE

In double tagging experiments, the electrons are detected in order to get the characteristics of the exchanged photons. Figure 4 describes the kinematics. The scattering angles θ_1 and θ_2 are assumed to be very small (for instance, lower than ten milliradians). Thus the q^2 of the photons are very small : they are quasi real and we can use the equivalent photon approximation.

$$q_1^2 = -t_1 \simeq -EE_1'\theta_1^2 - t_{1min} \qquad \omega_1 = E - E_1' \qquad t_{1min} = m^2\omega_1^2/EE_1'$$
$$q_2^2 = -t_2 \simeq -EE_2'\theta_2^2 - t_{2min} \qquad \omega_2 = E - E_2' \qquad t_{2min} = m^2\omega_2^2/EE_2'$$

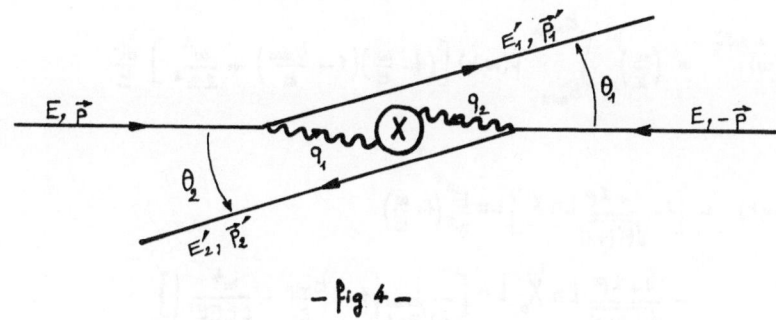

$$- Fig 4 -$$

Using the double equivalent photon approximation, the non corrected cross section is

$$\frac{d\bar{\sigma_o}}{dW} = \frac{2}{W} \sigma_{\gamma\gamma}(W) \int_{\omega_{1min}}^{\omega_{1max}} N_o(\omega_1) N_o\left(\frac{W^2}{4\omega_1}\right) \frac{d\omega_1}{\omega_1}$$

where

$$N_o(\omega_1) = \frac{\alpha}{\pi}\left[\left(1 - \frac{\omega_1}{E} + \frac{\omega_1^2}{2E^2}\right) Ln \frac{t_{1max}}{t_{1min}} - \left(1 - \frac{\omega_1}{E}\right)\left(1 - \frac{t_{1min}}{t_{1max}}\right)\right]$$

is the spectrum of equivalent photons and $\sigma_{\gamma\gamma}(W)$ is the cross section of the process $\gamma\gamma$ (real) $\longrightarrow$ X. Soft + virtual term is given by

$$\frac{d\sigma}{dW}^{virtual+soft} = \frac{4}{W}\sigma_{\gamma\gamma}(W) \int_{\omega_{1min}}^{\omega_{1max}} N^{vrt.+soft}(\omega_1) N_o\left(\frac{W^2}{4\omega_1}\right) \frac{d\omega_1}{\omega_1}$$

where $N^{soft+virtual}$ is the spectrum associated to the diagrams of figure 5.

$$- Fig 5 -$$

Hard term

$$\frac{d\sigma}{dW}^{hard} = \frac{\alpha^3 W^3}{16 \pi^3 E^4} \int_{\Lambda}^{\Delta E} Y(W, k_o) dk_o$$

contains an integration over the radiated photon energy from the cut-off Λ to the maximum value which is the resolution ΔE of the tagging system.

Computation is made in the experimental conditions of the DCI experiment[3]:

$$E = 0.8 \ GeV \ , \quad \Delta E = 14 \ MeV \ , \quad \theta_1, \theta_2 \leqslant 10 \ mrad \ , \quad \omega_i \in [0.2E, 0.5E]$$

For the cut-off Λ we chose the electron mass. We find that the total contribution is practically equal to the soft + virtual term because the hard term is rather small. As a consequence, δ depends weakly on the $\gamma\gamma$ process considered. The detailed expression of $N^{soft+virtual}$ is given by:

$$N^{virt. + soft}_{(\omega)} = \left(\frac{\alpha}{\pi}\right)^2 \int_{t_{min}}^{t_{max}} P(\omega,t)\left[\left(1-\frac{\omega}{E}\right)\left(1-\frac{t_{min}}{t}\right) + \frac{\omega^2}{2E^2}\right]\frac{dt}{t}$$

with

$$
\begin{aligned}
P(\omega,t) = &\left[1 - \frac{1+2\rho}{\sqrt{\rho(\rho+1)}}\,Ln\,X_0\right]Ln\frac{E^2}{m^2}\left(1-\frac{\omega}{E}\right)\\
&- \frac{1+2\rho}{\sqrt{\rho(\rho+1)}}\,Ln\,X_0\,Ln\left[\frac{1}{2\sqrt{\rho+1}}\left|1 - \frac{t_{min}}{t} - \frac{m^4}{2tEE'}\right|\right]\\
&+ \frac{1+2\rho}{2\sqrt{\rho(\rho+1)}}\left[\phi\left(\frac{\omega}{\omega-2\alpha E}\right) - \phi\left(\frac{\omega}{X_0(\omega-2\alpha E)}\right) + \phi\left(\frac{\omega X_0}{\omega+2\beta E}\right)\right.\\
&\left.- \phi\left(\frac{\omega}{\omega+2\beta E}\right) - \phi\left(\frac{X_0}{1+X_0}\right) + \phi\left(\frac{1}{1+X_0}\right)\right]\\
&- 2 + \left[\frac{1+2\rho}{\sqrt{\rho(\rho+1)}}\,Ln\sqrt{\rho+1} - \frac{2+3\rho}{2\sqrt{\rho(\rho+1)}} - \frac{2\rho-1}{3\rho^2}\sqrt{\rho(\rho+1)}\right]\\
&- \frac{2}{3}\left(\frac{5}{3}-\frac{1}{\rho}\right) - \frac{1+2\rho}{\sqrt{\rho(\rho+1)}}\left[\phi(-\beta) - \phi(\alpha) + \frac{1}{2}\phi(2\alpha) - \frac{1}{2}\phi(-2\beta)\right]
\end{aligned}
$$

where

$$\rho = \frac{t}{4m^2} \qquad \alpha = \sqrt{\rho(\rho+1)} - \rho \qquad \beta = \sqrt{\rho(\rho+1)} + \rho$$

$$X_0 = \frac{\beta}{\alpha} \qquad \phi(z) = -\int_0^z \frac{Ln|1-z|}{z}\,dz \qquad (\text{Dilog or Spence function})$$

Numerical results lead to a correction rate δ of about 10%. Figures 6a and 6b show the cross sections for the processes $\gamma\gamma \longrightarrow e^+e^-$, $\gamma\gamma \longrightarrow \mu^+\mu^-$, $\gamma\gamma \longrightarrow \pi^+\pi^-$ (in the frame of the Born-term model). Note the correction is negative.

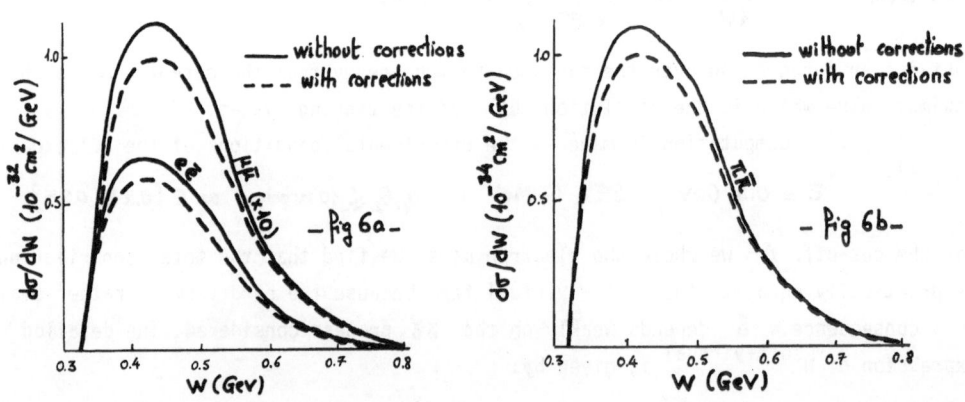

3- $\gamma\gamma$ PROCESSES WITHOUT TAGGING

In the case of experiments without tagging, one must integrate over the whole phase space of the scattered electrons and, for the inelastic corrections, of the radiated photon.

a) In a first time, we use the equivalent photon approximation applied to the non corrected electron vertex (figures 7a,7b,7c)

— fig 7a —

— fig 7b —

— fig 7c —

We get the following results:

-non corrected term:

$$\frac{d\sigma_0}{dW} = \int N_0(\omega) \frac{d\sigma_\gamma}{dW}(\omega,W) \frac{d\omega}{\omega}$$

-virtual corrections:

$$\frac{dC^{virt.}}{dW} = \int N_0(\omega) \frac{dC_\gamma^{virt.}}{dW}(\omega,W) \frac{d\omega}{\omega}$$

-inelastic corrections:

$$\frac{dC^{inel.}}{dW} = \int N_0(\omega) \frac{dC_\gamma^{inel.}}{dW}(\omega,W) \frac{d\omega}{\omega}$$

In the non corrected cross section, $d\sigma_\gamma(\omega, W)/dW$ is given by

$$\frac{d\sigma_\gamma}{dW}(\omega, W) = \frac{\alpha\, W^3}{16\pi\omega^2 E^2}\int\left[\sigma_T(S+2) + \sigma_L S + \frac{4m^2}{t_1}(\sigma_T S + \sigma_L(S+1))\right]\frac{dt_1}{t_1}$$

where

$$S = \frac{4s\,(t_{1max} - t_1)(t_1 - t_{1min})}{(t_1 + 4m^2)(W^2 + t_1)^2}$$

and where $\sigma_{T,L}$ (T = transverse, L = longitudinal) means the cross section of the process γ (virtual) + γ (real) $\longrightarrow$ X.

After cancellation of the infrared divergences,

$$\frac{dC_\gamma^{virt.}}{dW} + \frac{dC_\gamma^{ind.}}{dW} = \frac{dC_\gamma^{virt+soft}}{dW} + \frac{dC_\gamma^{hard}}{dW}$$

where

$$\frac{dC_\gamma^{virt+soft}}{dW}(\omega, W) = \frac{\alpha\,W^3}{16\pi\omega^2 E^2}\int\left[\delta_T(\sigma_T(S+\ell) + \sigma_L S) + \frac{4m^2}{t_1}\delta_L(\sigma_T S + \sigma_L(S+1))\right]\frac{dt_1}{t_1}$$

$$\frac{dC_\gamma^{hard}}{dW}(\omega, W) = \frac{\alpha\,W^3}{16\pi\omega^2 E^2}\int\frac{dt_1}{t_1^2}\int_{m_o^2}^{m_{1max}^2}dm^2\,\frac{m^2 - m^2}{4\pi^2\alpha}\left[\sigma_T^c\sigma_T(S+\ell) + (\sigma_T^c\sigma_L + \sigma_L^c\sigma_T)S^k + \sigma_L^c\sigma_L(S+1)\right]$$

with

$$S^k = \frac{4s\,t_1(t_{1max}^k - t_1)(t_1 - t_{1min}^k)}{\Lambda(W_i^2 m_i^2 - t_1)(W^2 + t_1)^2}$$

$$\delta_T = -\frac{2\alpha}{\pi}(I_F - J + K - \psi) \qquad \delta_L = -\frac{2\alpha}{\pi}(I_F + \rho J + K - \psi)$$

$$I_F = \ell + \frac{1+2\rho}{\ell\sqrt{\rho(\rho+1)}}\left[\left[Ln(\rho+1) - \frac{3\rho+2}{1+2\rho}\right]Ln\,X_o - \phi(\alpha) + \phi(-\beta) - \frac{1}{2}\phi(-2\beta) + \frac{1}{2}\phi(2\alpha)\right]$$

$$J = \frac{1}{\ell\sqrt{\rho(\rho+1)}}Ln\,X_o$$

$$K = \frac{2}{3\rho}\left[5\rho - 3 - (\rho - \frac{1}{2})\sqrt{\frac{\rho+1}{\rho}}\,Ln\,X_o\right]$$

$$\psi = \ell + \left(\ell\,Ln\frac{m^2}{m_o^2 - m^2} - 1\right)\left(1 - \frac{1+2\rho}{\ell\sqrt{\rho(\rho+1)}}Ln\,X\right) - \frac{1+2\rho}{4\sqrt{\rho(\rho+1)}}\left[\phi\left(\frac{2(\alpha+\beta)}{X_o}\right) - \phi(-2(\alpha+\beta)X_o)\right]$$

m_o is a cut-off (arbitrary small) and $\overline{\sigma}_{T,L}^c$ means the cross section of the process $e + \gamma$(virtual) $\longrightarrow$ $e + \gamma$(real) (Compton effect).

Preliminary results were given during the Workshop but, though analytical calculation was exact, numerical computation of the soft + virtual contribution was unfortunately wrong. We give here the right results, quite different from the previous ones, from independent computations by M.DEFRISE and J.SILVA[4]. Table 1 below gives the correction rate for different energies and various invariant masses when using the single equivalent photon approximation.

E = 1.5 Gev

W/E	$\delta(\pi^+\pi^-)$ (%)	$\delta(\mu^+\mu^-)$ (%)
0.20	0.10	0.14
0.25	0.02	0.03
0.30	-0.05	-0.06
0.35	-0.12	-0.13
0.40	-0.23	-0.23
0.45	-0.33	-0.31
0.50	-0.46	-0.40
0.55	-0.57	-0.46
0.60	-0.78	-0.63

_ Table 1 _

E = 4.5 Gev

W/E	$\delta(\pi^+\pi^-)$ (%)	$\delta(\mu^+\mu^-)$ (%)
0.10	0.45	0.45
0.15	0.28	0.30
0.20	0.09	0.13
0.25	0.08	0.00
0.30	0.06	-0.13
0.35	-0.01	-0.21
0.40	-0.09	-0.24
0.45	-0.18	-0.27
0.50	-0.34	-0.42
0.55	-0.47	-0.56
0.60	-0.54	-0.58

Our conclusion is that δ is extremely small and rather insensitive to the $\gamma\gamma$ process considered. Thus, we can use the double equivalent photon approximation.

b) Use of the double equivalent photon approximation. To apply this approximation, we simply set in the previous formulas,

$$\sigma_T(W, t_1) \simeq \sigma_T(W, 0) = \sigma_{\gamma\gamma}(W) \qquad \sigma_L(W, t_1) \simeq \sigma_L(W, 0) = 0$$

and we get

$$\delta = \frac{\int N_0(\omega)\frac{d\omega}{\omega^3}\int\frac{dt_1}{t_1}\left[\delta_T(\mathcal{S}+2)+\frac{4m^2}{t_1}\delta_L\mathcal{S}\right]+\int N_0(\omega)\frac{d\omega}{\omega^3}\int\frac{dt_1}{t_1^2}\int_{m_0^2}^{m_{max}^2}dm^2\frac{m-m^2}{4\pi\alpha}\left[\overline{\mathcal{C}}_T^c(\mathcal{S}+2)+\overline{\mathcal{C}}_L^b\mathcal{S}\right]}{\int N_0(\omega)\frac{d\omega}{\omega^3}\int\frac{dt_1}{t_1}\left[\mathcal{S}+2+\frac{4m^2}{t_1}\mathcal{S}\right]}$$

which is clearly independent of the $\gamma\gamma$ process. Numerical results are given in Table 2 for different energies and various invariant masses.

	E = 1.5 Gev	E = 4.5 Gev
W/E	δ (DEPA) (%)	δ (DEPA) (%)
0.10	0.21	0.36
0.15	0.17	0.32
0.20	0.10	0.20
0.25	0.05	0.15
0.30	-0.04	0.12
0.35	-0.15	0.03
0.40	-0.24	-0.13
0.45	-0.34	-0.17
0.50	-0.48	-0.32
0.55	-0.57	-0.44

_ Table 2 _

	E = 15 Gev	E = 70 Gev
W/E	δ (DEPA) (%)	δ (DEPA) (%)
0.10	0.49	0.84
0.15	0.47	0.74
0.20	0.43	0.61
0.25	0.29	0.55
0.30	0.19	0.39
0.35	0.11	0.33
0.40	0.01	0.19
0.45	-0.09	0.11
0.50	-0.26	-0.01
0.55	-0.37	-0.18
0.60	-0.47	-0.39

4-CONCLUSION

Computation of the correction rate has shown that:

--in double tagging experiments, the radiative corrections are rather small. The inelastic corrections are limited by the resolution of the tagging system and, thus,we have a very small contribution of the hard photons. Moreover, the use of the double equivalent photon approximation is completely justified in that case.

--in no tagging experiments, the radiative corrections are extremely small and even negligible. They cannot practically affect the cross section.

However, there are other types of $\gamma\gamma$ experiments, namely single-tagging experiments and measurements at finite angle, and the next work is to compute the corresponding radiative corrections to obtain their order of magnitude.

References

(1) G.COCHARD & S.ONG, Phys. Rev. D19, 810 (1979).

(2) S.ONG, Thèse de Troisième Cycle, Univ. of Paris VI (1976), unpublished.

(3) A.COURAU et al., Phys. Lett. 84 B, 145 (1979).

(4) M.DEFRISE, S.ONG and J.SILVA, submitted for publication to Phys. Rev. D. (1980). Numerical results relative to the radiative corrections in the no tagging case published in " A double equivalent-photon approximation including radiative corrections for photon-photon collision experiments without electron tagging" (LPTP 80/1, S.ONG et al.) and in "Update on photon-photon collisions" (LPC 80/06, N.ARTEAGA-ROMERO et al.) are wrong.

γ.γ PHYSICS AT D.C.I.

COLLABORATION :   L.P.C. CLERMONT-FD
L.A.L. ORSAY

M. BROSSARD, A. FALVARD, J. JOUSSET, B. MICHEL
G. MONTAROU, J.C. MONTRET, P. REICHSTADT

Laboratoire de Physique Corpusculaire
Université Clermont II - IN2P3
B.P. 45 - 63170 AUBIERE (France)

A. COURAU, J. HAISSINSKI

Laboratoire de l'Accélérateur Linéaire
Université Paris-Sud
Bâtiment 200
91405 ORSAY Cédex (France)

This talk reports the γ.γ experimental programm achieved the last year at D.C.I. (Orsay).

We present the zero degree tagging system and the results on the tagged experiment.

I. A ZERO DEGREE TAGGING SYSTEM FOR D.C.I.

1) GENERAL VIEW OF THE EXPERIMENT

The goal of our experimental study is to identify the X states produced in $\gamma + \gamma \rightarrow X$ reactions. Within our tagging acceptance, and in the energy range of D.C.I. ($\simeq$ 1 GeV/beam), the system X produced can be a e^+e^- pair, $\mu^+\mu^-$ or $\pi^+\pi^-$.

The identification is performed by the reconstruction of the masses of the detected particles by means of :

- measurement of kinematical parameters of particles produced in the central detector ;
- measurement of kinematical parameters of scattered electrons at very small angle : < 10 mrad.

This is realized by the tagging system.

2) CHARACTERISTICS OF D.C.I. (1)

The D.C.I. is constituted by two rings, one on top of the other, with a common interaction section. This structure allows to perform e^+e^-, e^+e^+ or e^-e^- collisions studies.

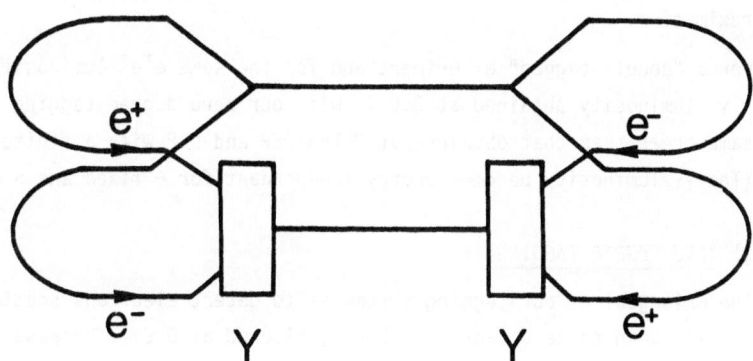

Figure 1

The reunion of two beams is realized by means of two bending magnets called Y magnets.

In this experiment, these magnets have a very important role : they are used as spectrometers.

The main characteristics of D.C.I. are summed up in the next list :

- Energy (for data taking) : 0.85 GeV
- Number of rings : 2
- Length of orbit : 94.6 m

- Number of experimental sections : 1
- Length of interaction region : 6 m
- Rotation frequency : 315 ns
- Number of bunches : 1 per beam and per ring
- Length of the bunches crossing region : 15 cm
- Section of the bunches crossing region : 10 mm²
- Intensity ⎫ for data taking   40 mA/beam
- Luminosity ⎭   2-3 10²⁹ cm⁻² s⁻¹.

With these characteristics, some essential conclusions for the γγ physics at D.C.I. are possible :

- zero degree tagging is possible by the structure machine and the duty cycle (leading to less than 10^{-2} Bremsstralhung hit per bunches crossing) ;

- with the zero degree tagging we use the dominant part of the quasi real photon spectrum.

Another advantage is that, in the analysis, we can use the equivalent photon approximation.

For a "double tagged" experiment and for the same e^+e^- luminosity, the integrated γγ luminosity obtained at D.C.I. with our zero degree tagging system is of the same order than that obtained at PETRA/PEP and LEP with a finite angle tagging. (The γγ luminosity becomes energy independant for θ fixed and > m/E).

3) *ZERO DEGREE TAGGING* (2)

The main goal of our tagging system is to detect electrons scattered at very small angles down to zero degree. This is allowed at D.C.I. because two magnetic fields (Y magnets) , on each side of interaction section are used as spectrometers (Fig. 2).

The magnetic field has been measured at the crossing of a 2 x 2 x 2 cm³ network. The analysis of the field map, has shown that, in the medium vertical magnet plane, the optics differs by less than 1 % from that of ideal magnet with a uniform field 72 cm long.

In the ideal magnet approximation, the equations which relate the momentum p and the vertical angle θ_v of the scattered electron to the intercept Y_d and slope θ'_v of its trajectory measured in a vertical frame located after Y magnet (Fig. 3) are :

$$Y_d = L tg\, \theta_v + \frac{1}{\sin \alpha_o}(1 - \cos \theta'_v)\, \frac{p}{p_o}$$

$$\sin \theta'_v = \sin \theta_v + \frac{P_o}{P} \sin \alpha_o$$

P_o is the beam energy and $\alpha_o = 10°$ is the beam deflection by the Y magnets.

Y_d, θ'_v are measured by our tagging system ; the horizontal angle θ_H comes directly from the measurement of the z_d coordinate by the tagging system.

4) ACCEPTANCE AND RESOLUTION

In order to detect electrons in an acceptance as extended as possible towards small $x = E_Y/E$ values, special vacuum chambers had to be built to implement our 4 tagging hodoscopes.

To minimize multiple Coulomb scattering these vacuum chambers have 50 μm titane windows on 400 cm^2 area.

The tagging system defined before, has an x acceptance such as : $0.2 < x < 0.5$ and, an angular acceptance $\theta < 10$ mrad with full azimuthal acceptance ϕ for x value above 0.33.

The memory time of the detector must be less than 315 ns (the beam's revolution time in D.C.I.).

The spatial resolution of detector must be less than 300 μm. This value allows an accuracy on the momentum of the order of 1 % and a resolution better than 1.5 mrad on the scattering angle θ.

These various requirements led us to choose drift chambers of the Charpak type.

5) DRIFT CHAMBERS HODOSCOPES

4 hodoscopes are in place on D.C.I. Each hodoscope contains three pairs of drift chambers. Each pair (Fig. 4) is constituted by two chambers, each chamber having 14 cells of different sizes.

The variation of the drift length is necessary to give the counting rates at a similar level in various cells. More precisely, each drift chamber is composed of :

- 4 (12 + 12) mm cells
- 1 (12 + 8) mm cells
- 4 (8 + 8) mm cells
- 1 (8 + 4) mm cells
- 4 (4 + 4) mm cells.

In the first and the last couple of chambers, all wires are horizontal and, we have shifted by 4 mm one drift plane to remove the up-down ambiguity.

In the central couple, the wires are tilted at an angle of ± 2°30' to measure the track projection on the horizontal plane.

Each hodoscope is completed by a scintillator, giving a fast signal used in the trigger logic.

These chambers are used with a mixed gas : 31 % isobutane and 69 % argon at the atmospheric pressure. For this mixed gas, the saturation of drift velocity is obtained when the electric field reaches 1 000 V/cm.

The sense wires signals are discriminated and amplified by Lecroy D.C. 201 hybrid circuits.

They are sent to a T.D.C. Lecroy and give the start of the conversion process.

All channels have a common stop signal provided by a pick-up electrode of the storage ring.

6) TRACK RECONSTRUCTION AND DETECTOR PERFORMANCES

The characteristics of our detector have been obtained by the study of beam-beam and beam-gas bremsstrahlung events.

A sample of 10^5 electron trajectories have been registered.

The following characteristics have been obtained :

Efficiency : its value, averaged over all cells is 0.996.

Resolution : the spatial resolution is obtained using the first and the last pairs of our chambers. In this condition, we limit the influence of multiple coulomb scattering.

In these conditions :

$$Z = \sum_{i=1}^{4} (y_{ci} - y_{mi})^2/\sigma^2$$

(y_{ci} = calculated value of the coordinate ; y_{mi} = measured value of the coordinate), has a probability distribution which can be identified with a χ^2 distribution law.

$$f(Z) = A\, e^{-Z/2}$$

Fig. 5 shows the distribution that we have obtained. It leads to $\underline{\sigma = 0.21\text{mm}}$.

Acceptance

We compare Fig. 6 the plot (p, θ_v) deduced from the analysis of beam-beam bremsstrahlung and beam-gas bremsstrahlung events with a Monte Carlo simulation.

Fig. 7 shows the angular distribution in the vertical plane and in the horizontal plane. The standard deviations are :

$$\sigma_{\theta_H} = 0.75 \text{ mrd}$$

$$\sigma_{\theta_V} = 1.5 \text{ mrd.}$$

II. FIRST PHOTON-PHOTON EXPERIMENTS AT D.C.I. : PAIRS PRODUCTION MEASUREMENTS WITH THE DM1 DETECTOR

Results presented here correspond to :

- data taken between March 1979 and September 1979
- DCI operated with two beams of opposite charge e^+e^- and one ring. (For September 1979 until December 1979, DCI has been operated with four beams $e^{\pm}e^{\mp}$ and two rings)
- energies varying between E_o = 750 MeV and E_o = 1 GeV
- an integrated luminosity of the order of 500 nb^{-1} (about 70 % of the total luminosity used for photon-photon experiments).

ee → ee $X^+ X^-$ processes have been studied in the following way :

- the pairs $X^+ X^-$ (mainly e^+e^-, $\mu^+\mu^-$ and $\pi^+\pi^-$ pairs) are measured by the magnetic detector DM1 at large angles (roughly > 45°), and for momenta greater than p_{min} = 75 MeV/c ;
- one of the two scattered electrons (case of "single tagged" events) or both scattered electrons (case of "double tagged" events) are measured at very small angles ($\theta < 10^{-2}$ radian) by the tagging system just described before.

1) THE DM1 APPARATUS

The DM1 magnetic detector has already been described in detail (3)[*]
Fig. 8 shows a general view of this detector. Let us recall that it consists of :

---

[*] People who participate in the DM1 experiment on DCI are :
J.L. Bertrand, J.C. Bizot, J. Buon, A. Cordier, B. Delcourt, P. Eschstruth, L. Fayard, J. Jeanjean, M. Jeanjean, F. Mané, J.C. Parvan, M. Ribes and F. Rumpf (L.A.L. Orsay).

- a solenoïdal magnet which provides an homogeneous magnetic field along the beam axis
- four cylindrical wire chambers, which provide both anodic and cathodic detection and give three coordinates of each hit of particles in the detector
- scintillation counters, used to reject cosmic background by time of flight measurements : in fact they have not been used in our analysis of photon-photon events
- resolution on the momentum determination is of the order of 1 % in the case of photon-photon events
- accuracy on the angles measurements is better than 1°
- an important feature of the DM1 detector is that it does not provide any particle identification, neither neutral detection.

2) *EVENTS SELECTION*

Candidates for photon-photon events were selected according to the following criteria :

a) events occur within a 60 ns gate following the bunches interaction time

b) they involve two and only two tracks of opposite charge detected by the DM1

c) the reconstructed vertex of the two tracks is distant from the center of the interaction region by less than 250 mm in a longitudinal view, and less than 6 mm in the transverse plane

d) the acollinearity angle ζ is larger than 4°, in order to eliminate some residual cosmic background

e) the acoplanarity angle of the two tracks   is less than 20°

f) the reconstructed q^2 of the photon which has not been tagged is less than q^2_{max} = 3 000 MeV^2.

Let us note that the last two criteria are related to the dominance of quasi-real photons in photon-photon processes.

3) *BACKGROUND*

The observed background to $\gamma\gamma \rightarrow X^+ X^-$ events consists of random coincidences between :

- annihilation events, beam-gas events or cosmic events in the DM1
- bremsstrahlung events in the tagging system.

The low rate of bremsstrahlung events, about 10^{-2} for each bunches crossing in our experimental conditions, insures a large background rejection.

- Better than 10^2 in the case of "single tagged" events
- better than 10^4 in the case of "double tagged" events. Therefore, the last three cuts (d, e, f) are not necessary in the latter case.

Let us note that, without any tagging, and with some more restrictive cuts, a photon-photon signal appears.

For that reason, we have performed a very short experiment in the following conditions.

4) e^+e^+ EXPERIMENT

To reduce background and to increase the photon-photon signal, mainly e^+e^- pairs production :

- DCI has been operated with two e^+ beams of 1.2 GeV each
- the magnetic field in the DM1 has been reduced from the usual value 8200 Gauss to 4000 Gauss only
- the cut (d) on the acollinearity angle became $\zeta > 8°$
- the cut (f) is replaced by a cut on the transverse momentum conservation :
$$\frac{|P_{t+} - P_{t-}|}{P_{t+} + P_{t-}} < 0.1$$

Results are given in the following table.

Total luminosity	$8.3 \pm 0.3 \text{ nb}^{-1}$
Total number of observed events	88
Background contamination	6 ± 2
Number of $\gamma\gamma$ events	82 ± 2
Number of expected $\gamma\gamma$ events	73.0 ± 2.7 ($65.6\ e^+e^- + 7.4\ \mu^+\mu^-$)
Experimental cross section[*]	9.3 ± 1.3 nb
Theoretical cross section[*]	8.3 nb (e^+e^- : 7.5 nb + $\mu^+\mu^-$: 0.8 nb)

[*] Of course, cross sections are measured or calculated within our peculiar acceptance and kinematical cuts.

We have also obtained angular and energy distributions for these events. For instance, we show you their invariant mass distribution on the Fig. 9. More details on this experiment may be found in the reference (4).

5) *KINEMATICAL ANALYSIS FOR TAGGED EVENTS*

Since the DM1 does not provide any particle identification, we must reconstruct the mass m of the particles of the pair $X^+ X^-$ in the following way :

- the DM1 measures the two momenta $\vec{p}_+$ and $\vec{p}_-$
- the tagging system measures :
  $\vec{p}_1$ in the case of "single tagged" events,
  $\vec{p}_1$ and $\vec{p}_2$ in the case of "double tagged" events.

a) Single tagged events

Assuming $\vec{p}_1 + \vec{p}_2 + \vec{p}_+ + \vec{p}_- = \vec{0}$ (1)

We deduce $\vec{p}_2$ and the squared mass q_2^2 of the virtual photon which has not been tagged (on which the cut (f) is applied).

Neglecting the electron mass, we also deduce the photon-photon energy :

$E = 2E_o - p_1 - p_2$ (2)

Then, the mass m is given by :

$E = (p_+^2 + m^2)^{1/2} + (p_-^2 + m^2)^{1/2}$ (3)

Let us note that radiative effects shift the measured mass m towards higher masses.

b) Double tagged events

The momentum conservation :

$\vec{p}_1 + \vec{p}_2 + \vec{p}_+ + \vec{p}_- + \vec{p}_{rad} = \vec{0}$ (4)

gives the resulting momentum $\vec{p}_{rad}$ of real photons which has been radiated by the incoming or outgoing particles involved in the photon-photon process. Then :

$E = 2E_o - p_1 - p_2 - p_{rad}$ (5)

m is also given by the formula (3).

Therefore radiative effects are reduced in the case of "double tagged" events, since they are taken into account in the formulas (4) and (5).

6) *RESULTS*

Results for "single tagged" events are given on the Fig. 10.

We have fitted our data with Monte-Carlo predictions for the 3 processes

($e^+ e^-$, $\mu^+\mu^-$ and $\pi^+\pi^-$) including radiative effect[*], in the following way :

a) <u>Assuming that we have $\pi^+\pi^-$ events</u>

The only free parameters of the fit are 3 amplitudes a_e, a_μ and a_π, one for each process. The best fit, represented by a solid line on the Fig. 10, gives the following results :

χ^2 = 17 for 17 degrees of freedom.

Confidence level of about 50 % :

a_e = 157 ± 12 e^+e^- events

a_μ = 68 ± 9 $\mu^+\mu^-$ events

a_π = 14 ± 5 $\pi^+\pi^-$ events.

b) <u>Assuming that we have no $\pi^+\pi^-$ events</u>

Now, there is only two free parameters a_e and a_μ. The best fit, represented by the dashed line on the Fig. 10, gives :

χ^2 = 26.5 for 18 degrees of freedom.

Confidence level of about 15 %.

a_e = 160 ± 12 e^+e^- events

a_μ = 70 ± 9 $\mu^+\mu^-$ events

a_π = 0 $\pi^+\pi^-$ events.

This method gives an evidence for the existence of $\pi^+\pi^-$ events.

The mass distribution obtained for "double tagged" events is shown on Fig. 11. Best fit gives also an evidence for the presence of $\pi^+\pi^-$ events. It is represented by the solid line, while the shaded area shows the $\pi^+\pi^-$ contribution. Results are :

a_e = 27 ± 6 e^+e^- events

a_μ = 27 ± 6 $\mu^+\mu^-$ events

a_π = 6 ± 3 $\pi^+\pi^-$ events.

The following table shows that the observed ratios between "single tagged" and "double tagged" events agree with the Monte-Carlo predictions whithin statistical errors.

[*] For each line, real photons have been drawn indepently of each other, except for the incoming electron (positron) and the correspondant electron (positron) scattered at small angle ; in this case, the photons radiated by these two lines can't be distinguished and interference terms have been taken into account.

If we add "single tagged" and "double tagged" events, we obtain :

a_e = 184 ± 14 e^+e^- events
a_μ = 95 ± 11 $\mu^+\mu^-$ events
a_π = 20 ± 6 $\pi^+\pi^-$ events.

Errors are statistical only.

R	Expected	Observed
$e^+ e^-$	0.14	0.15 ± 0.04
$\mu^+ \mu^-$	0.26	0.28 ± 0.07
$\pi^+ \pi^-$	0.40	0.30 ± 0.17

$$\text{Ratio R} = \frac{\text{Number of "double tagged" events}}{\text{Number of "double tagged" + "single tagged" events}}$$

7) PHOTON-PHOTON CROSS SECTIONS

Results are presented in the following table.

	Expected	Measured
$\dfrac{\sigma_{\mu\mu}}{\sigma_{ee}}$	0.55	0.52 ± 0.07
$\dfrac{\sigma_{\pi\pi}}{\sigma_{ee} + \sigma_{\mu\mu}}$	0.030	0.072 ± 0.022

Normalized to e^+e^- cross section, the measured $\mu^+\mu^-$ cross section agrees quite well with predictions (QED + Williams Weiszäcker approximation + Monte-Carlo calculation). Normalized to pure QED processes ($\gamma\gamma \rightarrow e^+e^-$ and $\gamma\gamma \rightarrow \mu^+\mu^-$) the measured $\pi^+\pi^-$ cross section near threshold is somewhat larger than the predicted one, where "Born terms" only have been taken into account.

These results are still very preliminary.

8) OTHER RESULTS AND CONCLUSION

We have also looked for K^+K^- events. Fig. 12 shows a mass spectrum for all events, the scale being increased. We find some events in the region of the kaon mass, but we cannot determine if they are K^+K^- events, or if they comes from a radiative tail of e^+e^-, $\mu^+\mu^-$ or $\pi^+\pi^-$ events.

Therefore there is no evidence for the existence of K^+K^- events.

Let us note that, for all events, the invariant masses of the produced pairs are small in comparison with the beams energy. They are practically included

- between 150 MeV and 600 MeV for e^+e^- events
- between 250 MeV and 600 MeV for $\mu^+\mu^-$ events
- between 325 MeV and 600 MeV for $\pi^+\pi^-$ events.

Finally there is no evidence for tagged events with other final states than e^+e^-, $\mu^+\mu^-$ and $\pi^+\pi^-$. In particular, we have not found any candidates for $\gamma\gamma \rightarrow \eta$, η', 3π, 4π ... processes.

One of the reasons is that tagging acceptance and beam energies chosen for data taking did not favour most of those channels.

Therefore, the first photon-photon experiments with the DM1 detector consist mainly of pairs production measurements at low energies.

We shall see that, for experiments using the DM2 apparatus, the tagging acceptance must be increased towards higher photon energies, so that the counting rates for processes mentioned above will be larger by one order of magnitude.

Moreover, since the DM2 provides a particle identification, we should have a better separation between $\pi^+\pi^-$ events and leptonic events.

III. NEW APPARATUS ON D.C.I.

1) CENTRAL DETECTOR AND TAGGING SYSTEM (5)

The new experimental set-up (Fig. 13) consists of a conventional solenoïd with a 2 meters diameter, and a 3 meters length, producing a 5 kgauss magnetic field.

The function of the detectors inside the coïl is to measure the charged particles momentum, to identify pions, kaons and protons.

For that, the following detectors are installed from the vacuum chamber :

- two cylindrical proportional wire chambers
- 13 cylindrical drift wire chambers
- a ring of 36 water Cerenkov counters
- a ring of 36 scintillation counters
- outside the coïl, there will be :
      . a photon detector
      . a muon identifier.

Momenta of charged particles are given with an expected precision of

1 at 2 %.

The photon detector measures the conversion point of the photon with a precision of 1°5, the direction of the photons with a precision of a few degrees and have a good detection efficiency for low energy photons.

The whole photon detector will be divided into 8 octants and consists of 120 scintillators and 2 900 proportional bitubes.

Typically, the muon identifier will covert a solid angle of 0.5 x 4π steradian. It will allow a clear separation between $\pi^{\pm}$ and $\mu^{\pm}$ events.

For this new central detector we built a new tagging system with a larger x acceptance. The upper limit is given by the thin Ti window of D.C.I. and the magnetic field of Y magnets used like spectrometers.

We can reach x = 0.7 instead of x = 0.5. The lower limit x = 0.2 is unchanged. For instance, for 1 GeV per beam, we can see with this tagging the resonant states : η, η', δ, S^*, ϵ, f, A_2 ... (Fig. 14).

In its conception, this tagging system is very similar to the first one. It is composed by four hodoscopes of drift chambers larger than in the first tagging system. It is presently in construction.

References

(1) The Orsay Storage Ring Group
Proceedings of the Eight International Conference on High Energy Accelerators, C.E.R.N. (1971)

(2) A zero degree tagging system used for photon-photon experiments at D.C.I.
M. BROSSARD, A. FALVARD, J. JOUSSET, B. MICHEL, G. MONTAROU, J.C. MONTRET, P. REICHSTADT, P.Y. BERTIN, G. FOURNIER, M.C. VIALATTE, J. BUON, A. COURAU, J. HAISSINSKI, J.P. MARX, R. SOUCHET
To be published in Nuclear Instruments and Methods

(3) J. JEANJEAN et al.
Nuclear Instruments and Methods, 117 (1974) 349

A. CORDIER et al.
Nuclear Instruments and Methods, 133 (1976) 237

(4) A. COURAU, A. CORDIER, B. DELCOURT, P. ESCHTRUTH, L. FAYARD, J. HAISSINSKI, F. MANE, B. MICHEL, A. FALVARD, J. JOUSSET, J.C. MONTRET
Physics Letters, 84 (1979) 145

(5) DM2 : un détecteur magnétique pour D.C.I.
J.F. AUGUSTIN et al.
L.A.L. 78/14.

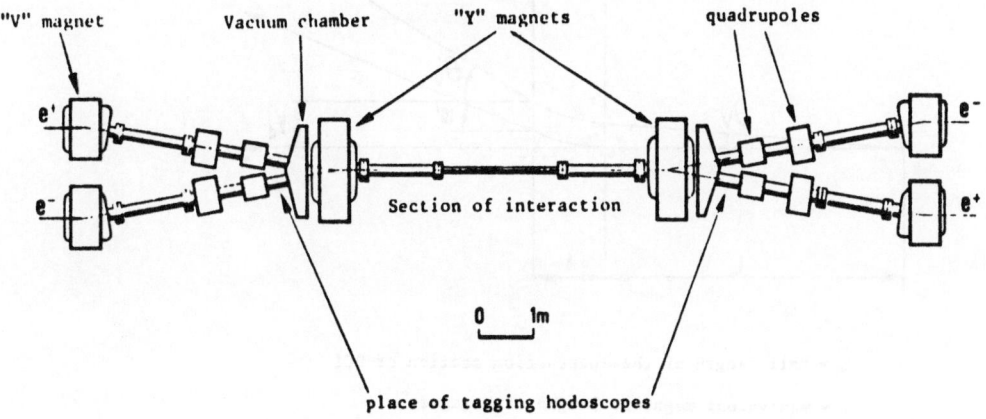

"V" magnet   Vacuum chamber   "Y" magnets   quadrupoles

e⁺

e⁻

Section of interaction

0   1m

place of tagging hodoscopes

Figure 2

Interaction region of D.C.I.

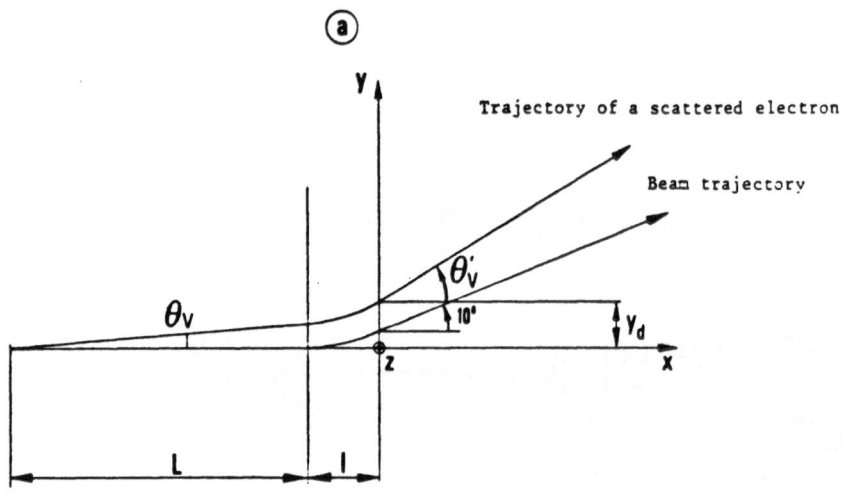

L = half length of the interaction section of DCI

l = equivalent magnetic length of Y magnet

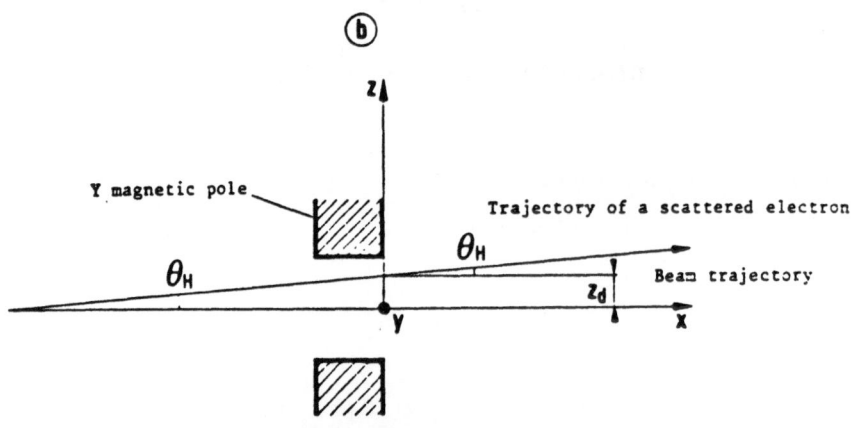

Figure 3

Trajectory projections on (a) the vertical plane,
(b) the horizontal plane

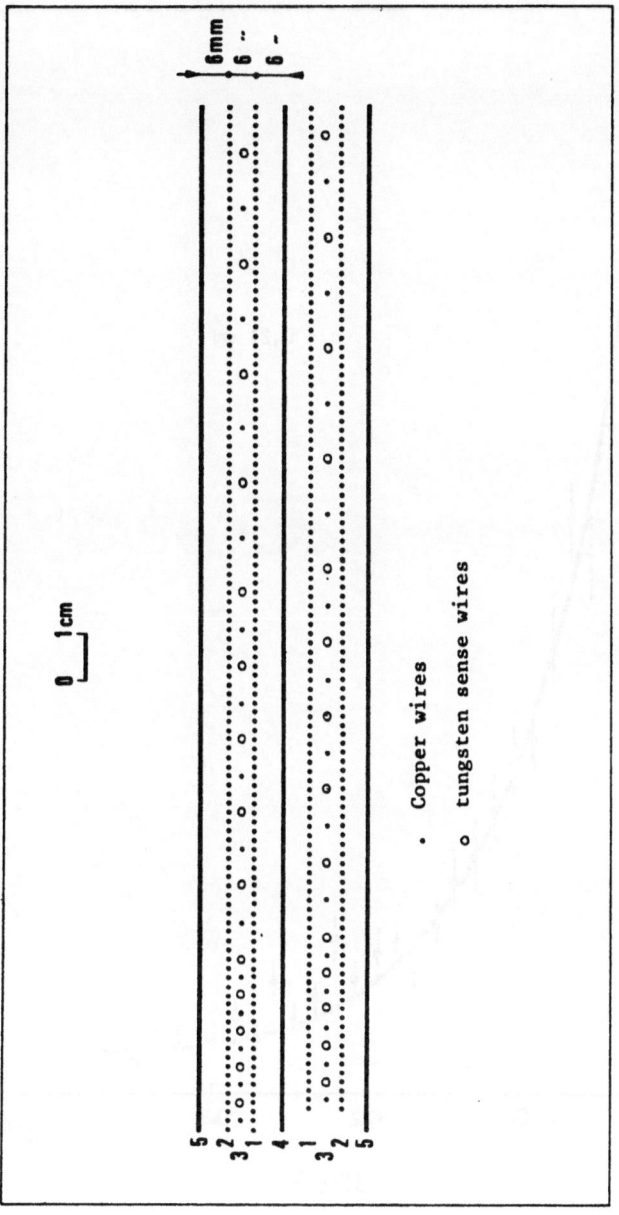

Figure 4

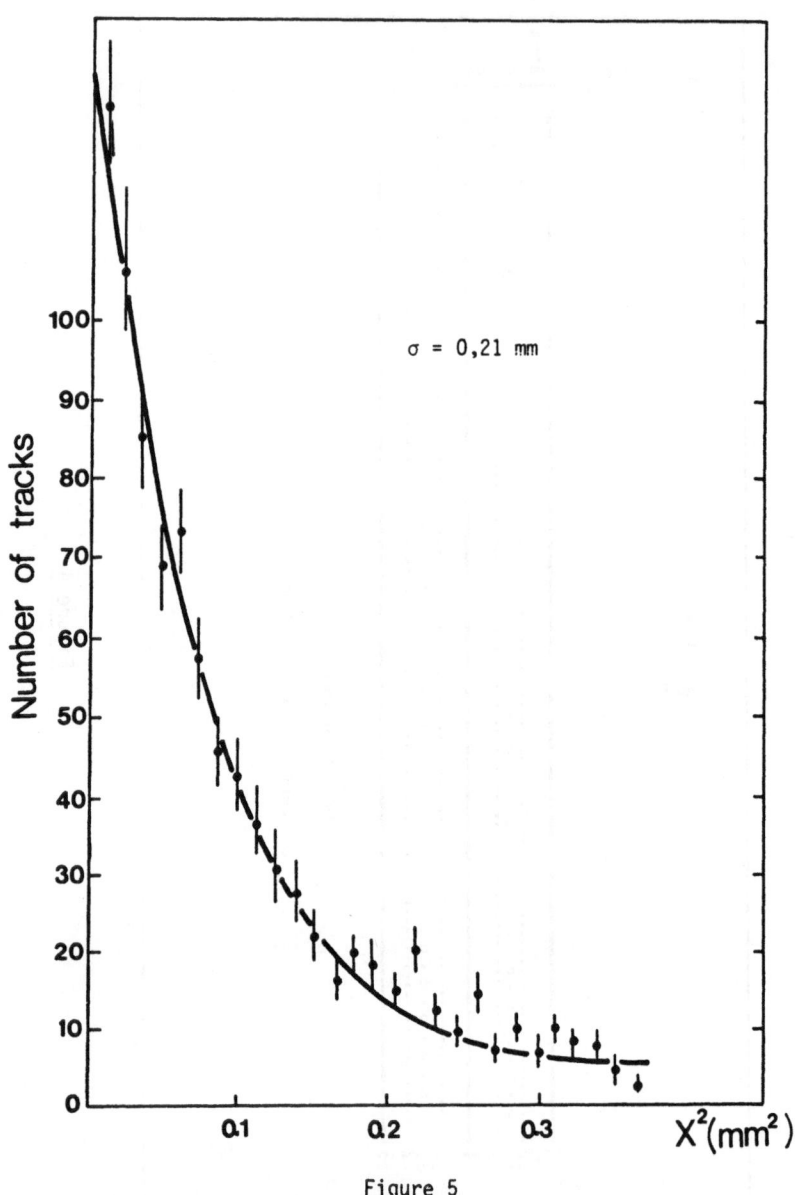

Figure 5

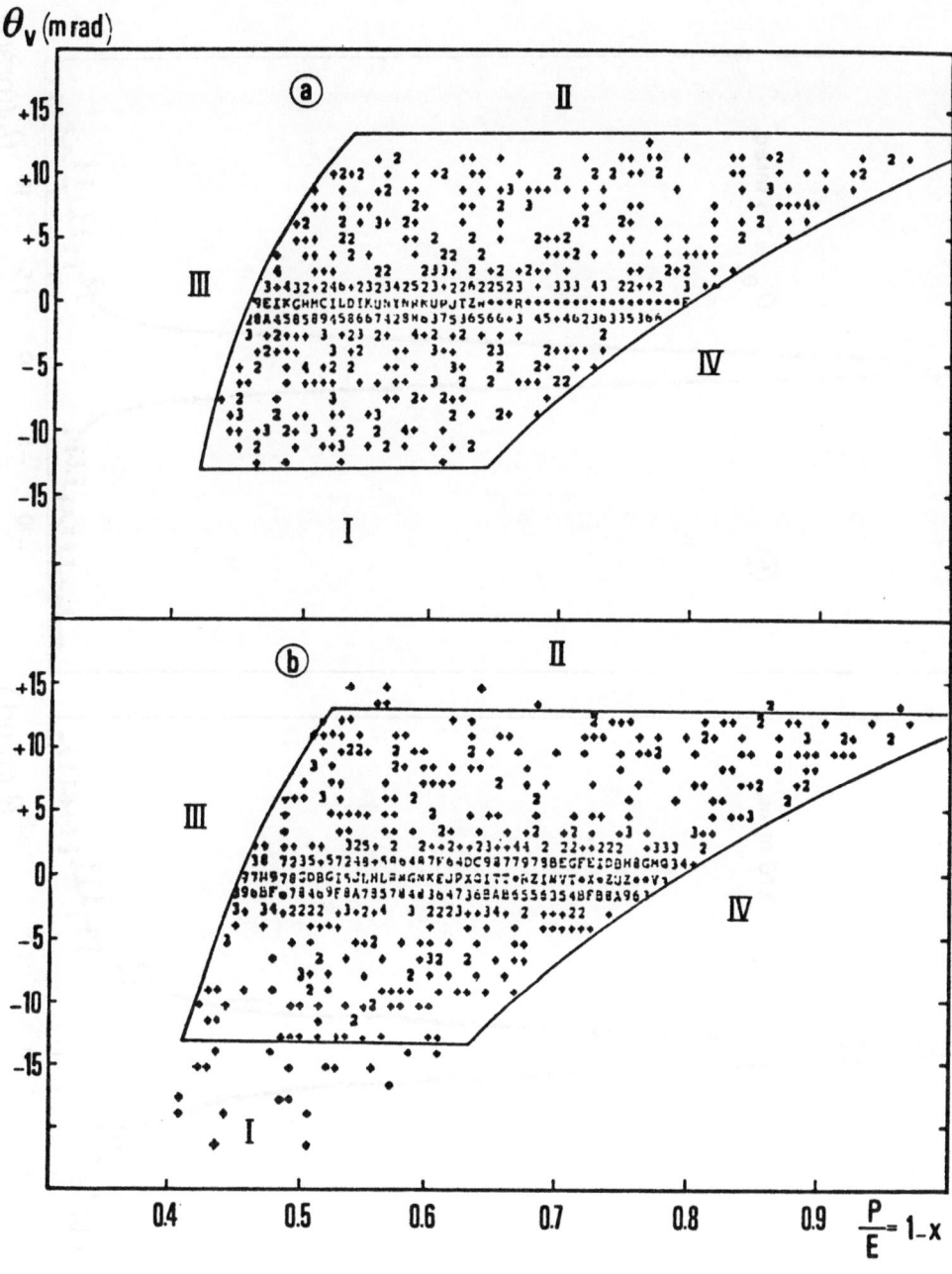

Figure 6

a
σθH = 0.75 mrad.
−8
−4
0
+4
+8
θH(mrad.)
b
σθV = 1.5 mrad.
−10
−5
0
+5
+10
θV(mrad.)
Number of tracks

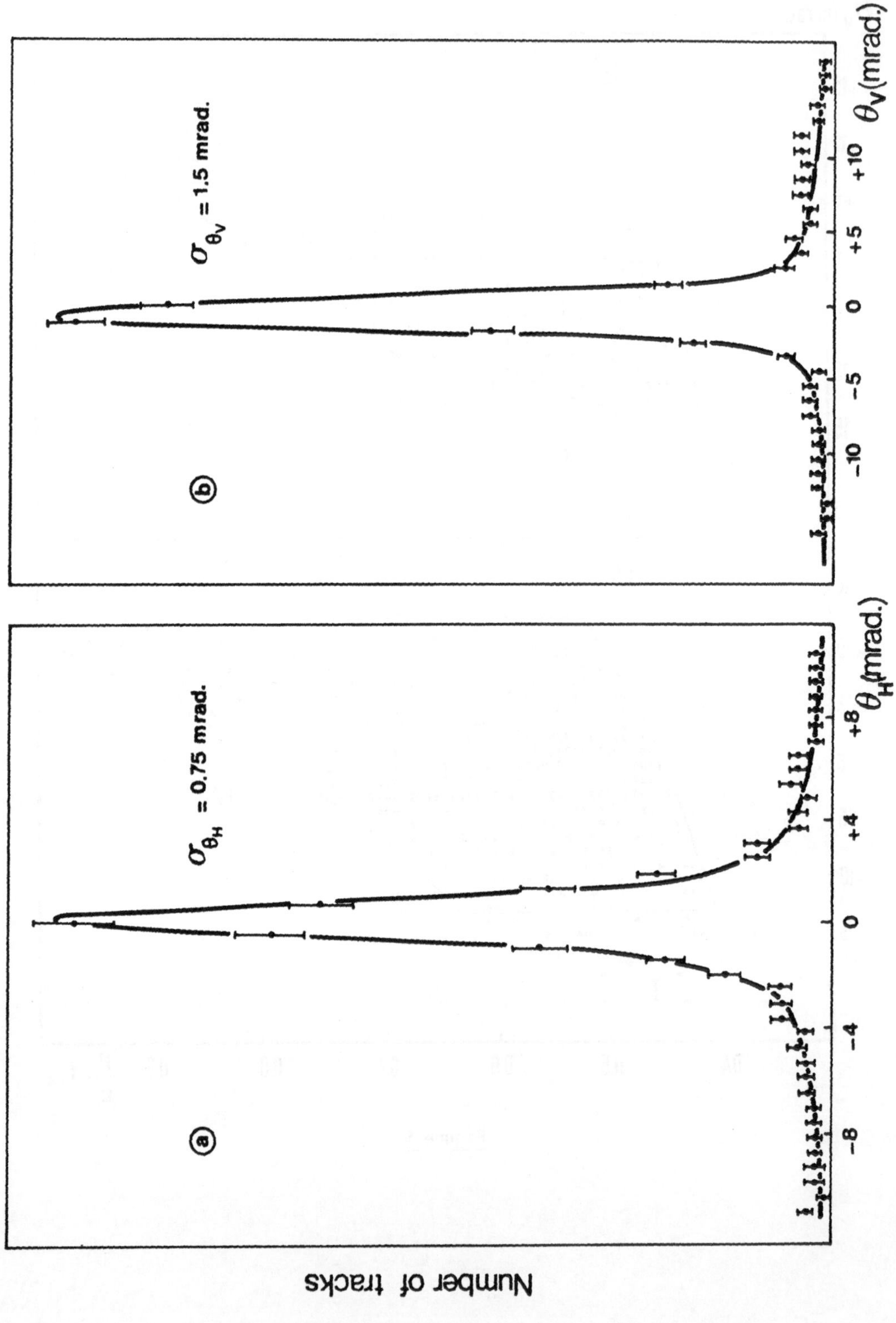

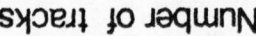

Figure 7

231

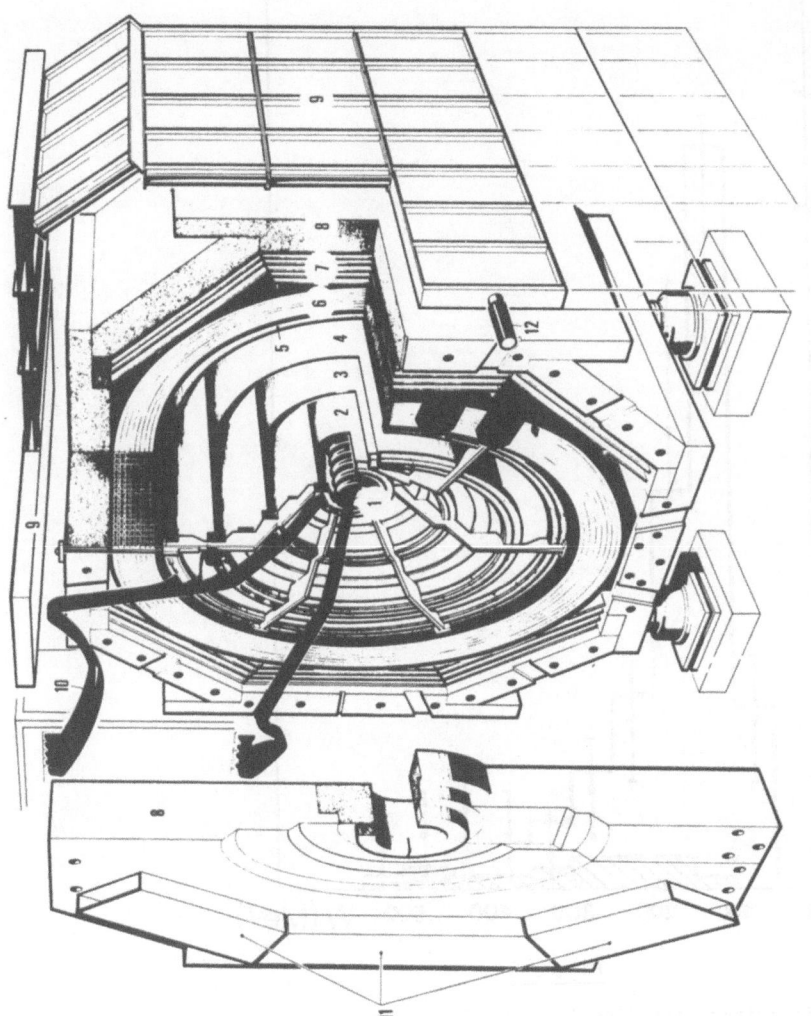

FIGURE 8

1) Vacuum chamber
2) 3) 4) 5) Cylindrical multiwire proportional chambers
6) Magnet solenoïd
7) Drift chambers
8) Magnet Yoke
9) Cosmic ray veto system
11) Electronic shielding
12) Photo-multiplier

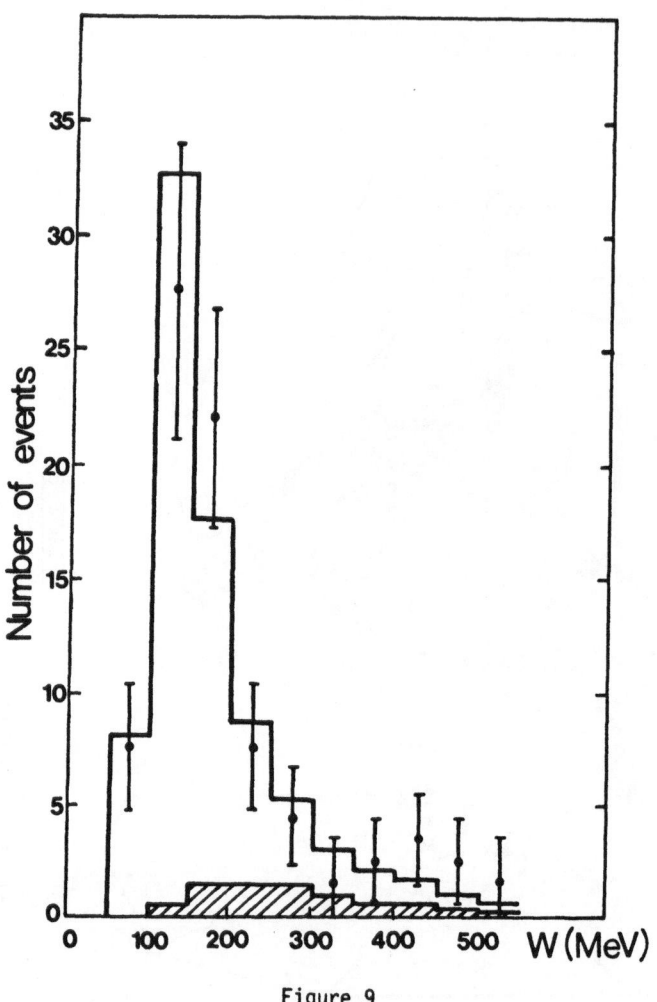

Figure 9

Invariant mass distribution of the lepton pairs,
computed as electron pairs.
The solid line represents the Monte-Carlo prediction.
The shaded area shows the muon contribution.

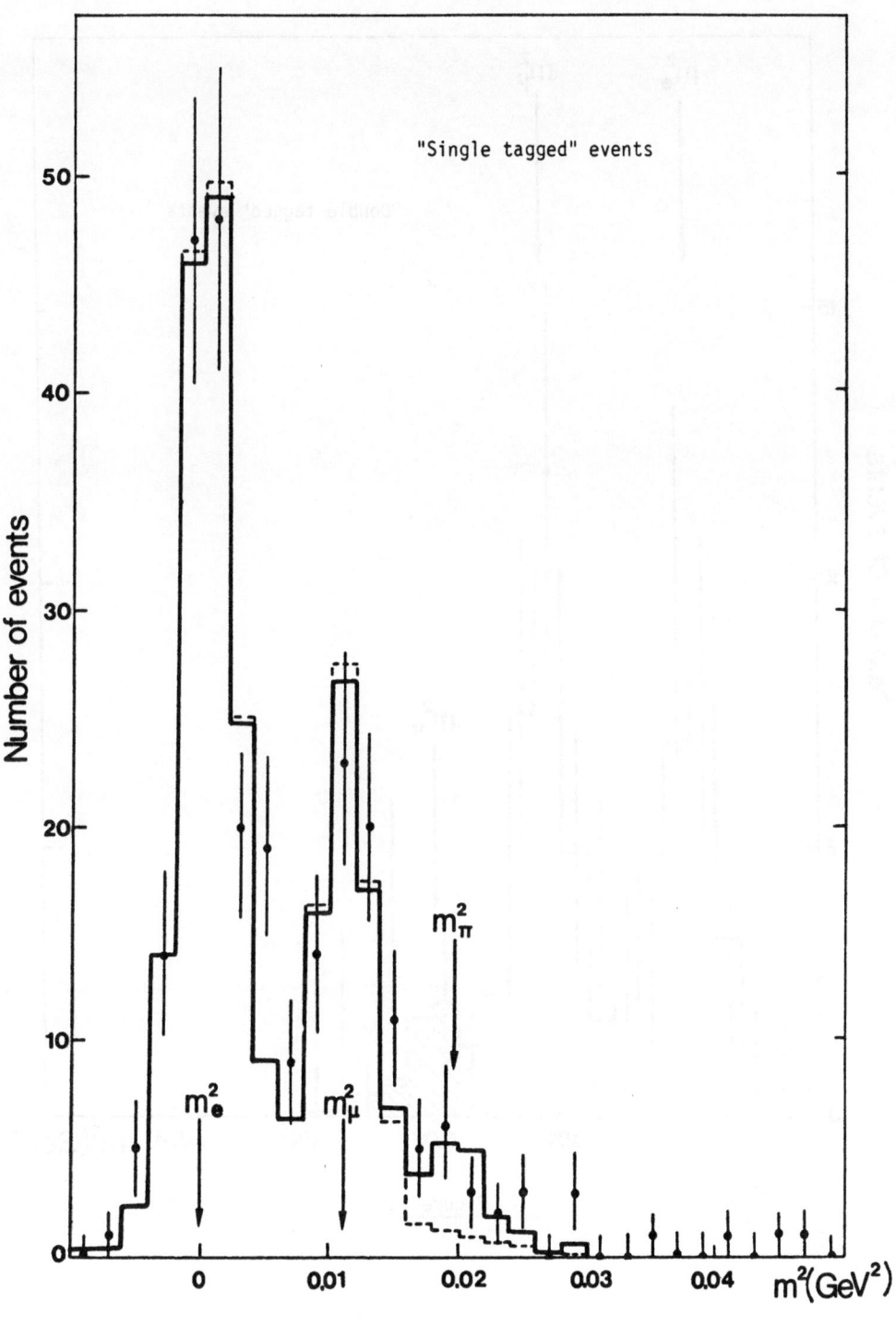

Figure 10

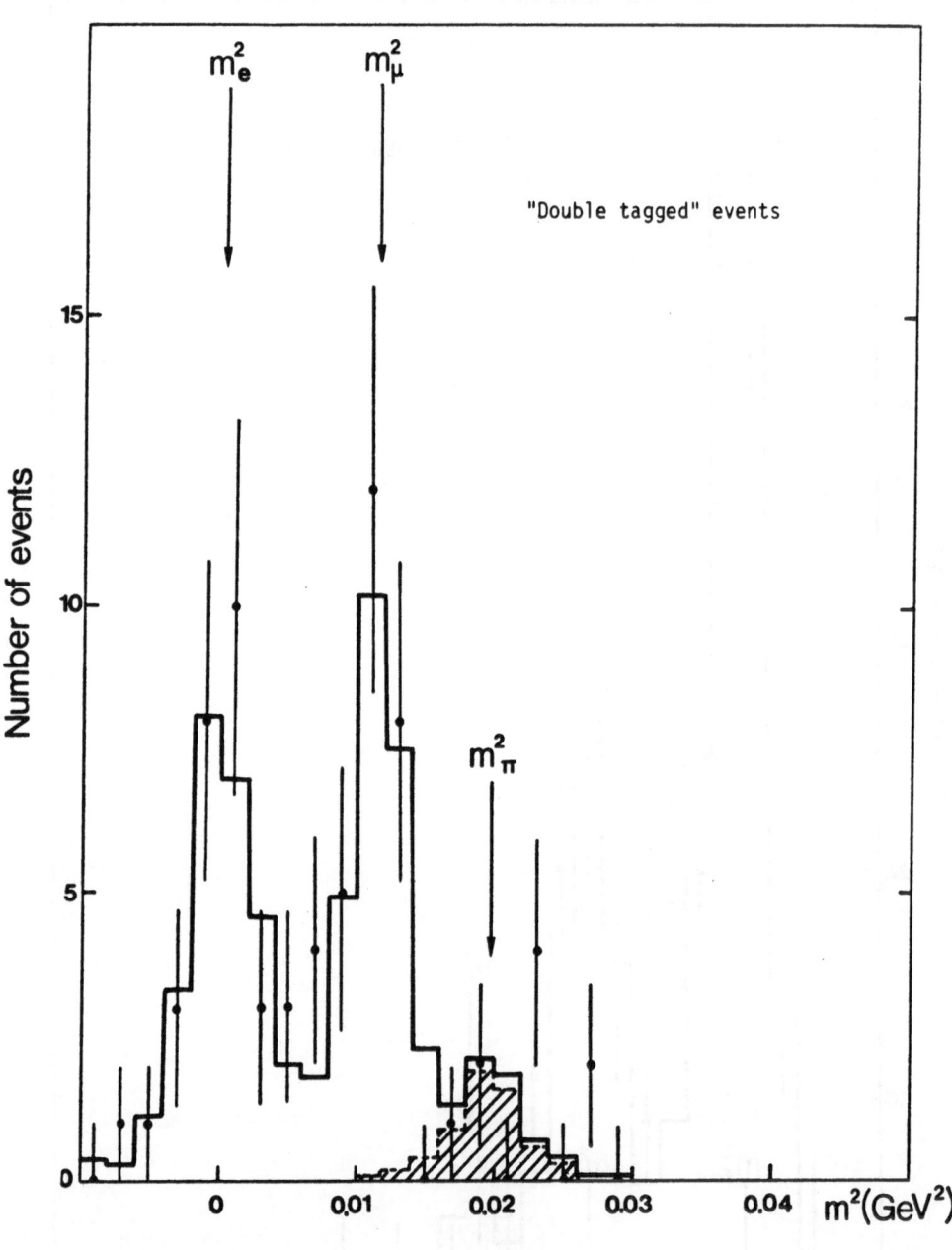

Figure 11

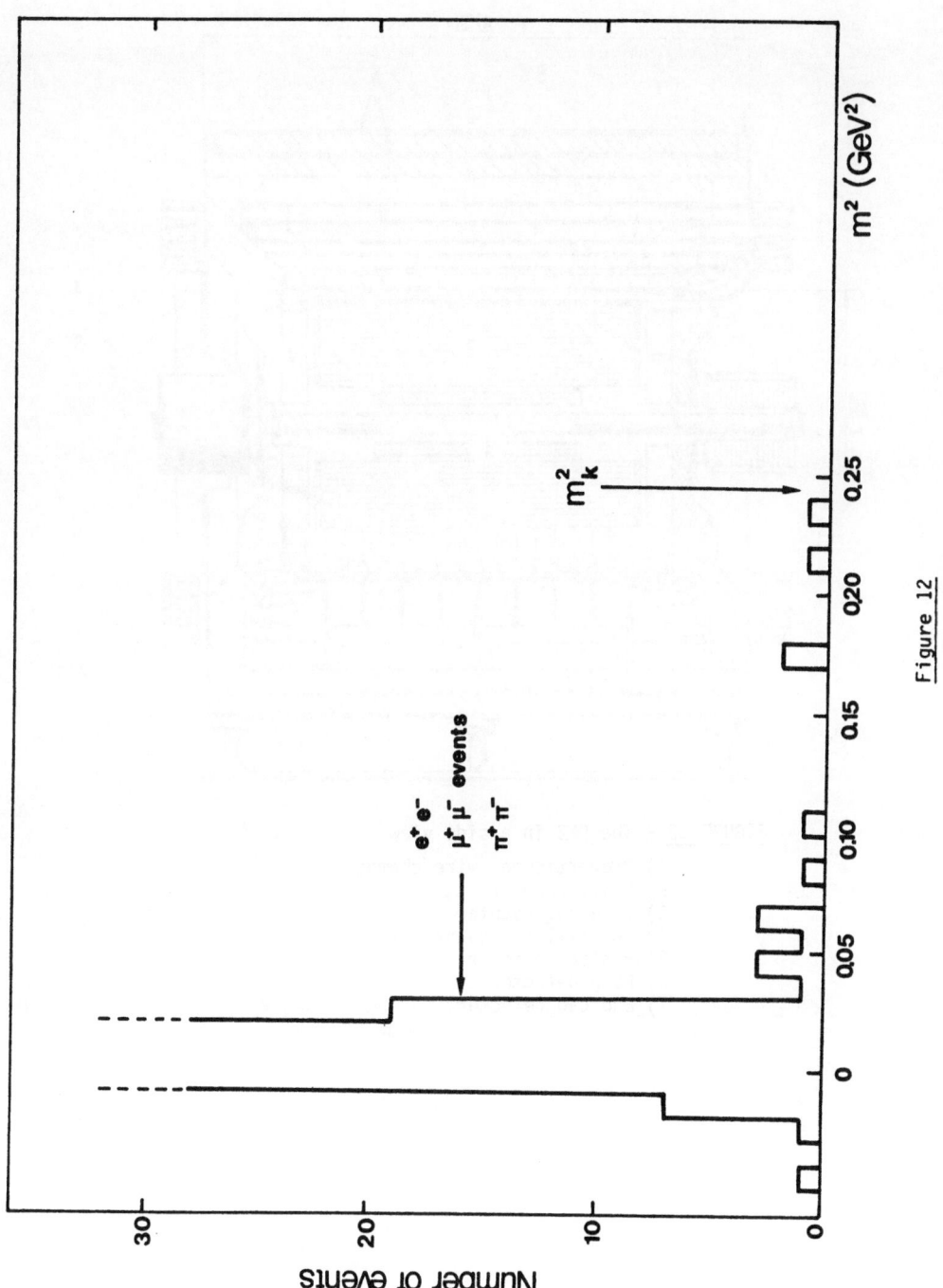

Figure 12

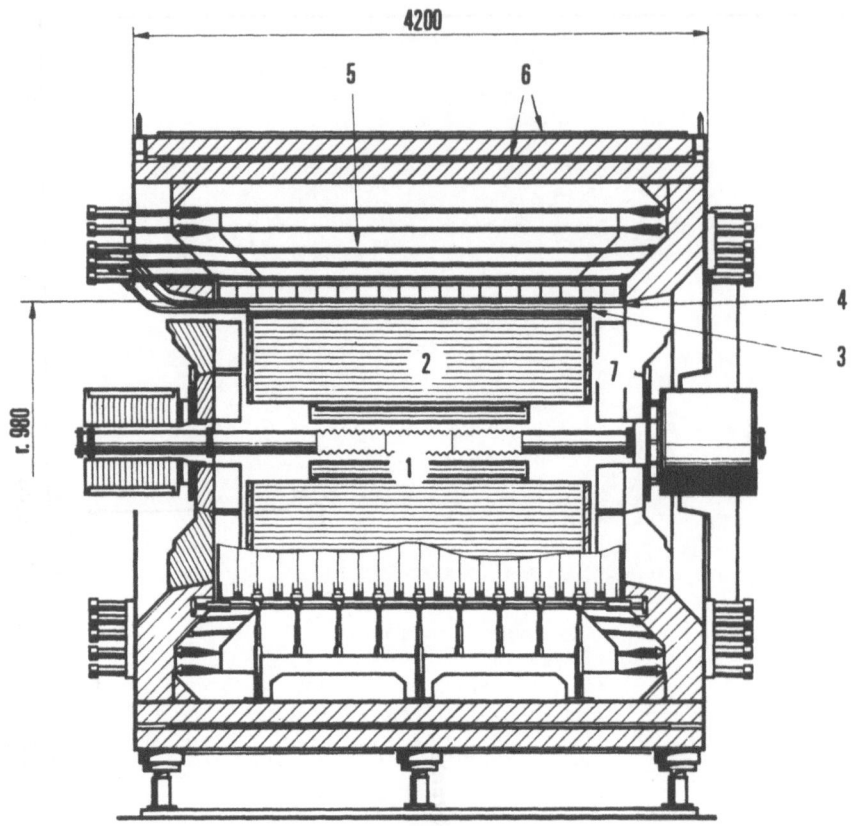

FIGURE 13 - The DM2 in a side view

1) Proportionnal wire chamber
2) Drift chamber
3) Cerenkov counter
4) Scintillator counters
5) Photon detector
6) Muon detector
7) End cap detector

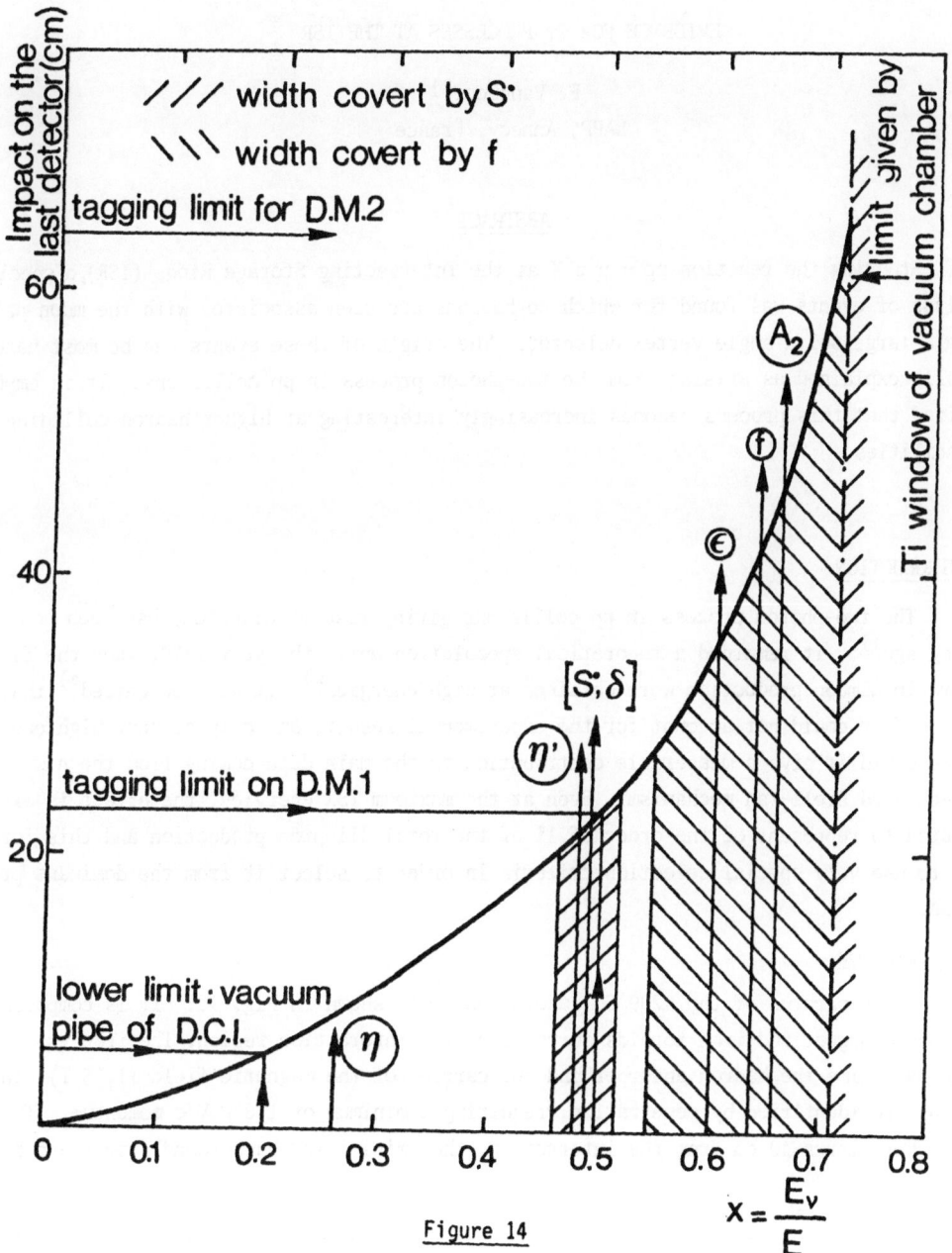

Figure 14

EVIDENCE FOR 2γ PROCESSES AT THE ISR

F. Vannucci[*]

LAPP, Annecy, France

ABSTRACT

Studying the reaction $pp \rightarrow \mu^+\mu^- X$ at the Intersecting Storage Rings (ISR), a special class of events was found for which no hadrons are seen associated with the muon pair in a large solid angle vertex detector. The origin of these events can be most naturally explained as arising from the two-photon process in pp collisions. It is emphasized that this process becomes increasingly interesting at higher hadron colliding facilities.

INTRODUCTION

The two-photon process in pp collisions giving rise to dileptons has been computed long ago[1]. It remained a theoretical speculation until the year 1970, when the first data in dimuon production were obtained at high energies[2]. It was recognized[3] that the effect could not account for the experimental result, and only at very high energies could it give a measurable contribution to the main data coming from the now celebrated Drell-Yan mechanism. Even at the maximum ISR energies, the effect is expected to represent of the order of 1% of the total dilepton production and this forces us to use very special selection criteria in order to select it from the dominant process.

THE DETECTOR

The apparatus of the R209 Collaboration[**] is shown in Fig. 1a. It is composed of seven magnetized iron toroids surrounding the interaction region. The iron is at the same time the hadron absorber and the carrier of the magnetic field (1.75 T). The muons are identified by penetration, requiring a minimum of 1.8 GeV/c momentum. The absorber starts 40 cm from the interaction, thus minimizing background from π and K decays.

---

[*] Visitor at CERN, Geneva, Switzerland.
[**] Experiment performed by the CERN-Harvard-LAPP-MIT-Naples-Pisa Collaboration:
D. Antreasyan, W. Atwood, R. Battiston, U. Becker, G. Bellettini, P.L. Braccini,
J.G. Branson, J. Burger, F. Carbonara, R. Carrara, R. Castaldi, V. Cavasinni,
F. Cervelli, M. Chen, G. Chiefari, T. Del Prete, E. Drago, M. Fujisaki, M. Hodous,
T. Lagerlund, P. Laurelli, O. Leistam, P.D. Luckey, M.M. Massai, R. Matsuda,
L. Merola, M. Morganti, M. Napolitano, H. Newman, D. Novikoff, J.A. Paradiso,
L. Perasso, K. Reibel, R. Rinzivillo, G. Sanguinetti, C. Sciacca, M. Steuer,
K. Strauch, S. Sugimoto, S.C.C. Ting, W. Toki, M. Valdata-Nappi, C. Vannini and
F. Visco.

Large-size drift chambers (up to 6.5×2.7 m^2) in between and around the toroids detect the muon trajectories. The multiple scattering in the iron limits the momentum resolution to $\Delta p/p \simeq 16\%$. Three layers of scintillator counters in coincidence form a candidate muon track. The trigger requires two tracks at an angle of $180° \pm 50°$ in the non-bending plane. With a luminosity of 10^{31} cm^{-2} s^{-1} the trigger rate is about 1 per second.

Immediately around the intersection a vertex detector[4] (see Fig. 1b) made of 136 modular drift chambers measures the direction of associated charged hadrons. It consists of a total of 544 wires with 500 μ spatial resolution for the transverse coordinates, and 272 delay lines which give the longitudinal positions with an accuracy of 5 mm. This inner detector covers the full 360° in the azimuthal angle, and loses only 9° of the polar angle. A charged track traverses 3 to 5 planes of chambers in this solid angle. There is no magnetic field, only the direction of the tracks can be measured.

HADRONLESS EVENTS

In the vertex detector a normal event looks as shown in Fig. 2a: the dimuon is produced with many charged tracks reconstructing the interaction vertex. Figure 2b shows an example of an event where the two muons extrapolated from the outside spectrometer are not accompanied by any hadron.

Defining a track as at least two hits pointing to the vertex, the efficiency of the inner detector was found to be greater than 95%. This was measured with a sample of muons well reconstructed in the outer spectrometer.

If one considers an event with n real hadronic tracks, given an inefficiency ε, the probability for this event to appear like a hadronless event is ε^n, and the probability for it to appear like an event with 1 hadron is $n(1-\varepsilon)\varepsilon^{n-1}$. The loss of tracks due to inefficiency populates more the sample of 2μ + 1 hadron than the sample of 2μ + 0 hadron. In order to evaluate the importance of the effect, one can compare the number of events appearing like 2μ + 0 hadron and 2μ + 1 hadron. This has been performed by eye-scan, the computer program having a low efficiency for reconstructing a vertex when few tracks are present. Because of electronic noise and back scattering at the surface of the absorber, extra hits which do not reconstruct a track may be visible. The comparison has been done for different levels of cleanliness, i.e. different numbers of extra hits. The result is given in Table 1. In all classes there is an excess of 2μ + 0 hadron with respect to 2μ + 1 hadron. The background can be evaluated by counting the number of like-sign events: 11 such dimuons are found to be compared with 167 opposite-sign events.

The full significance of these hadronless events is shown in Fig. 3, where the total multiplicity (including the two muons) is displayed. The bulk of the data shows a maximum around 12 charged tracks associated with a dimuon. This is the Drell-Yan contribution. This distribution, extrapolated to the low end of the plot, does not account for the measured number of hadronless events. The effect is as high as 10 standard deviations. The break in the multiplicity plot indicates that two different processes contribute and the origin of the hadronless events cannot be explained by the Drell-Yan mechanism.

DYNAMICS OF HADRONLESS DIMUONS

For further investigation, various kinematical characteristics are studied for the 103 events having less than 5 extra hits in the vertex detector.

Figure 4 shows the invariant mass of the hadronless events. The acceptance of the muon spectrometer eliminates events with mass below 3 GeV because of the minimum momentum required for each muon to traverse the iron, hence the events accumulate at the smallest allowed mass.

Figure 5 shows the dimuon transverse momentum distribution. The average $\langle p_t \rangle$ is much lower for these events than for the Drell-Yan events[5] for which $\langle p_t \rangle \simeq 1.7$ GeV/c. The distribution peaks at $p_t \simeq 700$ MeV/c, which reflects the effect of the multiple scattering.

Figure 6 shows the acoplanarity angle of the two muons $\phi_1 - \phi_2$, where the ϕ angle is the angle between the plane formed by one muon together with the beams and the plane of the ISR. This distribution shows that the two muons tend to be in the same plane while they are very acollinear.

The topology of the hadronless dimuons -- namely the fact that the hadronic component escapes at small angle -- the low mass, low p_t and coplanarity of the dimuon system point toward interpreting those events as originating from the two-photon process. The magnitude of the cross-section, as will be seen later, is also supporting this interpretation.

GUIDE TO 2γ PROCESSES IN HADRONIC COLLISIONS

The Feynman diagrams representing the 2γ process in pp collisions are shown in Fig. 7 for the elastic and the inelastic components.

The cross-section is given in the equivalent photon approximation by the expression[6]

$$\sigma = \iint \text{flux } \gamma_1 \text{ flux } \gamma_2 \, \sigma_{\gamma\gamma} \, ,$$

where

$$\sigma_{\gamma\gamma} = \frac{4\pi\alpha^2}{m_{\mu\mu}^2} \left(\ln \frac{m_{\mu\mu}^2}{m_\mu^2} - 1 \right)$$

is the elementary cross-section for the process $\gamma\gamma \to \mu^+\mu^-$. This cross-section is independent of the origin of the photons, and is the same in the 2γ process originating in e^+e^- reactions.

On the contrary, the flux of equivalent photons depends on the radiator. The dependence on the beam in leading log approximation is given by $(\ln E/m)^2$, where E and m are the energy and the mass of the radiating particle. For instance a 15 GeV beam of e^+ or e^- radiates 15 times more than the same energy beam of p or $\bar{p}$. Higher energies are, or will soon be, available in hadron colliding facilities. At higher energies the disadvantage due to the mass of the emitting particle is reduced (a factor of 8 reduction for a 70 GeV beam), but the form factors of the hadron drastically cut the rates. Figure 8 summarizes the comparison of cross-sections as a function of the dimuon mass for the PETRA/PEP, ISR, and $\bar{p}p$ collider machines. The rates go asymptotically as $(\ln s)^3$ for a given $m_{\mu\mu}$ and Fig. 8 shows that the $\bar{p}p$ collider overtakes the PETRA/PEP machines for final masses greater than 12 GeV. To be complete in this comparison one should remember that the luminosity gives a further advantage to the ISR or ISABELLE machines over the e^+e^- machines.

The cross-section obtained at the ISR, as a function of the invariant dimuon mass is shown in Fig. 9. The full line represents the cross-section computed from the two-photon elastic process [*]. Form factors for the proton have been included. The inelastic process, more difficult to compute, is expected to be of the same order as the elastic part. The agreement between the calculation and the experimental result is quite satisfactory, considering the difficulty of estimating the acceptance. The over-all normalization has an uncertainty of ±50%.

One difficulty of the 2γ process in pp or $\bar{p}p$ collisions is the competition from hadronic processes. The double pomeron exchange would likewise give particles scattered at small angle. This competition does not arise when one restricts the analysis to the electromagnetic process giving e^+e^- or $\mu^+\mu^-$ in the final state. On the other hand, tagging seems essential to study the more interesting hadronic events of the 2γ process.

CONCLUSION

We have given evidence for the production of dimuons which are created without associated hadrons in a large domain of rapidity around the central region. The topology of these events, their low transverse momentum, and the magnitude of the corresponding cross-section can be most naturally interpreted by the 2γ process. This effect which is still a small background at the ISR, compared with the dominant Drell-Yan mechanism, grows with energy and could trigger enough interest to be studied for itself, possibly at the $\bar{p}p$ collider, probably at the ISABELLE machine.

Acknowledgements

It is a pleasure to acknowledge the help of J.A. Paradiso, J.P. Revol, M. Steuer and J.A.M. Vermaseren in the course of this study.

---

[*] Computation done at CERN by J.A.M. Vermaseren.

REFERENCES

1) L.D. Landau and E.M. Lifshitz, Classical theory of fields (Pergamon, Oxford, 1975).

2) J. Christenson et al., Phys. Rev. Lett. 25, 1523 (1970).

3) M.S. Chen et al., Phys. Rev. D7, 11 (1973).

4) A. Bechini et al., Nucl. Instrum. Methods 156, 181 (1978).

5) D. Antreasyan et al., Measurement of dimuon production at the ISR, to be published.

6) V.M. Budnev et al., JETP Lett. 12, 238 (1970).
   K. Fujikawa, Nuovo Cimento 12A, 117 (1972).
   A. Soni, Phys. Rev. D8, 880 (1973).
   C. Carimalo et al., Phys. Rev. D7, 3485 (1978).
   R. Moore, Manchester University preprint, TH80/2 (1980).

Table 1

Total number of dimuon events
without and with hadrons

No. of hits	2μ + 0 hadron	2μ + 1 hadron
0	19	11
≤ 1	34	26
≤ 5	103	83
≤ 10	167	126

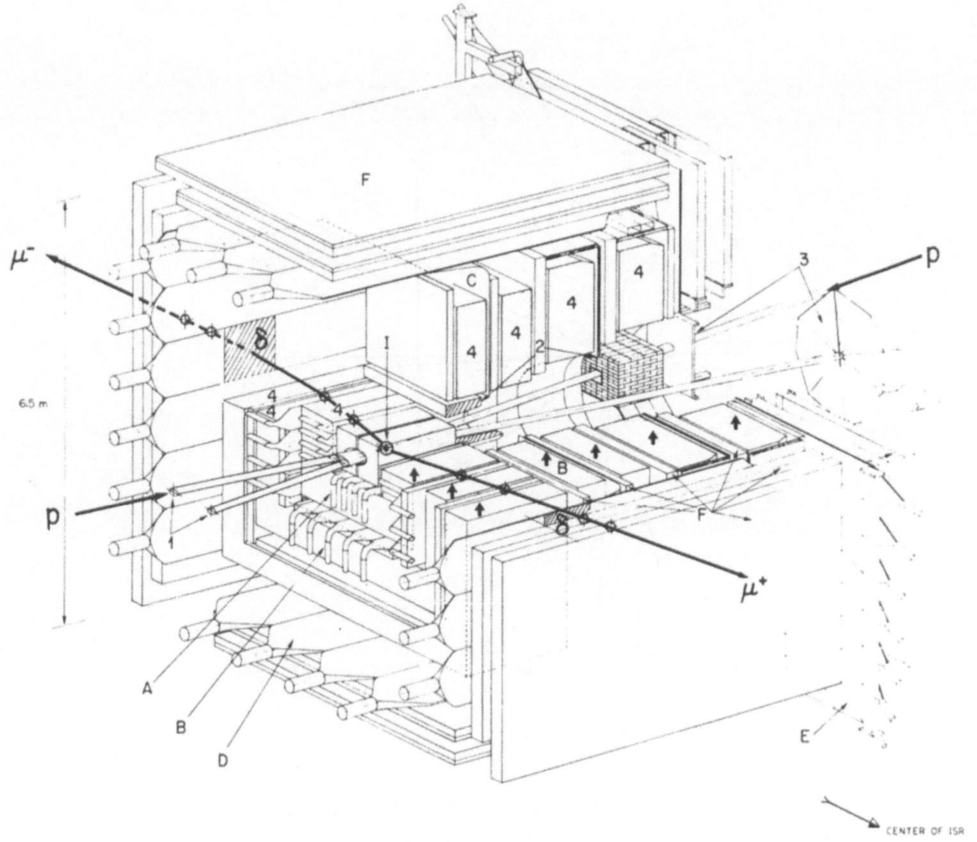

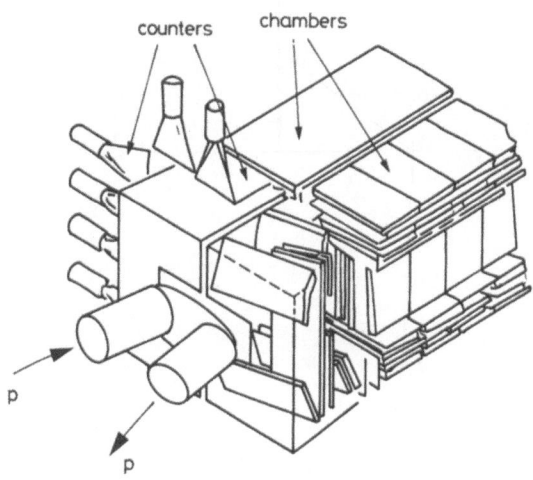

Fig. 1a Exploded view of the detector with a cut-out showing the internal structure

Fig. 1b View of the vertex detector surrounding the interaction point

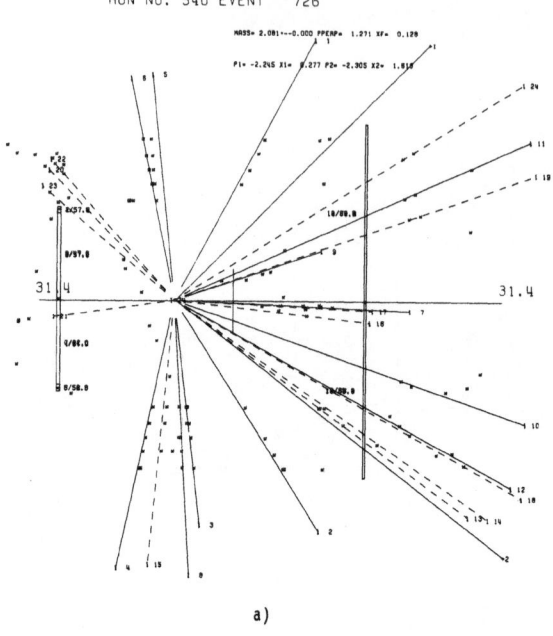

a)

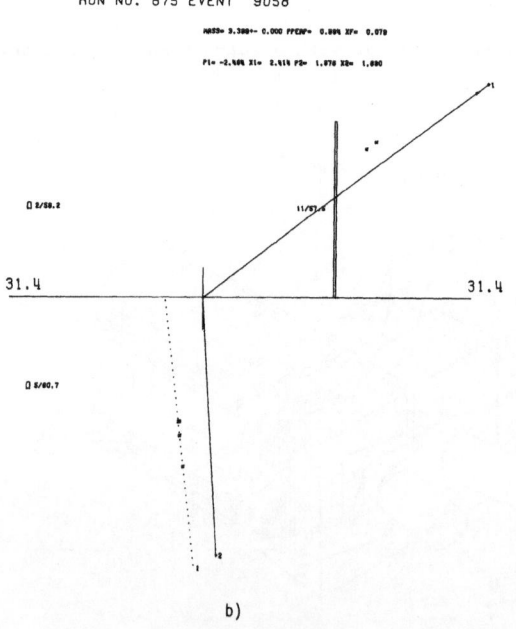

b)

Fig. 2 Dimuon events reconstructed in the vertex
detector a) with and b) without associated hadrons

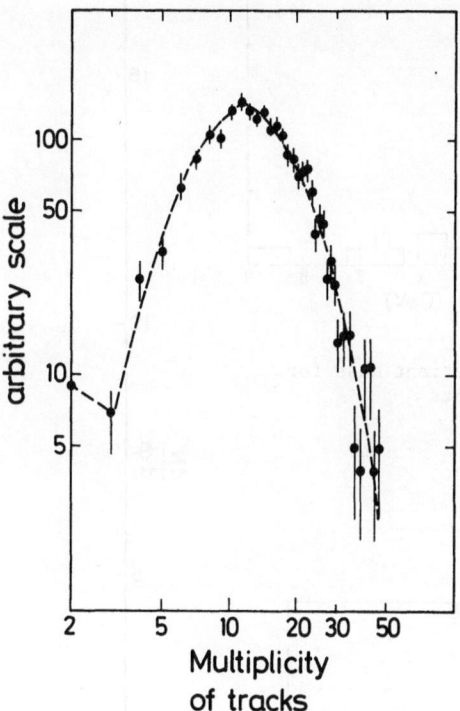

Fig. 3 Multiplicity of charged tracks associated with the dimuons. The two muons are included.

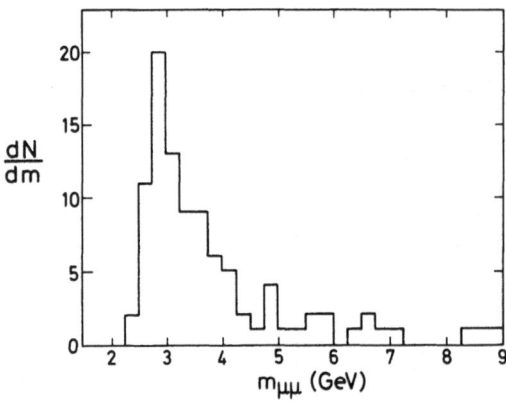

Fig. 4 Dimuon invariant mass for
the hadronless events

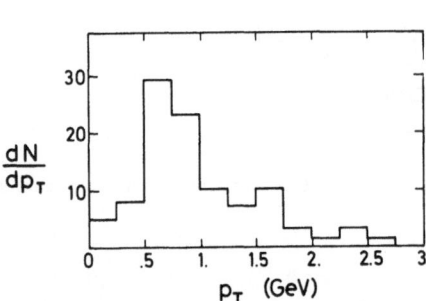

Fig. 5 Dimuon transverse momentum for
the hadronless events

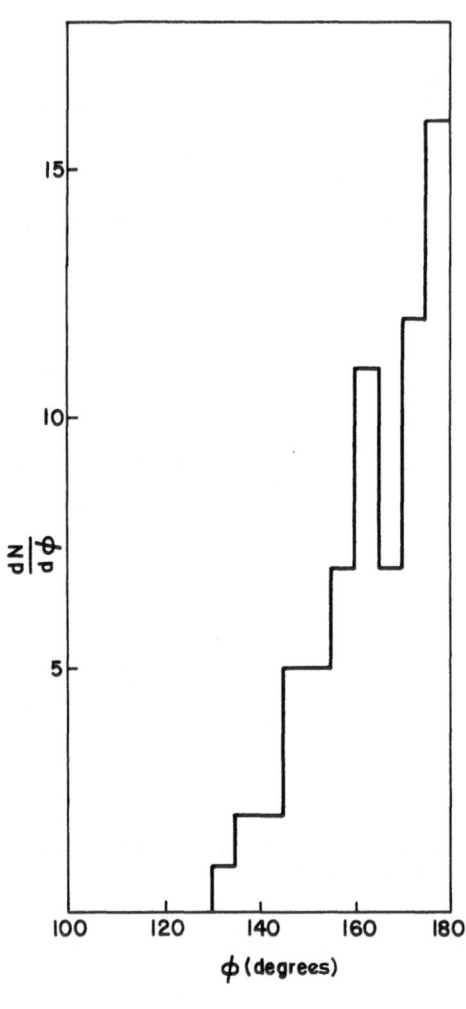

Fig. 6 Acoplanarity distribution of
the two muons for hadronless events

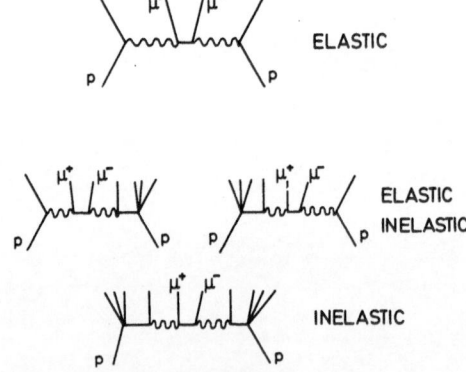

Fig. 7 Feynman diagrams for the two-photon process in pp collisions, with the elastic,
elastic-inelastic, and inelastic contributions.

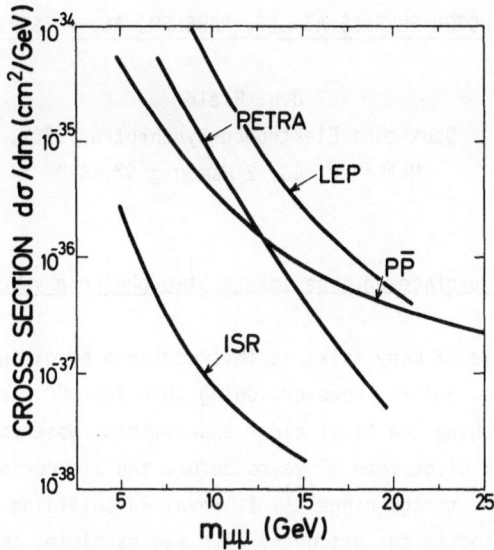

Fig. 8 Expectations for dimuon productions via the two-photon process at the ISR, at PETRA/PEP, at the p$\bar{\text{p}}$ collider, and at LEP.

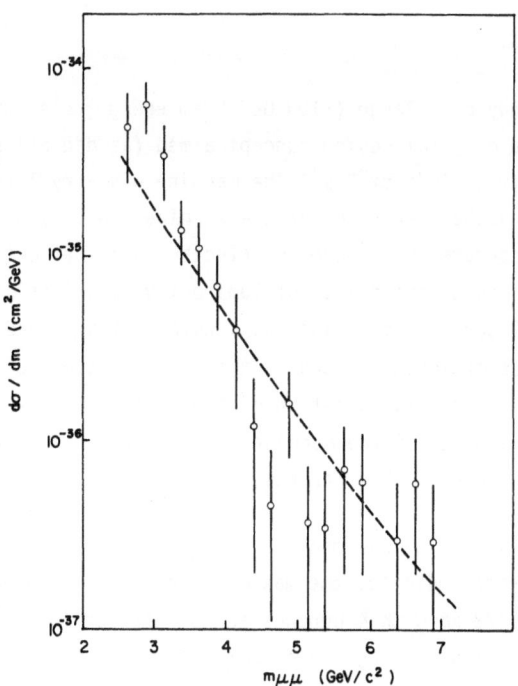

Fig. 9 Experimental cross-section obtained at the ISR. The dashed line represents the theoretical calculation for the elastic contribution.

TWO PHOTON PHYSICS AT LEP, PROBLEMS AND POSSIBILITIES

J.H. Field

Deutsches Elektronen-Synchrotron DESY,

Notkestr. 85, 2 Hamburg 52, BRD

Very brief historical preface on experimental two-photon physics

I shall follow the rule of many talks at this workshop by giving a few historical remarks at the beginning. Before, however, doing this for LEP I should like to set the
record straight concerning the first clear experimental observation of a two-photon
interaction. This took place some 40 years before the pioneering experiments at Novosibirsk using electron-storage rings[1]. It involved colliding a real with a virtual
photon (fig. 1), resulted in the discovery of a new particle, the positron, and earned
a Nobel Prize for the experimentalist, C. Anderson, who made the observation!

1. CHARACTERISTICS OF LEP

History of the project

The first detailed study of a large (~100 GeV beam energy) e^+e^- machine was made at
CERN in 1976 - 77. The original design concept aimed (as did all subsequent designs)
at a maximum luminosity of 10^{32} cm^{-2} s^{-1}. The machine was very large (circumference
51.5 km) and used 32 bunches per ring. As the machine had only 8 interaction regions,
it was necessary to separate the bunches by electrostatic deflection at the other
possible collision points to avoid loss of luminosity. The cost of the machine was
high (as compared to later versions) and had a number of unsolved technical problems.
One of the most serious of these related to the spacial tolerances necessary to separate and recombine the beams. The question of the site of the machine was not discussed. The technical aspects of the machine, and its physics justification respectively are given in two CERN yellow reports, ref. (2) and (3).

The second design study (LEP70) was made in 1978 and is described in ref. (4) (the
'Blue Book'). As the name indicates the maximum energy was reduced to 70 GeV, the
circumference was only 22 km with 4 bunches and 8 intersection regions (as in PETRA)
so that beam separation was no longer necessary. The possibility of placing such a

machine near the existing CERN site was studied. The cost of LEP70 which, from a technical viewpoint, seemed a feasible design, was roughly 1000 MSF. In 1978 further studies of the physics possibilities of such a machine were made and these were discussed at length during the Les Houches Summer Study in September 1978[5]. At the time a strong interest was expressed by the community of European physicists to go to significantly higher energies to enable the study not only of z^0 production which was the primary motivation for the first LEP proposal, but also of w^+w^- production to gain information on the basic trilinear couplings e. g. $z^0w^+w^-$ of the gauge theory of the electroweak interaction.

In response to these demands, and while further very detailed studies of the physics possibilities were being made under the auspices of ECFA, a further design study was completed in 1979 (see ref. (6), 'Pink Book'). The basic design was the same as LEP70, but scaled up in size (30 km circumference) and energy (86 GeV with conventional R.F.). More careful studies of component design resulted in a relatively small increase in the overall cost to 1275 MSF. If, at a later date, superconducting R.F. cavities became available, the maximum beam energy would be increased to 130 GeV. On the low energy side useful luminosities are expected down to energies of 22 GeV per beam, minimising the 'gap' between PEP/PETRA and LEP. As for LEP70 a site adjacent to CERN was specifically considered in detail in the design study. The earliest date at which LEP might be accepted as a project is 1981, which would give the first beams, according to the 'Pink Book' estimates, in 1988. All the LEP parameters and performance details given below are taken from the most recent LEP design study, ref. (6).

Selected LEP parameters

Table 1 gives a list of some parameters of LEP which may be of interest to an experimental physicist.

In physical dimensions LEP is almost exactly ten times larger than PETRA. The time between bunch crossings is large, 25.5 µs, allowing time for quite complicated trigger decisions to be taken without loss of useful luminosity. For a given instantaneous luminosity however, the duty cycle is ten times worse than PETRA, exacerbating the problems of occupation of tagging counters by background, and accidental trigger rates.

Two types of experimental interaction regions are planned: 4 with ±5 m and 4 with ±10 m free space for experimental detectors. The nominal machine luminosity refers to the ±5 m region, the ±10 m one having half of this. Taking into account the importance of good forward acceptance for 2-photon physics the conceptual designs of two-photon detectors so far made have all assumed a ±10 m interaction region. The location of the experimental interaction regions are indicated in fig. 2. 3 of the interaction regions are located deep under the Jura and access shafts will have to be bored horizontally (indicated on fig. 2 by broken lines). Fig. 3 shows a conceptual 2γ-detector design (taken from ref. (5)) located in one of the deep underground halls. The fractional energy spread of the beams, the bunch length, filling time and beam lifetime are not markedly different from PETRA values. The beam lifetime quoted in table 1 is dominated by the beam-beam bremsstrahlung lifetime at the assumed machine vacuum of $3 \cdot x \; 10^{-9}$ Torr (beam gas bremsstrahlung lifetime = 20 h).

Energy and power options

Table 2 shows the maximum energies and luminosities attainable in LEP for given investments in R.F. power. The first 4 stages: 1/6, 1/3, 1, 4/3 refer to the R.F. power relative to the nominal design figure of 96 MW. The construction costs are not markedly different for these different options, unlike the R.F. power costs. Stage 1/6 should reach the z^0 (assuming the correctness of the Weinberg-Salam Theory), while stage 1 should produce w^+w^- pairs with conventional R.F. Stage 4/3 in which a modest increase in beam energy is accompanied by a rather large increase in R.F.-power is a 'contingency' option should the superconducting R.F.-cavities required by stage 2 prove to be unrealisable.

Injector

The injection system for LEP will consist of:

- a 200 MeV electron linac
- a conversion target
- a 600 MeV positron linac
- a 600 MeV positron accumulation ring
- a 22 GeV injector synchrotron

As shown in fig. 4 the injector will be situated close to the present location of the ISR at CERN and will make use of the existing infrastructure for control rooms and laboratory space.

The luminosity of LEP

Fig. 5 shows the luminosity as a function of beam energy in a ±5 m interaction region for the 5 different stages given in table 2. The luminosity $\mathcal{L}$ is related to other essential machine parameters by the relation:

$$\mathscr{L}\ (\mathrm{cm}^{-2}\ \mathrm{s}^{-1}) \ = \ \frac{1.23 \times 10^{36} \times \Delta Q \times P_b\ (\mathrm{MW}) \times \rho\ (\mathrm{km})}{E\ (\mathrm{GeV})^3 \times \beta_y^*\ (\mathrm{m})}$$

where P_b = R.F. power (both beams)

ρ = machine radius

β_y^* = vertical β-function at intersection

ΔQ = beam-beam tune shift limit

The curves given in fig. 5 assume a value for ΔQ of 0.06, but to date PETRA has been able to reach only ≈ 1/2 of this figure. Should similar restrictions be found also on LEP some of the lost luminosity could be regained by reducing β_y^* from its design value of 10 cm. This would involve 'mini-β insertions' with quadrupoles nearer to the interaction point than in the planned interaction regions, and so possible restrictions on the space available for experimental detectors. Such modifications are likely to have more drastic effects on dedicated two-photon detectors, where forward acceptance is at a premium, than in the central detector aimed at studying electroweak interaction effects. If, however, it is possible to design in the 'mini-β' quadrupoles at an early stage useful acceptances in the forward direction still seen to be attainable, ref. (7).

e-p collisions at LEP

The LEP ring is situated conveniently close to the SPS (fig. 2) and the circumference has been chosen to be a rational multiple of that of the SPS:

$$\frac{\text{circumference of LEP}}{\text{circumference of SPS}} = \frac{22}{5}$$

A transfer tunnel could be rather easily constructed to give an ep collider with up to 130 GeV electrons colliding with 270 GeV protons.

Future use of the LEP tunnel

Fig. 5 shows a cross section of the LEP tunnel. As can be seen the space is more than adequate to accomodate the LEP magnets. In ref. (6) it is stated that '...., there is enough room left for the installation of a second machine at a later stage, should the need arise'.

2. ENERGY VARIATION OF 2γ PROCESSES

Cross sections with point-like photon couplings

LEP gives a roughly 5-fold increase in beam energy as compared to PEP and PETRA. The energy dependence of some typical cross sections and tagging efficiencies for the process:

$$e^+ e^- \to e^+ e^- \tau^+ \tau^-$$

are given in table 3. A process with a relatively large mass final state and high transverse momentum (several hundred MeV) of final state particles is chosen to avoid a strong dependence of the acceptance on the rapidity of the produced system. The kinematics and angular distribution of $\tau^+\tau^-$ production in 2γ collisions is expected to be almost the same as high p_T-$q\bar{q}$ production, refs. (8), (9) and (10), so the energy dependences given in table 3 should apply to this process also. The tagging efficiency quoted in table 3 is:

$$\mathcal{E} = \frac{\text{virtual photon flux in } \theta_{min} < \theta < \theta_{max}}{\text{virtual photon flux in } \quad 0 < \theta < \pi}$$

where θ is the lab. scattering angle of one (or both) electrons for single (double) tagging, respectively. The energy scaling laws given are true for $\theta_{min} \gg m_e/E$ (E = beam energy, m_e = electron mass) and not too small values of the $\gamma\gamma$ effective mass $W_{\gamma\gamma}$; $W_{\gamma\gamma}/2 E \gtrsim 0.1$ say. A typical experimental cut requires a minimum transverse momentum for one or more decay products, and so is approximately the same as the $p_T(\tau) \geq p_T^{min}$ cut.

Table 4 shows values of $\sigma(E = 75 \text{ GeV})/\sigma(E = 15 \text{ GeV})$ for untagged, singly tagged and doubly tagged cross sections with the different cuts given in table 3. Only the cross section with scaling cuts shows a dramatic energy variation. The energy variation of the tagged relative to the untagged cross section is most marked for the total cross section (i. e. no detection of final state particles from the $\gamma\gamma$ system required) and this difference becomes more pronounced for smaller values of $W_{\gamma\gamma}$.

An interesting consequence of the rapid $\alpha(\frac{\ln E}{E})^2$ energy dependence of the cross section with scaling cuts relates to the problem of looking for a new heavy lepton L under a background of τ production from the two photon process. As the momenta of the decay products of L in the one-photon process increase with the beam energy, momentum cuts can be scaled with the beam energy without loss of acceptance. If this is done, then:

$$\frac{\sigma^{cut}(e^+e^- \to e^+e^-\tau^+\tau^-)}{\sigma^{cut}(e^+e^- \to L^+L^-)} \quad \alpha(\ln E)^2$$

a slow increase with energy, so $2\gamma \, \tau^+\tau^-$ production is not expected to pose a serious problem in searches for high mass heavy leptons, contrary to some earlier expectations, refs. (11) and (12). The importance of energy scaling cuts in this case was pointed out by Gutbrod and Rek, ref. (13).

Vector meson propagator effects

The largest part of the hadronic $\gamma\gamma$ cross section is expected to result from hadronic couplings of vector meson components of the virtual photon (fig. 7). If the $\gamma\gamma$ total cross section is measured via tagging of the scattered electron, very strong suppression of the observed cross section is predicted at LEP energies due to propagator effects. If very small values of $W_{\gamma\gamma}$ are avoided, say $W_{\gamma\gamma}/2E \gtrsim 0.2$, the virtual photon flux at angle θ of the scattered electron is approximately:

$$\alpha \, d\theta^2/\theta^2$$

and a simple analytical expression can be given for the suppression factor S in the coupling of the virtual photon due to the vector-meson propagator:

$$\frac{1}{(1 - q^2/m_V^2)^2}$$

Defining

$$S \equiv \sigma(m_V)/\sigma(\infty)$$

$$S(x, \theta_{max}, \theta_{min}) = \frac{1}{\ln(\theta_{max}^2/\theta_{min}^2)} \left[\ln \frac{\theta_{max}^2}{\theta_{min}^2} \cdot \frac{P(x, \theta_{min})}{P(x, \theta_{max})} + \frac{1}{P(x, \theta_{max})} - \frac{1}{P(x, \theta_{min})} \right]$$

where $P(x, \theta) = 1 + \left[\frac{(1-x) \, E\theta}{m_V} \right]^2$

$$x = E_\gamma/E$$

and the scattered electron is detected in the angular range:

$$\theta_{min} < \theta < \theta_{max}$$

Fig. 8 shows S as a function of x for $\sigma_{min} = 20$ mrad, $\sigma_{max} = 200$ mrad at E = 15 GeV, 75 GeV and with $m_V = 0.773$ GeV (ρ) and $m_V = 3.1$ GeV (J/ψ).

For small values of x, S decreases for processes with a ρ propagator by a factor of 45 when the beam energy is increased from 15 to 75 GeV.

This type of analysis has been generalised by Donnachie, ref. (14) to include the effect of a whole Veneziano spectrum of vector mesons. The results for the case of double tagging in the angular range $10 < \theta < 150$ mrad at a beam energy of 70 GeV are presented in fig. 9. The left-hand plot shows the suppression factor for a given value of $z \equiv W_{\gamma\gamma}/2E$ while the right-hand plot shows the corresponding double tagging efficiency. n is the number of terms in the Veneziano spectrum. S is increased by a factor of roughly 2 in this model.

This strong suppression of diffractively coupled processes at LEP when tagging is used may be expected to improve the signal/background for some of the more interesting $\gamma\gamma$ interactions. Some examples are:

- Heavy-lepton production
- Heavy-flavour production (here is a high mass propagator J/ψ, Υ etc., so the suppression is less severe (see fig. 8)
- Processes with high p_T quark and/or gluon jets
- Deep inelastic $\gamma\gamma$ scattering

Resonance production

To compare resonance production on PETRA/LEP and LEP in two-photon reactions the direct production (fig. 10) of $J^{PC} = 0^{-+}$ $c\bar{c}$ and $b\bar{b}$ states, η_c, η_b with the parameters ref. (12):

	M (GeV)	$\Gamma_{\gamma\gamma}$ (keV)	σ_{tot} (15 GeV) pb	σ_{tot} (75 GeV) pb
η_c	2.8	10	116	265
η_b	9.2	20	2	9.4

is considered. A double tagging experiment in the angular range $20 < \theta < 200$ mrad is assumed. The double tagging efficiency in this angular range for E = 15 GeV is shown in fig. 11 as a function of $z = W_{\gamma\gamma}/2E$, ref. (15). The positions of the η_c, η_b at 15 GeV, 75 GeV are indicated. With an integrated luminosity of 10^{38} cm^{-2} (≈ 600 hours at the design luminosity in a ± 10 m intersection region) and allowing for the $(\ln E/m_e)^{-2}$ energy scaling of the double tagging efficiency, ref. (15), the following tagging efficiencies and numbers of events are found at the two energies:

	15 GeV		75 GeV	
	ϵ_{DT}	No. Events	ϵ_{DT}	No. Events
η_c	0.03	348	0.01	262
η_b	0.04	8	0.020	18

So for the η_c, because of reduced tagging efficiency, fewer events are expected in LEP than in PETRA. The experiment described is an extremely idealized one, and the practical requirement from the experimental trigger that at least one particle from the decay of the resonance is detected will reduce further the number of events observed. Clearly such an experiment is only possible at all if the design luminosity of the machine is achieved (say within a factor 2). One concludes that LEP is far from an ideal machine for resonance-searches in the $\gamma\gamma$ system. As stressed in his talk at this conference by Courau, resonance searches are done much more efficiently in lower energy machines where 0^0 tagging is possible.

The only advantage that LEP has for this type of measurement is in signal to background. The propagator suppression factor calculated above comes in squared in a double tagging experiment, so diffractive backgrounds should be several hundred times lower on LEP than on PEP/PETRA for the range of tagging angles considered above.

3. TAGGING AT LEP

In a series of recent papers[16],[17],[18] Kessler and collaborators have stressed the importance of tagging the scattered electrons at the smallest possible angles both so as to maximise the tagging efficiency and also so as to be able to use simple, factorisable Weizsäcker-Williams type formulae in the interpretation of the results. It was found, however, by comparing such approximate formulae with the exact QED result for μ-pair production that the angular restriction necessary to give good agreement between the exact and approximate calculations is much less severe in the case of single tagging[18] than for double tagging[16].

In any case tagging is impossible at LEP at the design luminosity at angles below $\simeq 1$ mrad because of occupation of the tagging counters by beam-beam bremsstrahlung background[19].

Rather detailed studies have been made of different tagging schemes for the ± 5 m interaction region at LEP where the main aim has been to obtain a sample of identified two-γ-production events for background estimates in the one-γ or one-z^0 processes[20,21,22].

Two opposing philosophies have been explored, both involving modification of the design of the interaction region quadrupoles:

  (i) The aperture of the quadrupole is increased by using a 'Panofsky' design and tagging counters are placed downstream of the quadrupoles.

(ii) 'Slim' superconducting quadrupoles are used with a smaller aperture than
the standard ones and tagging counters are placed in front of the quadrupoles. The space liberated in the forward direction could then be used to
increase the acceptance of the forward detector towards smaller angles.
The extra longitudinal space is available for particle identification
(time-of-flight or 'ISIS' type dE/dX measurements) or hadron calorimetry.

Results[21] for the tagging efficiency in downstream detectors given by the 'Standard', 'Panofsky' and 'Slim' quadrupoles are shown in fig. 12. The abcissa is the
scattered electron energy divided by the beam energy. Clearly a useful gain in tagging efficiency is obtainable by using the wide aperture quadrupoles.

Fig. 13 shows the regions of nonzero acceptance as a function of the polar and azimuthal angles of the scattered electron relative to the beam, for various scattered
electron energies and the 3 different types of quadrupole. In no case does the acceptance in polar angle extend below 2 mrad, and the acceptance biases as a function
of both angle and energy are quite severe, making the calculation of tagging efficiency complicated. The complications become even more marked when radiative corrections to the scattered electrons are taken into account. If tagging is done in front
of the quadrupoles using detectors looking directly at the interaction point with
negligible transverse field, as in the typical tagging systems of detectors at PETRA
or PEP, the radiative correction to the scattered electron can be almost neglected
as the radiated photon will in most cases be detected and measured in the same shower
counter as the scattered electron. This will clearly not be the case when tagging is
done through quadrupole fields, so the full radiative correction must be calculated
and applied.

On balance it seems that the slim quadrupole philosophy is preferable for a dedicated
two-γ detector particularly as the use of a ±10 m interaction region enables detectors in front of the quadrupoles and outside the beampipe to look at rather small
angles. The minimum vertical tagging angle could be as small as 4 mrad, the horizontal one, which must be larger to avoid problems with synchrotron-radiation background[23], could be 10 mrad.

4. PHYSICS POSSIBILITIES

Perhaps the best reason for looking at two-photon physics at LEP, from the view point
of the experimentalist, is simply that a new domain of energy in photon-photon collisions will be opened up and anything (particular what is not forseen by our present
day theories) may show up. However having said this nothing remains to be done but

to wait for 10 years or so till the first data from LEP arrives. From the view point of theory of course much more can be said and there is space here only to touch on the essential points. For details I refer the reader to the comprehensive discussions given in the talks of Gunion and Walsh.

The processes which seem, at the time of writing, to be of the greatest interest in the sense that there are many quite hard theoretical predictions to be either confirmed or ruled out by experiment, are those involving the point-like couplings of photons to hadronić systems. There are basically two different types of such processes where:

(i) a photon is far from mass shell
(ii) an internal propagator (quark or gluon) is far from mass shell

Fig. 14 shows Feynman diagrams for some typical processes of these two types.

In fig. 14a a high Q^2 photon interacts in a point-like manner with a quark constituent of a low Q^2 photon. The final state contains a high p_T jet resulting from the fragmentation of the struck quark and a 'beampipe jet' resulting from fragmentation of the remaining constituents of the photon. This is quite analoguous to the target fragmentation jet in deep inelastic lepton scattering. The process shown in fig. 14b has 2 high Q^2 photons and the final state contains 2 high p_T jets. Examples of type (ii) processes are shown in fig. 14c,d. In this case the photons are quasi-real ($Q^2 \approx 0$) but an internal quark propagator (boxed with a broken line) carries large transverse momentum. In the process of fig. 14c two high p_T quark jets are produced with no particles in the forward or backward directions. The first tentative evidence of the observation of such events in the PLUTO detector at PETRA is presentated in the talk of C. Berger at this conference. In fig. 14d a quark constituent of the lower photon undergoes a hard collision with the upper photon: $\gamma q \rightarrow qg$ to produce a high p_T quark and gluon recoiling from each other. As in fig. 14a the remaining constituents of the lower photon form a 'beampipe jet'. Note that high p_T jets are found in processes of both types (i) and (ii). Theoretical calculations of type(i) processes[24-28] were first done some 7 years ago. Although the type (ii) processes were first mentioned in the literature as long ago as 1971[29] the first detailed calculations were carried out in 1978[8,9,10].

Fig. 15 shows the F_2, F_L structure functions of the photon, measured in the process of fig. 14a as calculated in the parton model[24] and also to first order in QCD[28] when single gluon emission is taken into account. In both cases the structure functions show features which are qualitatively quite different to the structure functions of hadrons.

- The F_2-structure function increases as $\ln Q^2$, but in shape shows exact Bjorken scaling:

$$F_2(x,Q^2) = \ln Q^2 f_2(x)$$

- The F_L-structure function scales exactly:

$$F_L(x,Q) = f_L(x)$$

Actually there is another contribution to the photon-structure function important at small x, coming from the vector-meson components of the photon-wave function, and the simple scaling behaviour above does not survive to the next order in QCD.

These points are discussed in more detail in Walsh's talk. In any case LEP should be able to measure the photon-structure function up to Q^2 values as high as 20 $(GeV/c)^2$ given an integrated luminosity of $\sim 10^{38}$ cm^{-2}. This is shown in table 5 where event rates for <u>double</u> deep inelastic scattering (Q^2 of both photons > 1 $(GeV/c)^2$) as in fig. 14b are given. These rates are taken from the work of Cottingham and Landshoff[30] using the basic quark-box diagram of fig. 14b without QCD corrections.

Processes of type (ii) lead to a hierarchy of jet processes as one goes from the zeroth order (in QCD) process of fig. 14c to 1st and 2nd order in QCD. All processes have 2 high p_T jets but there may be 0, 1, 2 'beampipe jets' leading to overall 2, 3, 4 jet processes. Some examples of higher order processes are shown in fig. 16. Fig. 16a,b have 3 jets in the final state corresponding to the subprocesses $\gamma q \rightarrow gq$, $\gamma g \rightarrow q\bar{q}$. In fig. 16c constituents of each photon interact via the hard subprocess $gq \rightarrow gq$ leading to 4 final state jets. Fig. 16d is an example of a higher order two-jet process, resulting in 2 gluon jets. The rate has been calculated[31,32] to be about 20 % that of the leading $g\bar{g}$ process.(fig. 14c) with essentially identical differential cross sections.

A rough estimate of the rate of the leading $q\bar{q}$ two-jet process at a given p_T is given by the formula:

$$x_T^{>N} = \left(\frac{300 \mathcal{L}}{N E^2}\right)^{1/3}$$

Here $x_T^{>N}$ is the value of $x_T \equiv p_T^{jet}/E$ to have at least N events in a bin of width 0.05 in x_T. $\mathcal{L}$ is in units of 10^{38} cm^{-2} and E in GeV. For example taking E = 100, $\mathcal{L}$ = 1, N = 10 one obtains $x_T^{>N}$ = 0.14 corresponding to a p_T of 14 GeV. Even taking N = 500, $p_T \gtrsim 4$ GeV so the jet structure in the final state should be already evident.

Fig. 17 shows[33] the x_T dependence of the high p_T jets in the 2, 3, 4 jet processes for configurations where both jets are at zero rapidity in the lab (90° to the beam axis). It can be seen that, although at large x_T the two-jet process dominates, at small values of $x_T \lesssim 0.2$ where counting rates are high all processes have comparable cross sections, while for small values of x_T the four-jet processes dominate. Also shown in fig. 17 (the line labelled '2 brems') is the contribution to the two-jet cross section from the process $e^+e^- \rightarrow q\bar{q}$ where both the incoming electron and positron radiate a hard collinear photon. If the 2, 3, 4 jet events are topologically identified this background is no problem for the 3, 4 jet classes. For the two-jet events it becomes quantitatively important only at large values of x_T where in any case the event rate is low. This background can either be removed by requiring single tagging, or vetoed by detecting the forward γ rays. If LEP runs at its design luminosity occupation problems from photons from beam-beam bremsstrahlung may preclude the latter possibility.

A recent paper[34] has calculated the energy and angular distributions to be expected for the beampipe jet in the three-jet reactions corresponding to the hard scattering subprocesses $\gamma q \rightarrow gq$, $\gamma g \rightarrow q\bar{q}$. Results for the scaled energy x_B of the beampipe jet are shown in figs. 18, 19 for various kinematic configurations. The Born approximation, taking account only of the valence quark in the photon structure, and neglecting gluon radiation, already gives a quite accurate estimation. The two subprocesses show quite different behaviour, $\gamma q \rightarrow gq$ dominating at small rapidities (y_1, y_2) of the high p_T jets, and small values of x_B, while $\gamma g \rightarrow q\bar{q}$ becomes dominant for large rapidities and large values of x_B (see fig. 19).

Fig. 20 shows the p_T of the valence quark in the beampipe jet for different values of x_B. The angular distribution of the jet shrinks rapidly as its energy increases.

The two subprocesses which contribute to the three-jet events, as well as having different distributions of kinematical variables may also be distinguished using charge correlations[34]. If the charges of the two high p_T quanta in figs. 16a,b are denoted by Q_1, Q_2 and that of the beampipe jet (which may contain 1 or more elementary QCD quanta) by Q_B then two charge correlations may be measured:

$$C_1 = \langle Q_1 Q_2 \rangle$$
$$C_2 = \langle (Q_1 + Q_2) Q_B \rangle$$

C_1 and C_2 have very different behaviour for the two subprocesses. For the process in fig. 16a $C_1 = 0$ (the gluon is neutral) and $C_2 \simeq e_q^4$. For fig. 16b $C_2 = 0$ as the beampipe jet is neutral and $C_1 \simeq e_q^2$ (the quark-gluon coupling is flavour independent). Averaging over u, d, s, c quarks one then finds:

Subprocess	C_1	C_2
$\gamma g \rightarrow q\bar{q}$	$-17/45$	0
$\gamma q \rightarrow gq$	0	$-65/153$

If the charges of the quarks are replaced by the average charge of the hadrons in the jet above some low rapidity cut off the non zero charge correlations are expected to be reduced by some 20 %. C_1 should be easily measurable experimentally, whereas C_2 will be much more difficult, requiring the detection of all charged particles in the beampipe jet. It is interesting to note that charge correlations may also be useful statistically to separate the two-jet processes of fig. 14c and 16d which are otherwise kinematically almost indistinguishable. C_1 should have the value -65/153 for $e^+e^- \rightarrow q\bar{q} \; e^+e^-$ (4 quark flavours) and zero for $e^+e^- \rightarrow gg \; e^+e^-$. It might be interesting to study parameters sensitive to the angular spread of the jet fragments as a function of the charge correlation C_1.

The relative cross sections and the various differential distributions of the different multijet processes provide an almost parameter free test of QCD predictions (only the small x V.D.M. part of the photon-structure function is, at the time of writing, uncalculable from simple QCD perturbation theory). It is hard to think of any other type of process which could test the basic QCD assumptions in such a thorough way. The prospect then exists that tests could be made of the candidate theory of the strong interactions in two-photon reactions, comparable in importance to the tests of the electroweak interaction in the e^+e^- annihilation process that give the primary physics motivation for building LEP.

5. DETECTORS FOR TWO-PHOTON PHYSICS

General requirements

The physics of the two-photon process leads to important constraints on detector design:

(i)   In general the produced system has a Lorentz boost in the laboratory frame resulting in forward and backward collimation of the final state particles. For good overall acceptance it is then very important to detect these particles at the smallest possible angles. The angular range $20 < \theta < 300$ mrad is particularly important[15].

(ii)   Many of the potentially interesting processes discussed in section 4 are expected to have particles at small angles to the beams resulting from the beampipe jets. In order to make a detailed study of these processes they must be topologically classified as 2, 3, 4 jet events. The detection of at least one hadron from each beam pipe jet is necessary to do this. Again, good acceptance at small angles is necessary.

(iii)   In general the hadron momentum spectra from two-photon collisions are
        expected to be softer than in e^+e^- annihilation, even for forward pro-
        duction angles. This can be seen from figs. 21, 22, 23 which show a Mon-
        te-Carlo calculation[35] of inclusive hadron momentum spectra in diffe-
        rent angular regions for, respectively, e^+e^- annihilation, a high p_T
        two-photon process (that shown in fig. 14c), and the dominant low p_T
        two-photon process. The angular regions chosen correspond roughly to
        those imposed on a two-photon detector by technical constraints:

        $0 < \theta < 10$ mrad   beampipe
        $10 < \theta < 100$ mrad   tagging region
        $100 < \theta < 300$ mrad   forward detector
        $300 < \theta < \pi/2$ mrad   central detector

        The consequence for detector design of these soft spectra of the final
        state particles is that relatively small magnetic field integrals are
        sufficient in both the central and forward regions to give adequate mo-
        mentum resolution.

(iv)   The momentum spectra and angular distributions of final state particles
        change only slowly with beam energy, and are in fact almost independent
        of the mass of the produced two-photon system, as the larger average Lo-
        rentz boost of low mass systems compensates for the smaller C.M. momenta
        of the produced particles[35]. The dimensions and magnetic field inte-
        grals of two-photon detectors are then expected to scale only slowly
        with energy for a given total energy resolution - quite different to the
        situation for detectors intended to detect e^+e^- annihilation processes.

Design philosophy of forward detectors

The design of forward detectors for two-photon physics has been discussed in a number of ECFA/LEP reports during the last two years[36,37,38]. The design presented at
the Les Houches Summer Study[12] (fig. 24) was based on the two-photon detector
discussed in the PEP75 Study[39]. A strong magnetic field (1T) was used so as to
match the resolution in the electron momentum obtained by magnetic deflection with
the calorimetric measurements in the NaI counters. Optimum resolution in the electron energy is only important when double tagging is used to search for resonances
by the missing mass technique. As discussed in section 4 LEP is in any case not well
suited to such measurements because of low tagging efficiences. All later designs
proposed much weaker fields $\simeq 0.1$ T in the forward spectrometer magnets.

In an effort to increase the acceptance at small angles both for scattered electrons and the final state hadrons, a number of different designs were considered using un-shielded dipole fields in the forward detectors. Such a scheme has already been used in the MD-1 detector[40] on the VEPP-4 storage ring at Novosibirsk. The main disad-vantage of such detectors is that the acceptance function becomes a complicated func-tion of azimuth for low momenta and small production angles. Also, because tagging is done in a magnetic field, radiative corrections to the scattered electrons become important, as discussed in section 3. However, by applying suitable cuts on the ang-les and energies of the scattered electrons an azimuthally symmetric acceptance function can be regained[37]. The main advantage of the dipole magnets over the 'belt-buckle' septums used in both PEP9[41] and the updated PLUTO detector[7] is a uniform $\int B \, dL$ over the whole acceptance and a simple field configuration giving fa-ster particle tracking and momentum determination.

3 different 'self compensating' forward detector field configurations are shown in figs. 26, 27 and 28. In fig. 25 is shown, for comparison, the field configuration of the Les Houches design of fig. 24.

All detectors are taken to have a solenoid with a conventional coil as central de-tector. As solenoid compensation using skew quadrupoles should be possible at LEP[42] the compensating solenoid used in fig. 25 is dropped in the later designs.

The 'double dipole' arrangement shown in fig. 26 is a technically feasible realisa-tion of the forward 'split field magnet' detector discussed in some detail in ref. (37). Each dipole field is realised using a lumped solenoid with coils tilted so as to minimise the obstruction of the acceptance[38]. A particle traversing the entire double dipole has its production angle preserved and undergoes a lateral displace-ment of trajectory proportional to 1/momentum. This opens the possibility of fast trigger decisions based on the angle and momentum of scattered particles using simple processors. Because of the relatively small overall displacement however, particles produced at zero degrees are detected only for momenta below ~ 1 GeV/c.

The design shown in fig. 27 is still essentially a 'double dipole'. The down-stream magnet is reduced in dimensions, increased in field and relegated to the role of compensator only, all spectrometer functions being carried out by the upstream mag-net. The overall acceptance at smaller angles will be larger than for fig. 26 and 0° acceptance will extend to higher momenta $\simeq$ 3 - 4 GeV. However because of this back-ground problems may be severe, the simple tracking properties of the double dipole are lost and more severe cuts will be needed to ensure uniform azimuthal accep-tance[37]. Radiative corrections will be more severe too as the radiated photon and the parent scattered particle are more widely separated in space than for the double dipole. An advantage of the single spectrometer magnet is that the whole acceptance can be quite clear of obstructions. More detailed considerations on the actual design of the forward dipole magnets are found in ref. (38).

Finally, in fig. 28 is shown a design due to T. Taylor[38] which uses the same magnet arrangement as in fig. 27 but with a different ratio of $\int B \, dL$ between the spectrometer and compensating dipoles so that the beam is always undeflected at the interaction point. If the central magnet is a solenoid it is probably preferable to cross the axial field at a small angle as in fig. 28 than laterally off axis by several millimeters as in figs. 26, 27. Note that in fig. 28 the maximum displacement of the beams at 80 GeV is less than 1 mm over a 4 m long central solenoid.

Thorough discussion of all the machine physics aspects of these detectors, including the effects of the synchrotron radiation generated within the interaction region, are needed before any could be seriously proposed as a detector for LEP. Also Monte-Carlo studies of a few 'bench mark' processes are needed to see how they compare, one against the other in acceptance and resolution.

I present no conclusions; in any case it is too early yet for conclusions about LEP. This is a workshop, not a conference, so it seems more fitting to just stop in the middle of the work in progress, Much work has been done on studying two-photon physics at LEP, but much more remains to be done.

ACKNOWLEDGEMENTS

This talk has been largely a distillation of the work of many members of various ECFA/LEP working groups during 1978-79. I should like to thank M. Jacob for his enthusiastic organisation of and hard work for the LEP Summer Study of 1978 and A. Zichichi for his inspired and inspiring chairmanship of the later, wider, ECFA/LEP Study Group.

I have to thank the following for essential contributions to the work described above: M. Abud, R. Baldini-Ferioli, C. Carimalo, W.N. Cottingham, A. Courau, M. Davier, A. Donnachie, I. Duerdoth, M. Defrise, F. Erne, R. Del Fabbro, R. Gatto, K. Kajantie, P. Kessler, P.V. Landshoff, C.H. Llewellyn-Smith, T. Massam, G.P. Murtas, J. Parisi, J. Reignier and C.A. Savoy.

FIGURE CAPTIONS

1. The first two-photon process. Discovery of the positron in Bethe-Heitler pair production.

2. General Layout of the LEP-storage ring.

3. Plan of deep underground hall P3 and P5 with the two-photon experiment proposed at the LEP Summer Study.

4. General Layout of the LEP injector.

5. Luminosity at LEP in a ±5 m interaction region.

6. Cross section of the LEP tunnel.

7. Virtual photon coupling to hadrons via a vector-meson propagator.

8. Cross section suppression by propagator effects at E = 15, 75 GeV. Solid line: coupling via ρ, dashed line: coupling via J/ψ.

9. Generalised vector meson dominance predictions for propagator suppression and double tagging efficiency. n = number of terms in Veneziano spectrum. E = 70 GeV, $10 < \theta < 150$ mrad.

10. Diagram for resonance production in two-photon collisions.

11. Double tagging efficiency versus scaled two-photon mass. E = 15 GeV, $20 < \theta < 200$ mrad.

12. Single tagging efficiency versus scaled scattered electron energy for different quadrupole designs.

13. Regions of non zero acceptance versus polar (θ) and azimuthal (ϕ) angles of the scattered electron for different quadrupole designs. Dashed line: E' = 0.3 E, solid line: E' = 0.6 E, dot-dashed line: E' = 0.9 E.

14. Feynman diagrams involving a point-like photon quark coupling. See text for discussion.

15. The F_2 anf F_L photon structure functions as calculated in the simple parton model (dashed lines) and to first order in QCD (solid lines).

16. Feynman diagrams resulting in the production of 2, 3 and 4 jets in the final state in photon-photon collisions. See text for discussion.

17. $x_T \equiv q_T/E$ dependence of different 2, 3, 4 jet processes. y_1, y_2 are the Lab rapidities of the high p_T jets. q_T = jet transverse momentum.

18. Scaled beampipe jet energy distributions in 3 jet processes, with different kinematic conditions. $y_1 = y_2 = 0$.

19. Scaled beampipe jet energy distributions in 3 jet processes. See fig. 18 for definitions of curves.

20. Transverse momentum distributions of the valence quark within the beampipe jet for the 3 jet process: $e^+e^- \rightarrow e^+e^-$ qg + BPJ.

21. Inclusive hadron momentum spectra in $e^+e^- \rightarrow q\bar{q}$ (1γ annihilation) E = 70 GeV. See text for definitions of angular cuts.

22. Inclusive hadron momentum spectra in $e^+e^- \rightarrow e^+e^-$ $q\bar{q}$ (high p_T 2γ process) E = 70 GeV. See text for definitions of angular cuts.

23. Inclusive hadron momentum spectra in $e^+e^- \rightarrow e^+e^- X$ (low p_T 2γ process) E = 70 GeV. See text for definitions of angular cuts.

24. Plan view, one quadrant of 2γ detector presented at the Les Houches Summer Study, showing instrumentation.

25. Magnetic field configuration of 2γ detector of fig. 24.

26. Magnetic field configuration of 'Double Dipole' 2γ detector.

27. Magnetic field configuration of 'Compensated Dipole' 2γ detector.

28. Magnetic field configuration of a 'Compensated Dipole' 2γ detector with the beams undisplaced at the interaction point.

266

Table 1   Selected parameters of LEP

Circumference	30.108 km
Number of bunches	4
Number of intersections	8
Time between bunch crossings	25.5 µs
Free space for experiments	± 5 m (4)
	±10 m (4)
R.M.S. energy spread	1.24×10^{-3}
R.M.S. bunch length	63 mm
Vertical beam size, ±10 m intersection	0.025 mm
Horizontal beam size, ±10 m intersection	0.508 mm
Beam lifetime	5.8 hours
Filling time	15 min

Table 2   Energy and power options for LEP

Stage	1/6	1/3	1	4/3	2	
Energy	49.4	62.3	86.1	92.9	130	GeV
Luminosity	0.39	0.62	1.07	1.15	1.04	$\times 10^{32}$ cm^{-2} s^{-1}
R.F. power	16	32	96	128	96	MW

Note: Luminosity for ±5 m I.R., ±10 m has 2 times less.
Stage 2 only with superconducting R.F.

Table 3   Energy dependences of cross sections and tagging efficiency for
$e^+e^- \rightarrow e^+e^-\tau^+\tau^-$

Cross section	Energy dependence
σ_{tot}	$\left(\ln \frac{2E}{m_e}\right)^2 \ln \frac{2E}{m_\tau}$
Cut: $p_T(\tau) \geq p_T^{min}$	$\left(\ln \frac{2E}{m_e}\right)^2$
Cut: $x_T(\tau) \geq x_T^{min}$ $\quad x_T \equiv p_T(\tau)/E$ (scaling p_T cut)	$\left(\ln \frac{2E}{m_e}\right)^2 \frac{1}{E^2}$
*Single tagging efficiency	$\left(\ln \frac{2E}{m_e}\right)^{-1}$
*Double tagging efficiency	$\left(\ln \frac{2E}{m_e}\right)^{-2}$

---

*For angles $\theta \gg m_e/E$ and $\frac{m(\tau^+\tau^-)}{2E} \gtrsim 0.1$

Table 4   Cross section ratios $\sigma(E = 75 \text{ GeV})/\sigma(E = 15 \text{ GeV})$ for
$e^+e^- \rightarrow e^+e^-\tau^+\tau^-$

Cross section	Total	Cut: $p_T \geq p_T^{min}$	Cut: $x_T \geq x_T^{min}$
Untagged	2.1	1.3	0.052
Single tagged	≈ 0.9	1.1	0.045
Double tagged	≈ 0.8	1.0	0.039

Table 5   Event rates for double deep inelastic γγ scattering
           in the quark-parton model

Numbers are quoted for an integrated luminosity of 10^{38} cm^{-2}.
No acceptance corrections.

$-q_1^2$ (GeV/c)^2	1 - 3	3 - 5	5 - 8	8 - 12	12 - 20
$-q_2^2$ (GeV/c)^2					
1 - 3	1090	280	180	110	90
3 - 5		110	80	50	40
5 - 8			60	40	30
8 - 12				40	30
12 - 20					40

REFERENCES

1) V.E. Balakin, A.D. Bukin, E.V. Pakhtusova, V.A. Sidorov and A.G. Khabakhpashev, Physics Lett. 34B (1971) 663
2) J.R.J. Bennett et al., CERN 77-14 (1977)
3) L. Camilleri et al., CERN 76-18 (1976)
4) The LEP Study Group, CERN/ISR-LEP/78-17
5) M. Jacob (ed.), Proc. LEP Summer Study, CERN 79-01 (1979)
6) The LEP Study Group, CERN/ISR-LEP/79-39
7) Presentation at this workshop of the modified PLUTO detector, specialised to two-photon physics, by C. Berger
8) S.J. Brodsky, T.A. de Grand, J.F. Gunion and J.H. Weis, Phys. Rev. Lett. 41 (1978) 672 and Phys. Rev. D19 (1978) 1418
9) C. Llewellyn-Smith, Phys. Lett. 79B (1978) 83
10) K. Kajantie, Physica Scripta 29 (1979) 230 and University of Helsinki preprint HU-TFT-79-5 (1979), Acta Phys. Austriaca, Supp. XXI (1979) 663
11) R. Bhattacharya, J. Smith and C. Grammer, Phys. Rev. D15 (1977) 3267 J. Smith, J.A.M. Vermaseren and C. Grammer, Phys. Rev. D15 (1977) 3283
12) J.H. Field in ref. (5) p. 563
13) F. Gutbrod and Z.J. Rek, Z. Physik C (1979) 171
14) A. Donnachie, ECFA/LEP Report SSG/10/5/Mar. 79 (1979)
15) J.H. Field, DESY 79/78, to be published in Nucl. Phys.
16) C. Carimalo, P. Kessler and J. Parisi, Phys. Rev. 20 (1979) 1057
17) C. Carimalo, P. Kessler and J. Parisi, Phys. Rev. 20 (1979) 2170
18) C. Carimalo, P. Kessler and J. Pairsi, Phys. Rev. 21 (1980) 669
19) J.H. Field, ECFA/LEP Report SSG/10/16/ June 79 (1979)
20) B. Richter and P. Strolin, ECFA/LEP 48 (14/9/78) (1978)
21) R. Del Fabbro and G. P. Murtas, ECFA/LEP Report SSG/6/2/March 79 (1979) and Frascati Preprint LNF-79/23 (R) (1979)
22) R. Del Fabbro and G.P. Murtas, ECFA/LEP Report SSG/6/6/May 79 (1979) and Frascati Preprint LNF-79/40 (R) (1979)
23) Ref. (6) p. 189
24) T.F. Walsh and P. Zerwas, Phys. Lett. 44B (1973) 195
25) R.P. Worden, Phys. Lett. 51B (1974) 57
26) R.L. Kingsley, Nucl. Phys. B60 (1973) 45
27) M.A. Ahmed and G.G. Ross, Phys. Lett. 59B (1975) 369
28) E. Witten, Nucl. Phys. B120 (1977) 189
29) S. Berman, J. Bjorken and J. Kogut, Phys. Rev. D4 (1971) 3388
30) W.N. Cottingham and P.V. Landshoff, ECFA/LEP Report SSG/10/14/June 79 (1979)

31) K. Kajantie and R. Raitio, Phys. Lett. 87B (1979) 133
32) R.N. Cahn and J.F. Gunion, Phys. Rev. D20 (1979) 2253
33) K. Kajantie and R. Raitio, Nucl. Phys. B159 (1979) 528
34) J.H. Field, E. Pietarinen and K. Kajantie, DESY 79/85, to be published in
    Nucl. Phys.
35) M. Davier, ECFA/LEP 26. See also ref. (12)
36) J.H. Field, ECFA/LEP 28 (1978)
37) I. Duerdoth, J.H. Field and M. Steur, ECFA/LEP Report SSG/10/1/Feb. 79 (1979)
38) T. Taylor, ECFA/LEP Report SSG/13/6/Sept. 79 (1979)
39) Proceedings of the 1975 PEP Summer Study LBL-4800, SLAC-190
40) 'Magnetic Detector MD-1' Novosibirsk Preprint 77-75 (1977)
41) See talk of D. Burke at this workshop
42) Ref. (6) p. 8

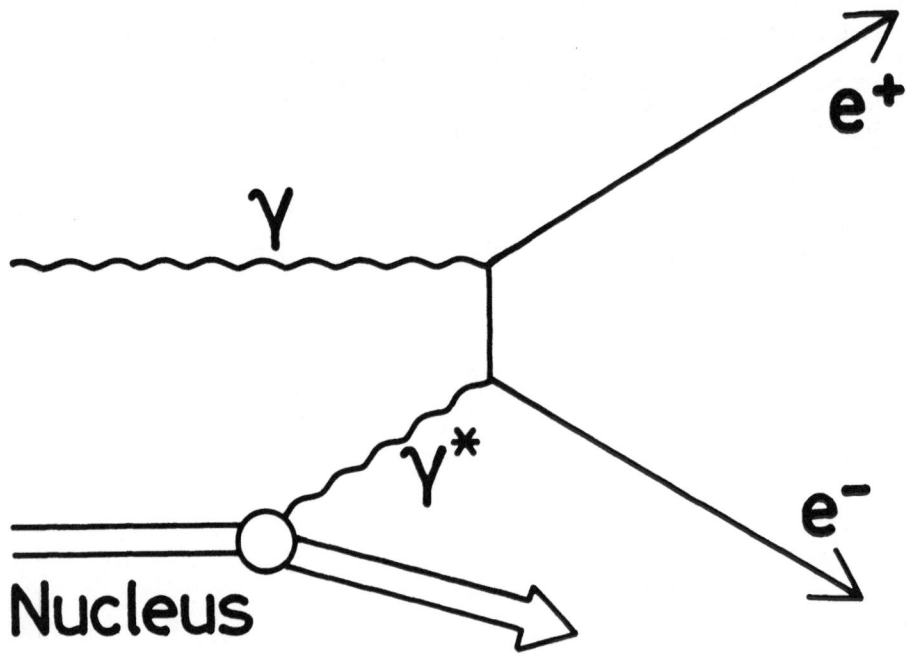

Fig. 1

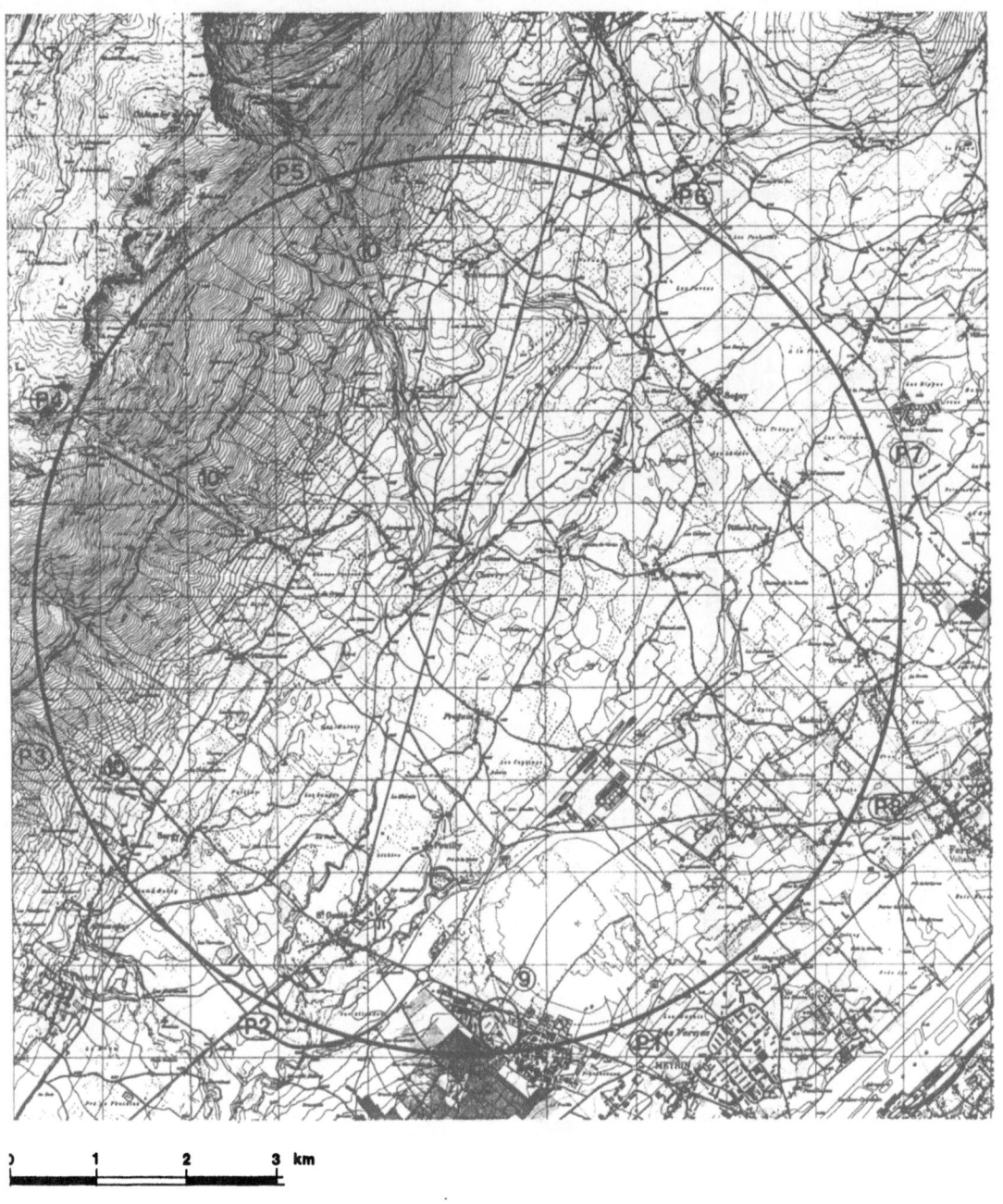

| | 1 | 2 | 3 km |

P 1,7,8 UNDERGROUND EXPERIMENTAL HALLS
P,2,6 SURFACE EXPERIMENTAL HALLS
P 3,4,5 DEEP UNDERGROUND EXPERIMENTAL HALLS
9 INJECTOR SYNCHROTRON
10 ACCESS TUNNELS TO THE DEEP UNDERGROUND EXPERIMENTAL HALLS

Fig. 2

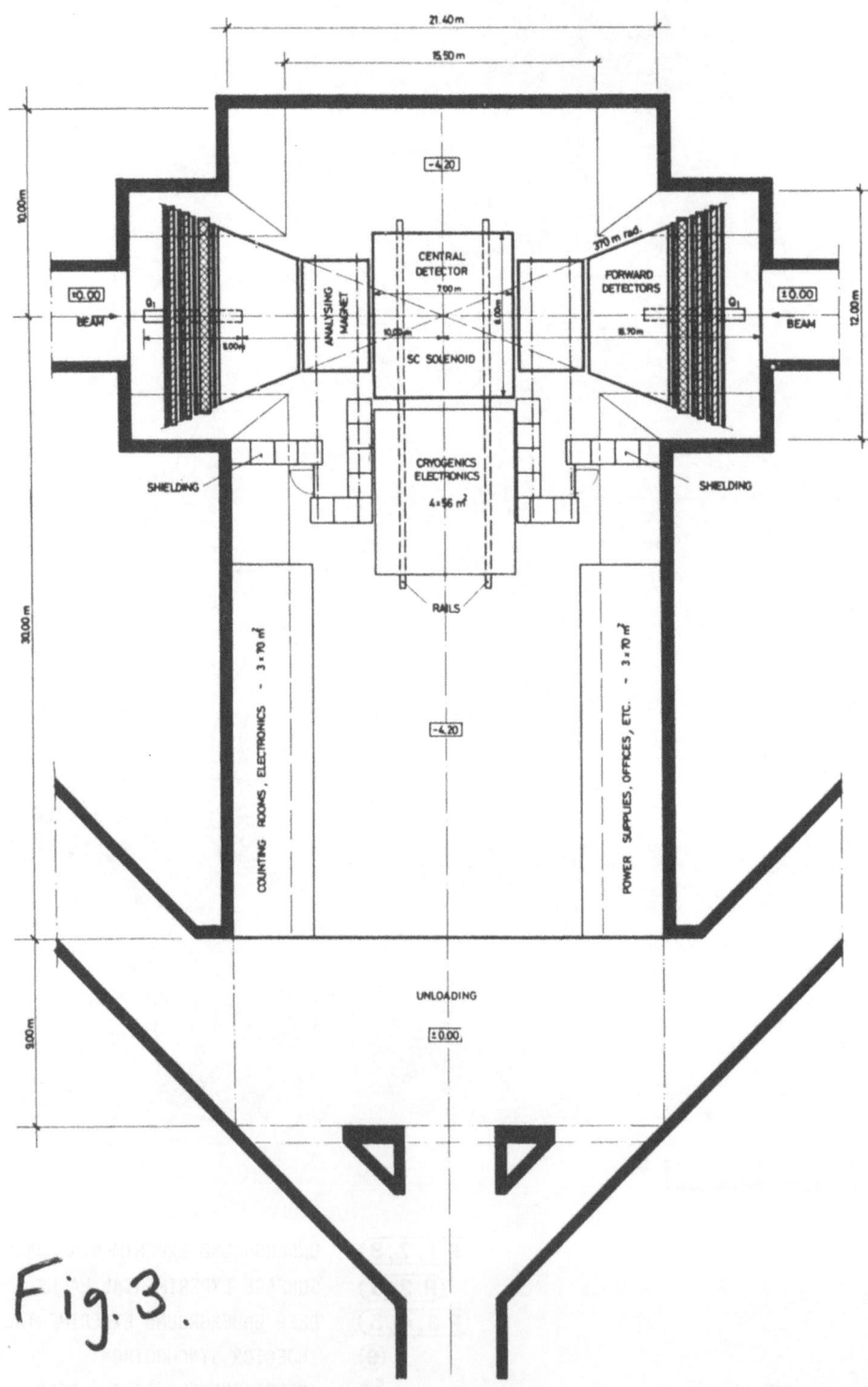

Fig. 3.

Plan of deep underground hall P3 and P5 with the two-photon experiment proposed at the LEP Summer Study

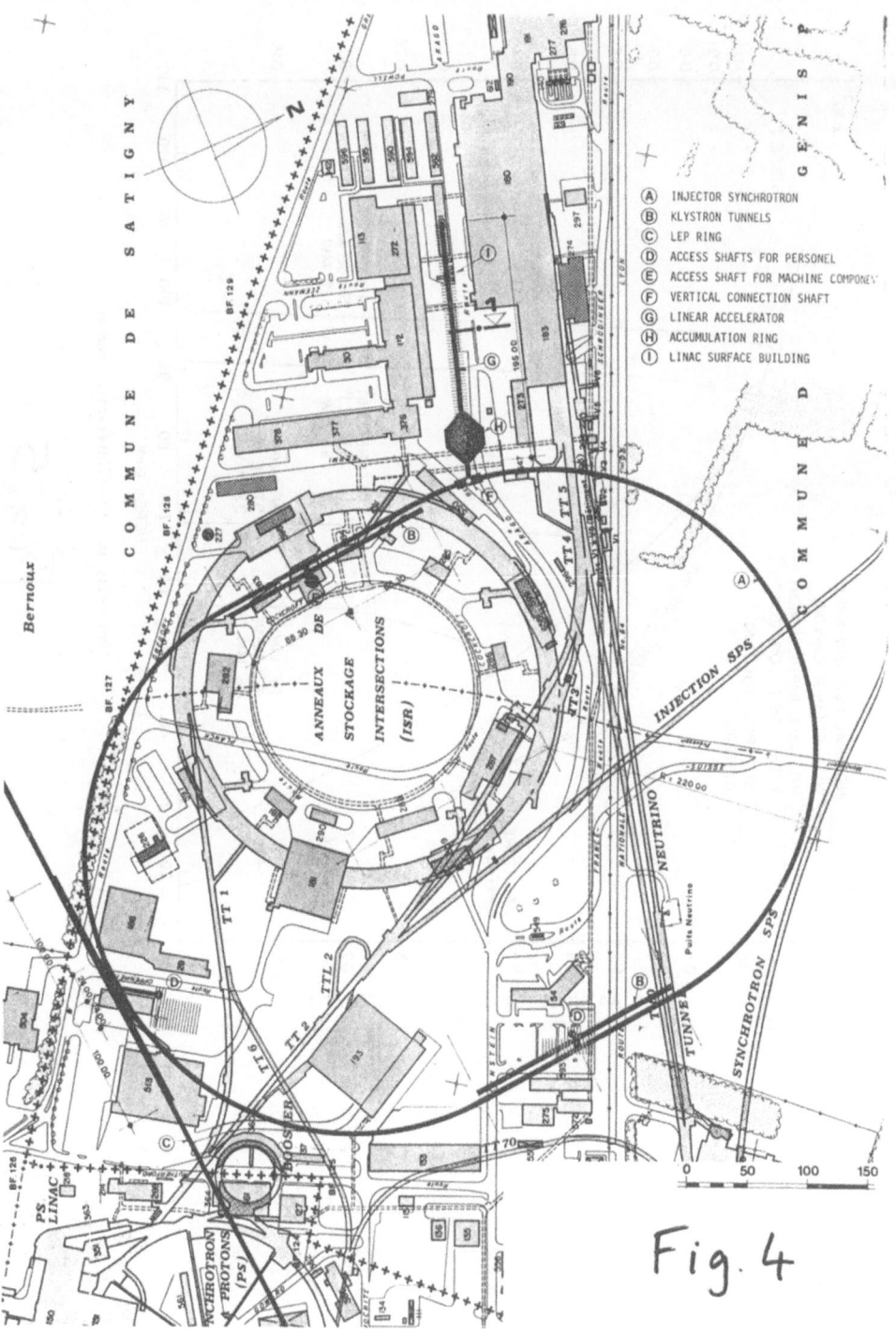

A INJECTOR SYNCHROTRON
B KLYSTRON TUNNELS
C LEP RING
D ACCESS SHAFTS FOR PERSONEL
E ACCESS SHAFT FOR MACHINE COMPONENTS
F VERTICAL CONNECTION SHAFT
G LINEAR ACCELERATOR
H ACCUMULATION RING
I LINAC SURFACE BUILDING

Fig. 4

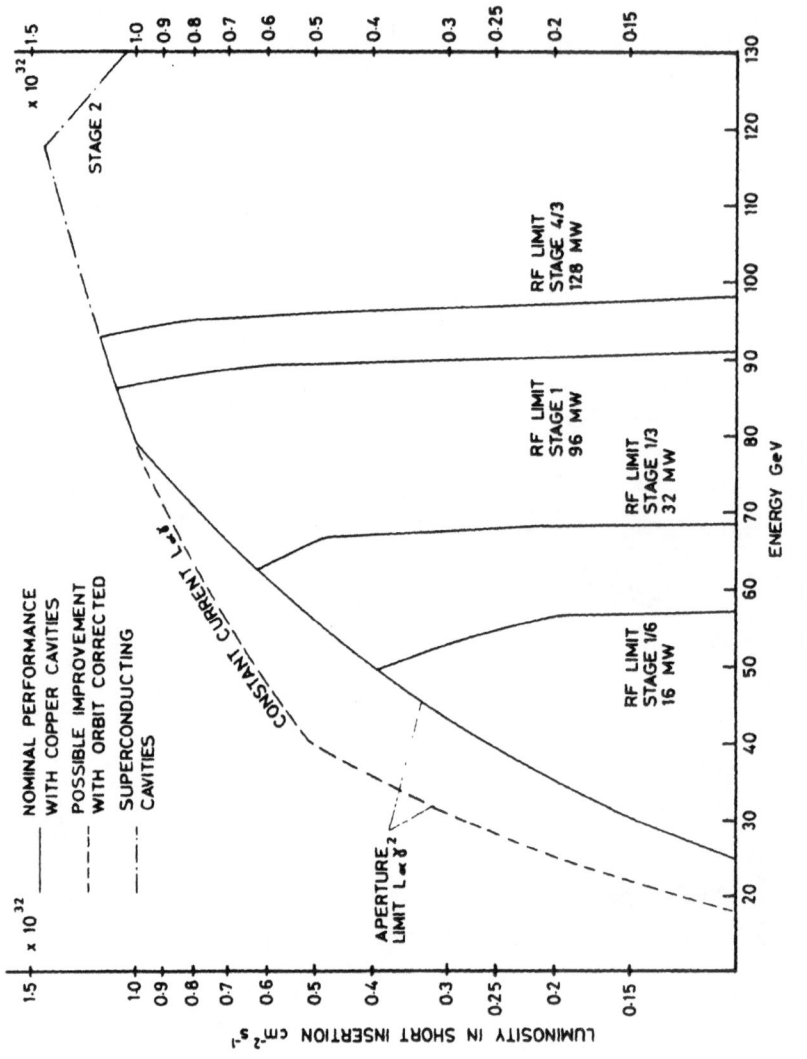

Luminosity in a ±5 m interaction region

Fig. 5

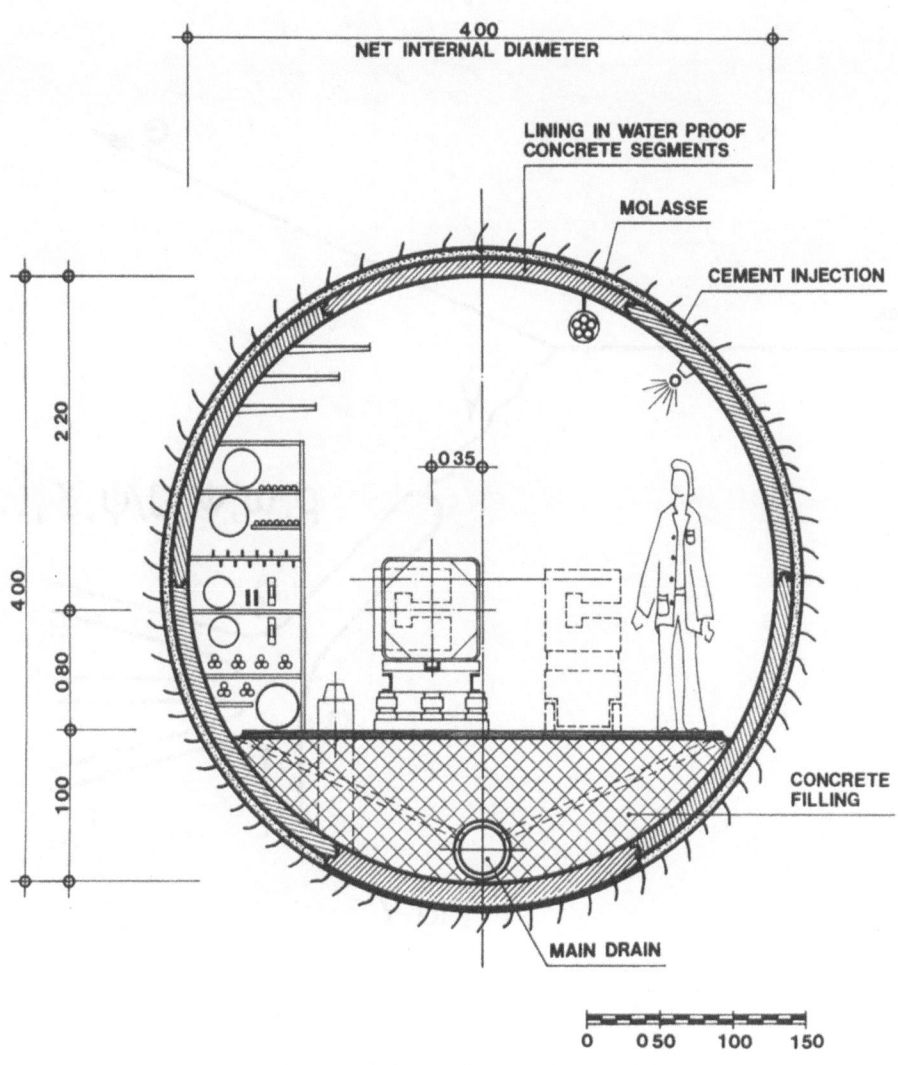

Main tunnel - Normal cross-section

Fig. 6

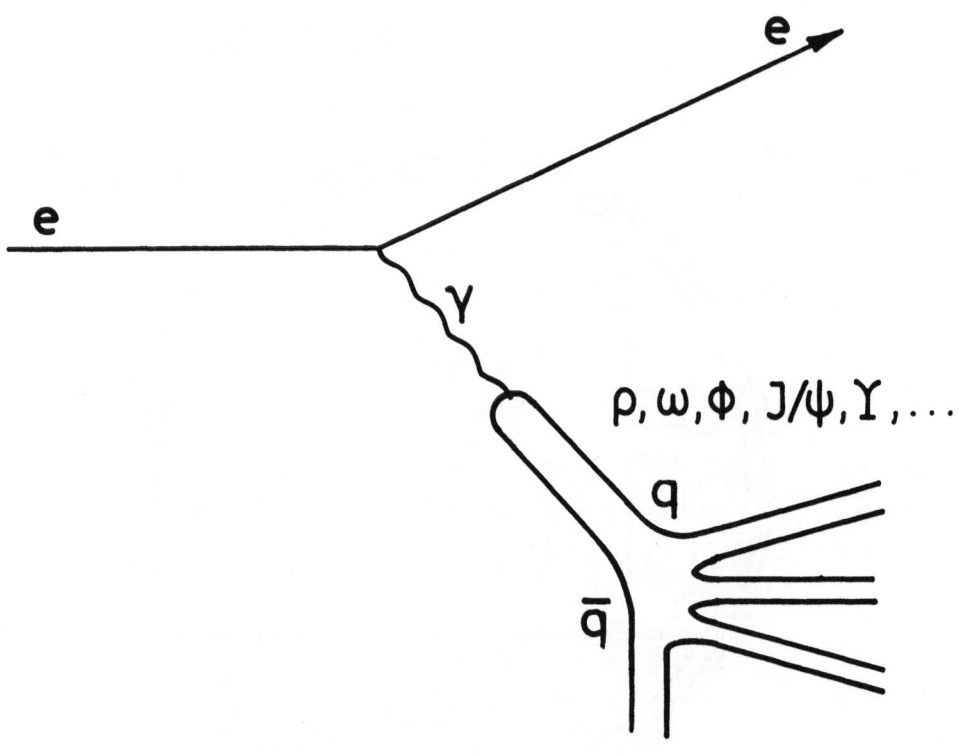

Fig. 7

Fig. 8

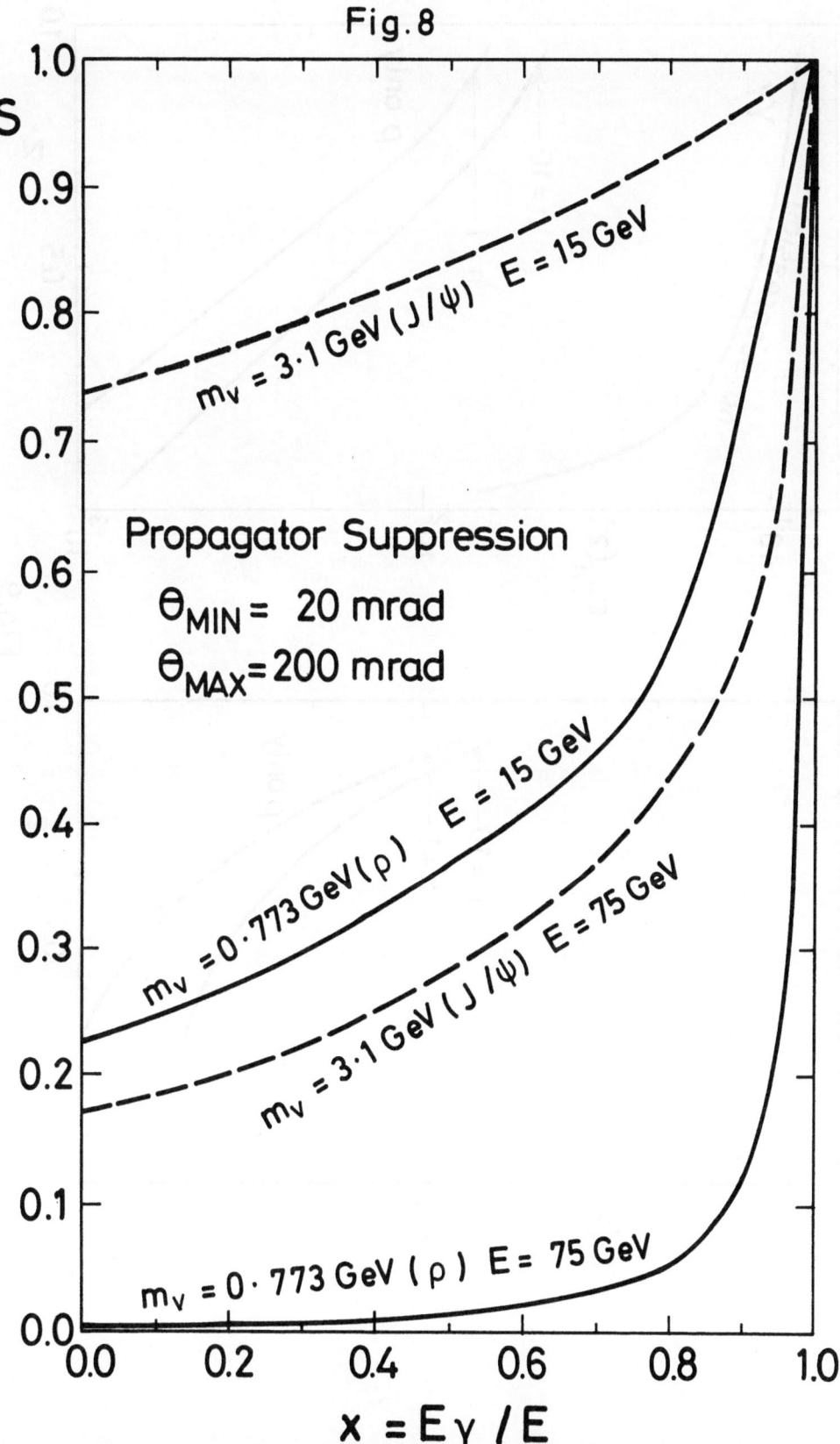

S

Propagator Suppression

$\theta_{MIN} = 20$ mrad

$\theta_{MAX} = 200$ mrad

$m_V = 3 \cdot 1$ GeV (J/ψ) E = 15 GeV

$m_V = 0 \cdot 773$ GeV (ρ) E = 15 GeV

$m_V = 3 \cdot 1$ GeV (J/ψ) E = 75 GeV

$m_V = 0 \cdot 773$ GeV (ρ) E = 75 GeV

$x = E_\gamma / E$

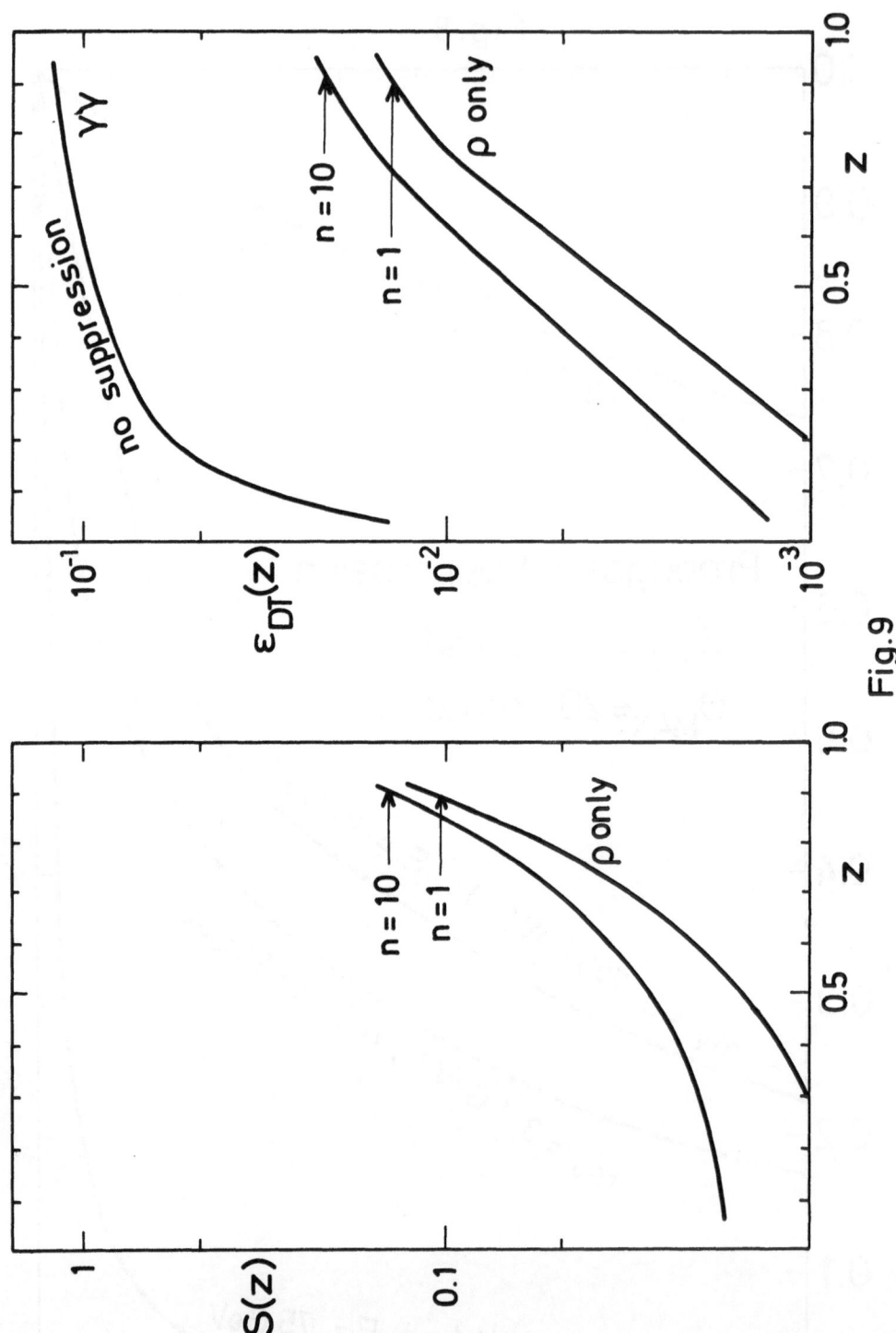

Fig.9

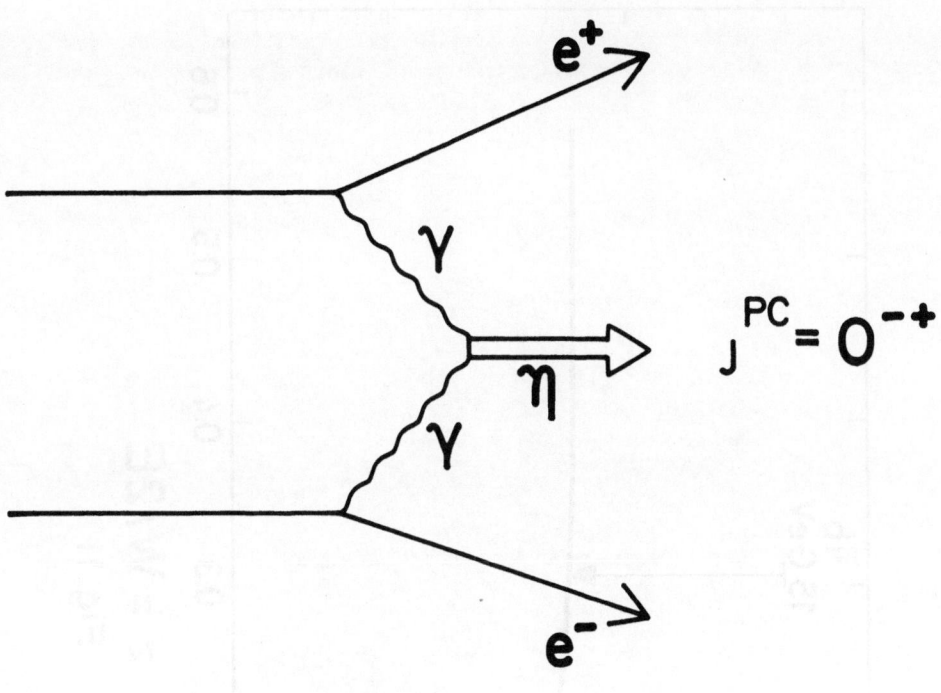

Fig. 10

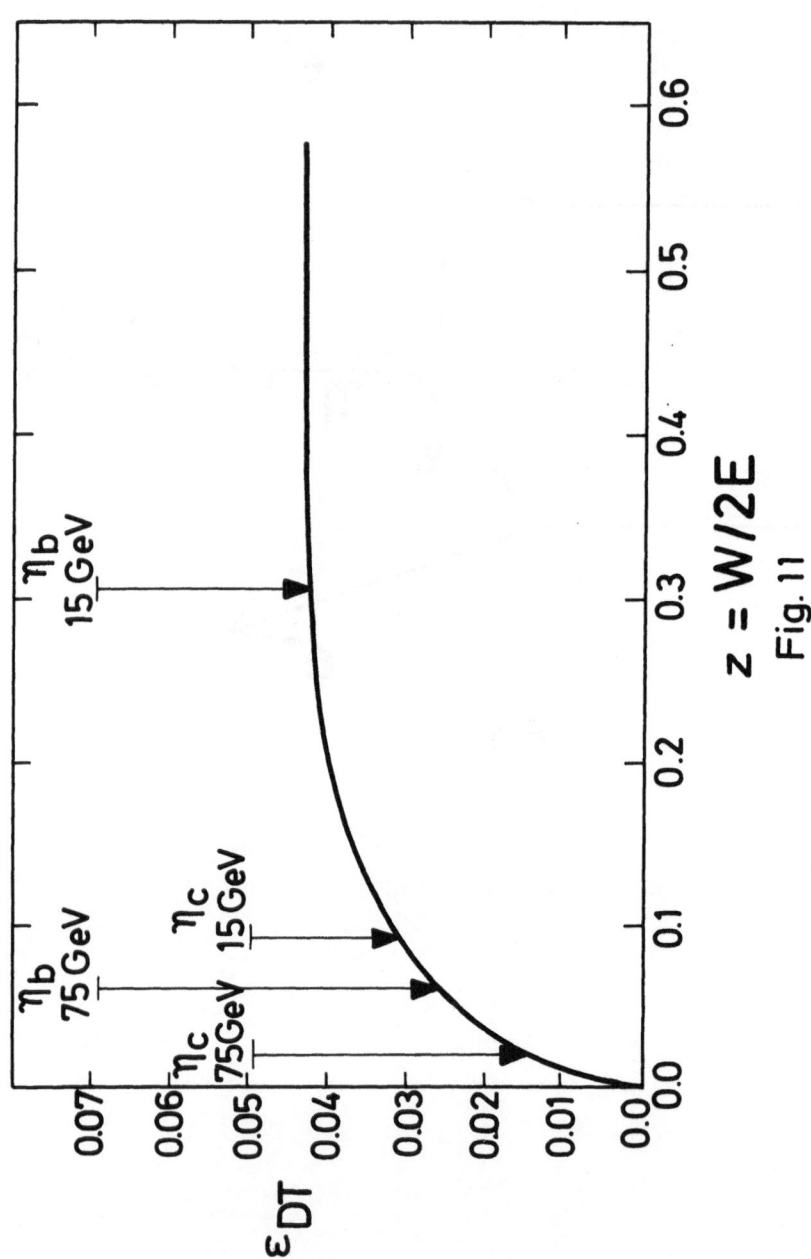

z = W/2E

Fig. 11

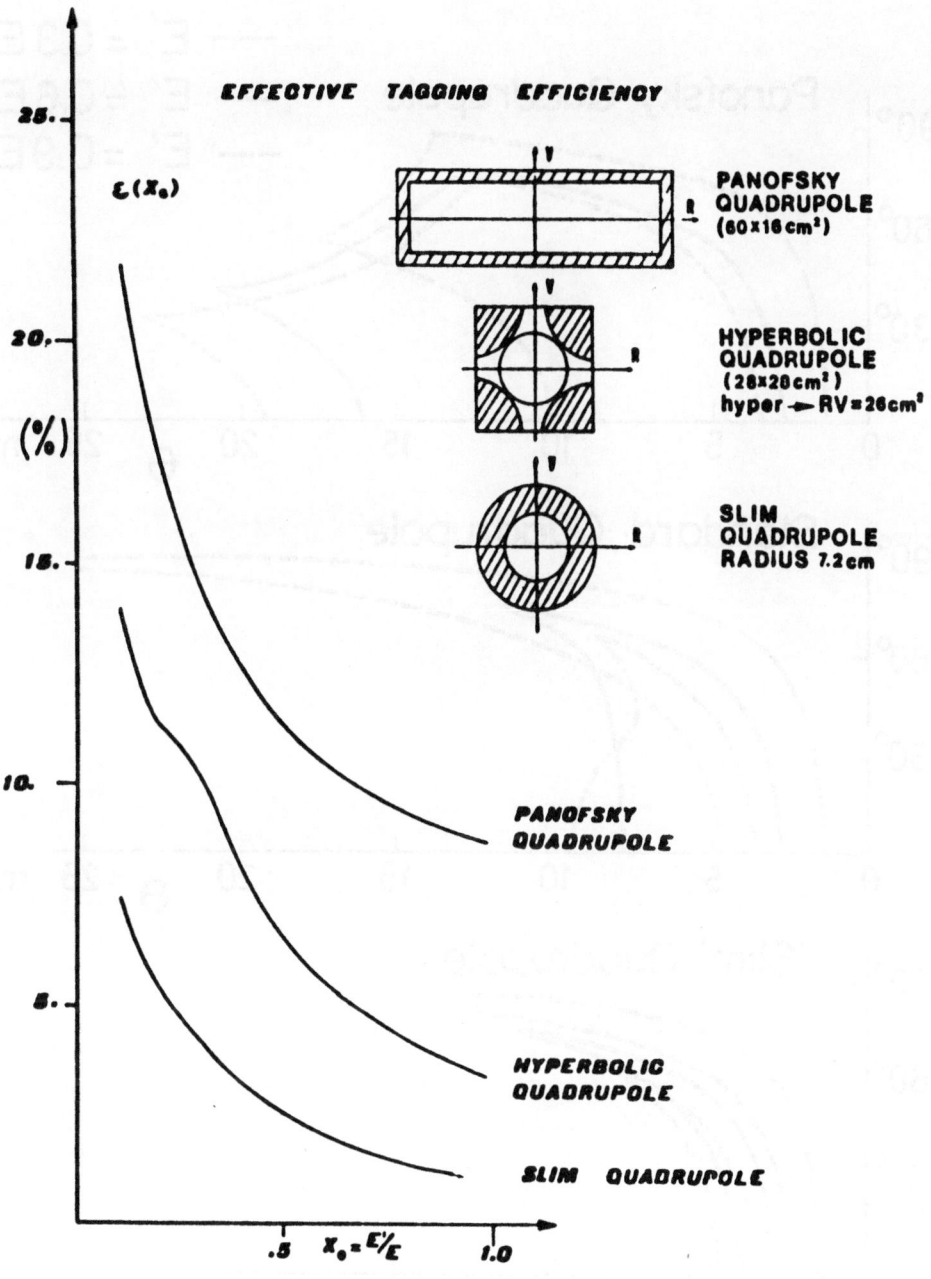

Fig.12

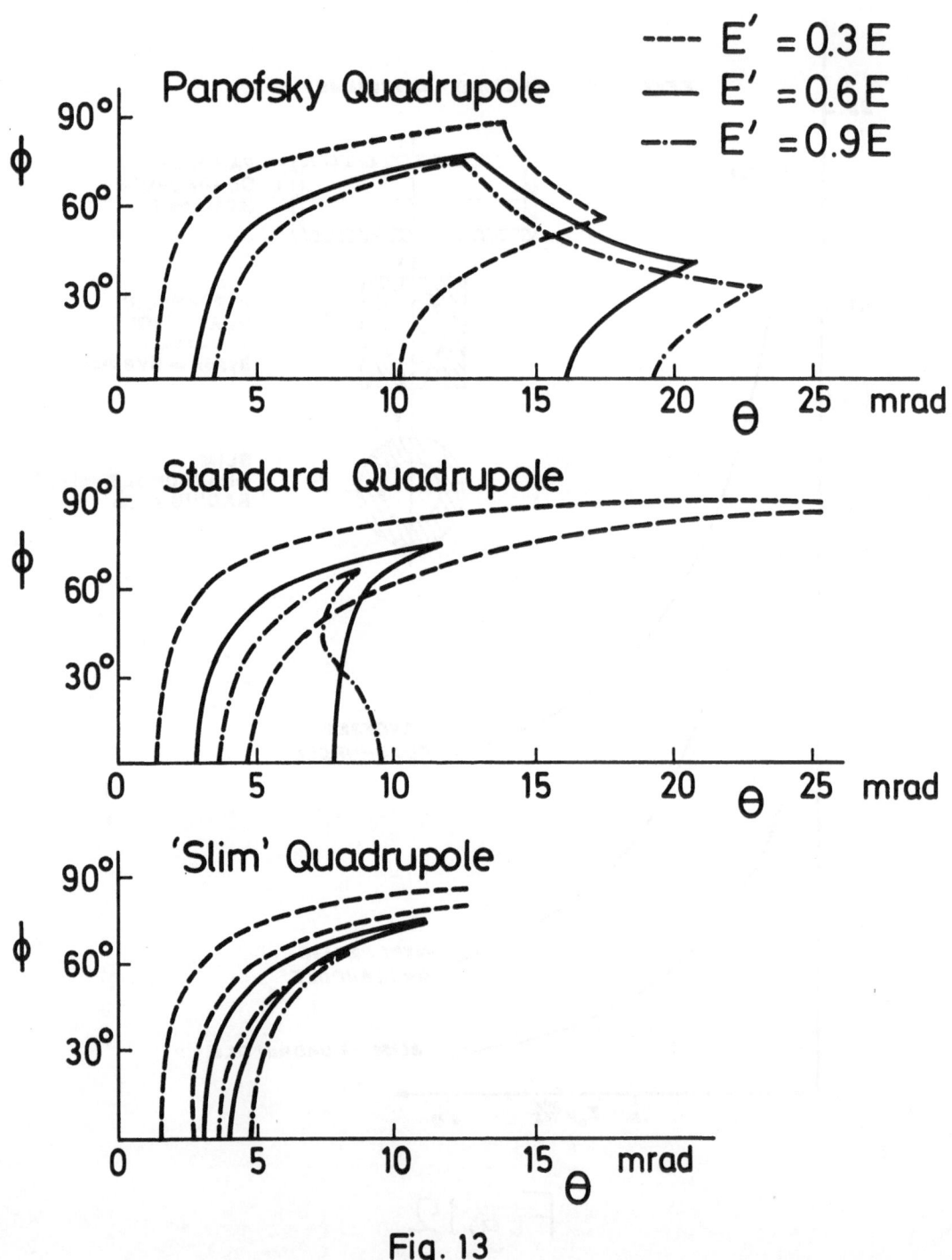

Fig. 13

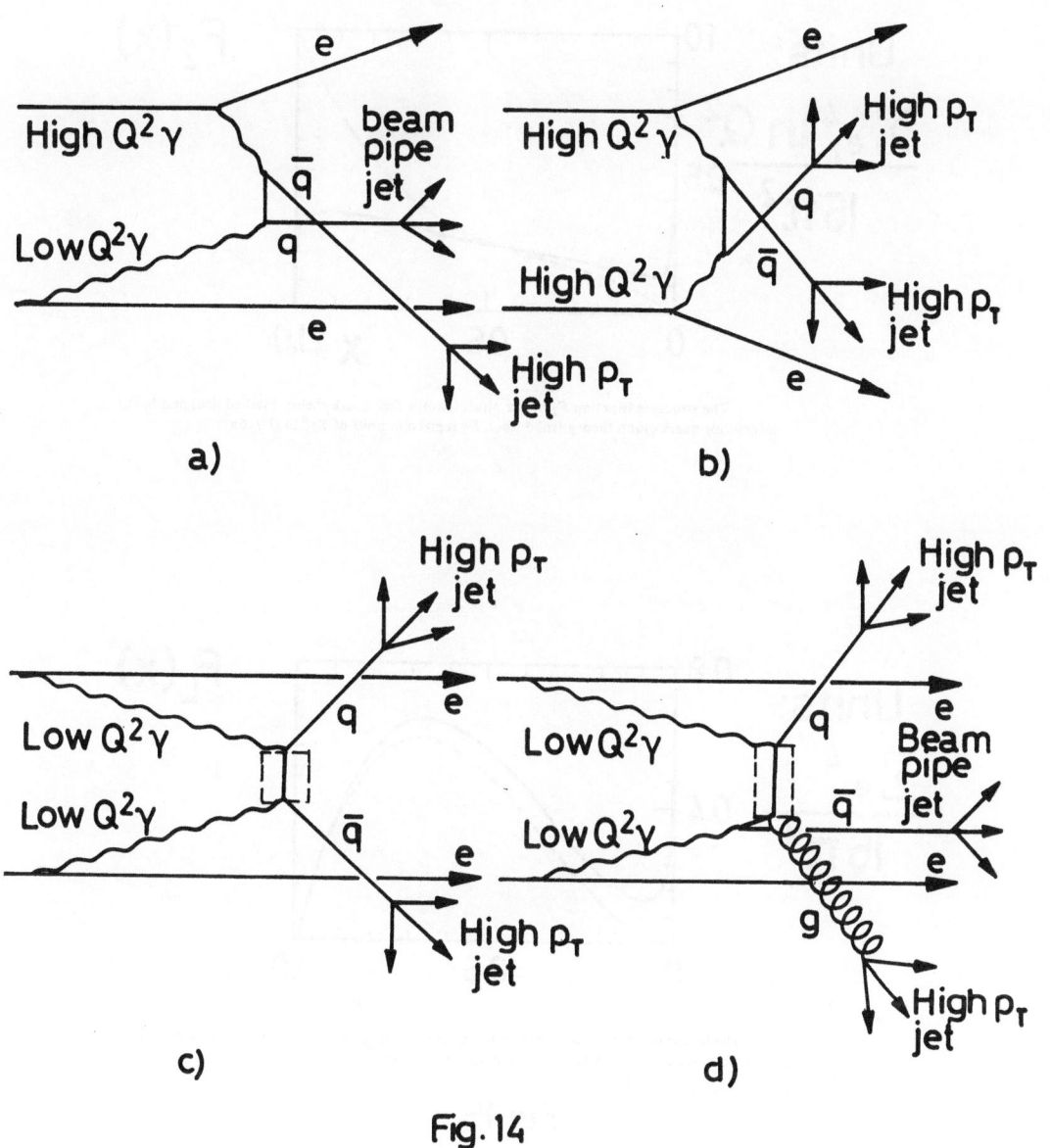

Fig. 14

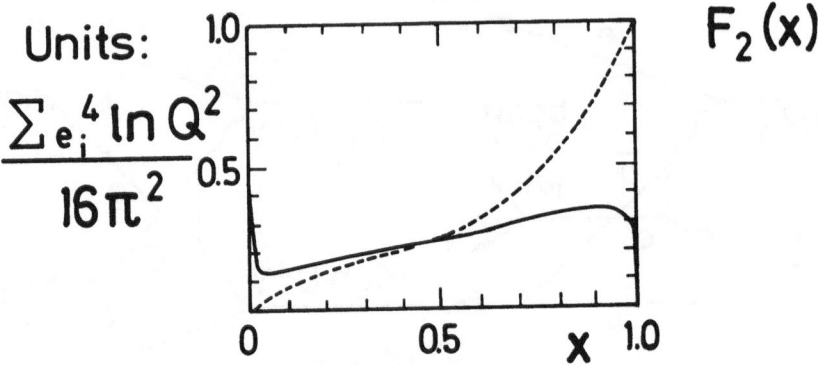

The structure function F_2 of the photon in the free quark theory (dashed line) and in the interacting quark-gluon theory (solid line). F_2 is given in units of $\Sigma e_i^4 \ln Q^2/16\pi^2$.

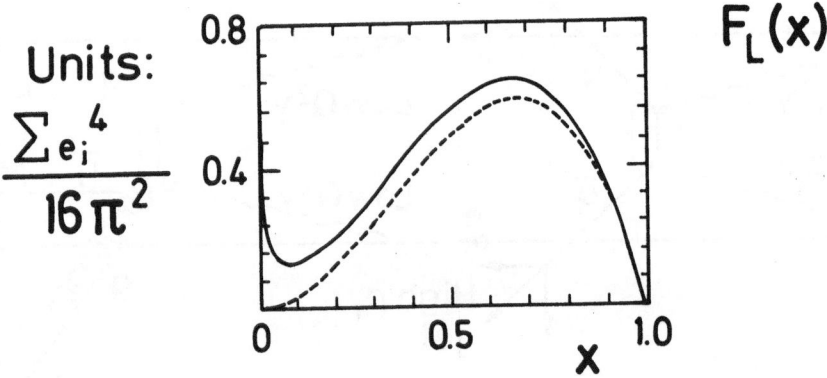

The longitudinal structure function of the photon in the free quark theory (dashed line) and in the interacting theory (solid line), in units of $\Sigma e_i^4/16\pi^2$.

Fig 15

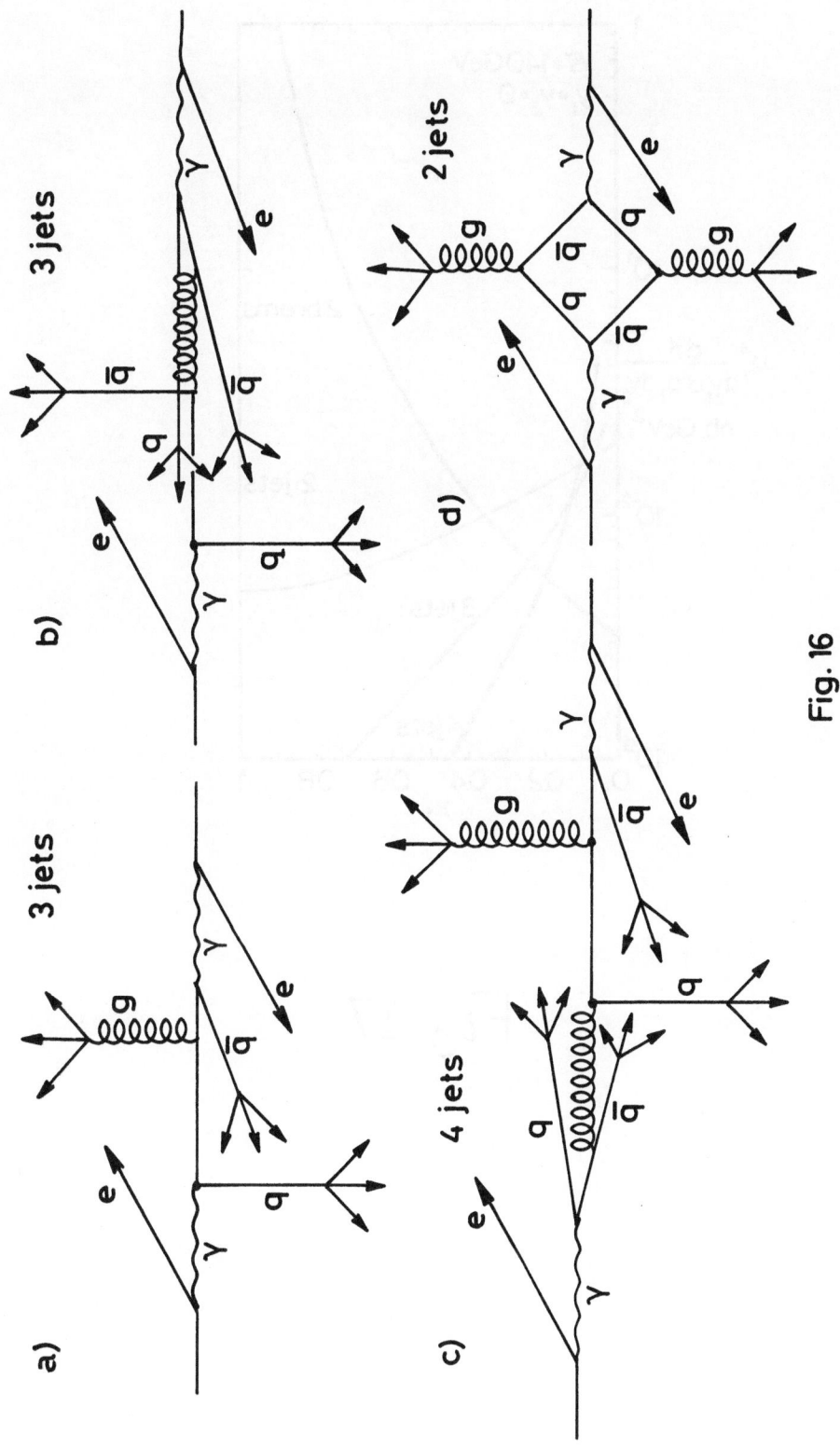

Fig. 16

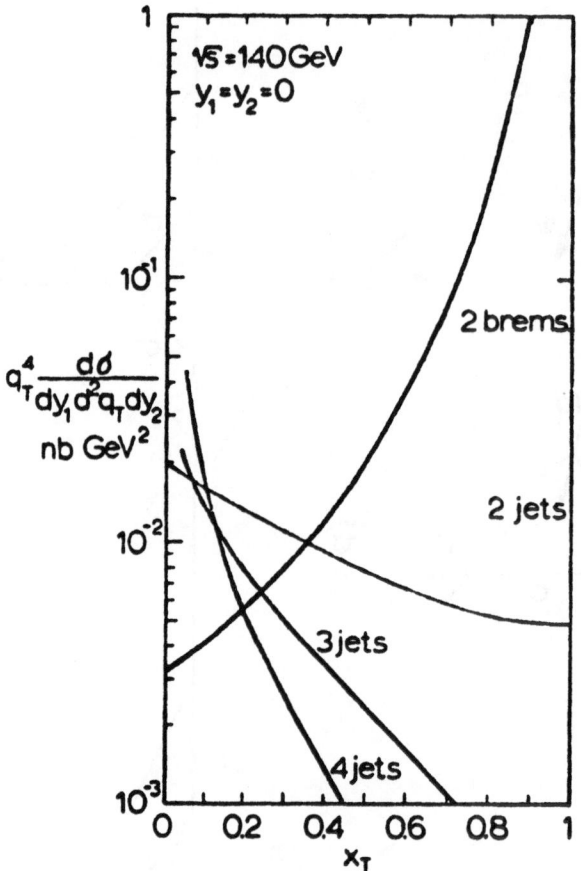

Fig.17

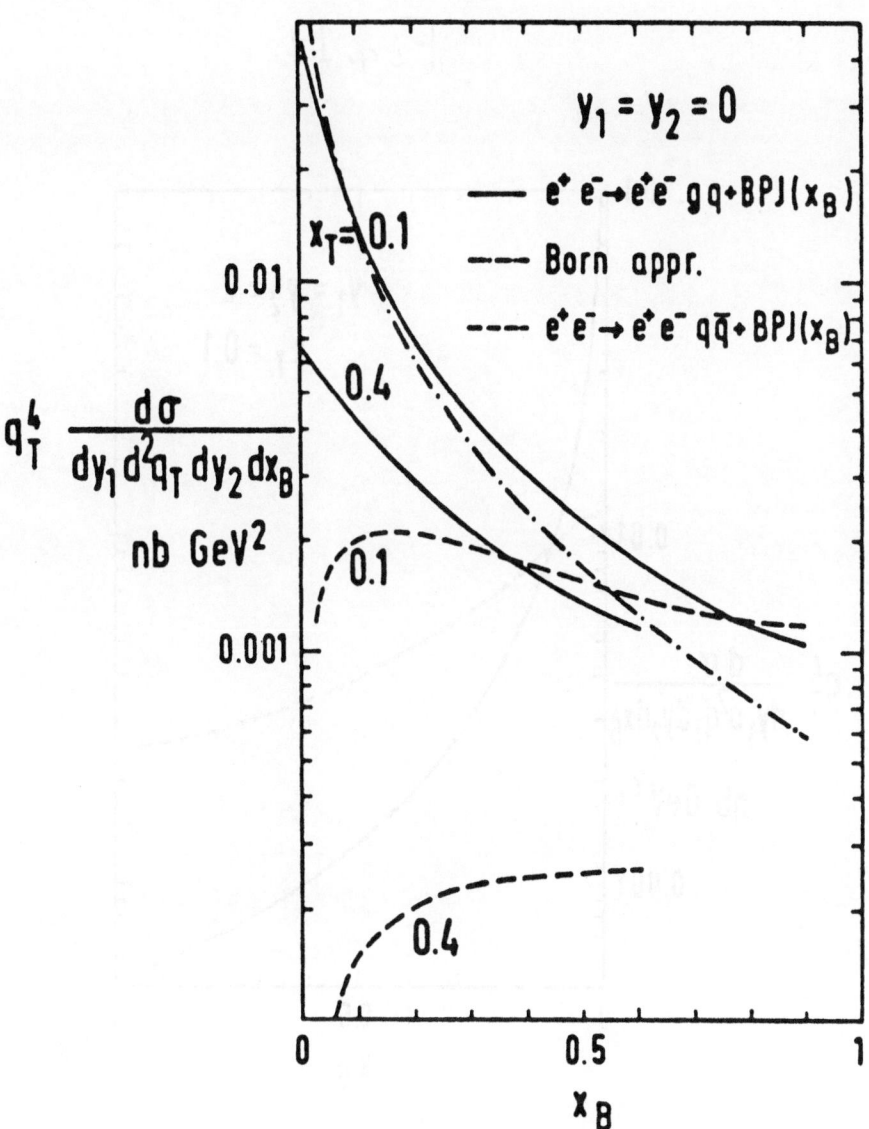

Fig. 18

Fig. 19

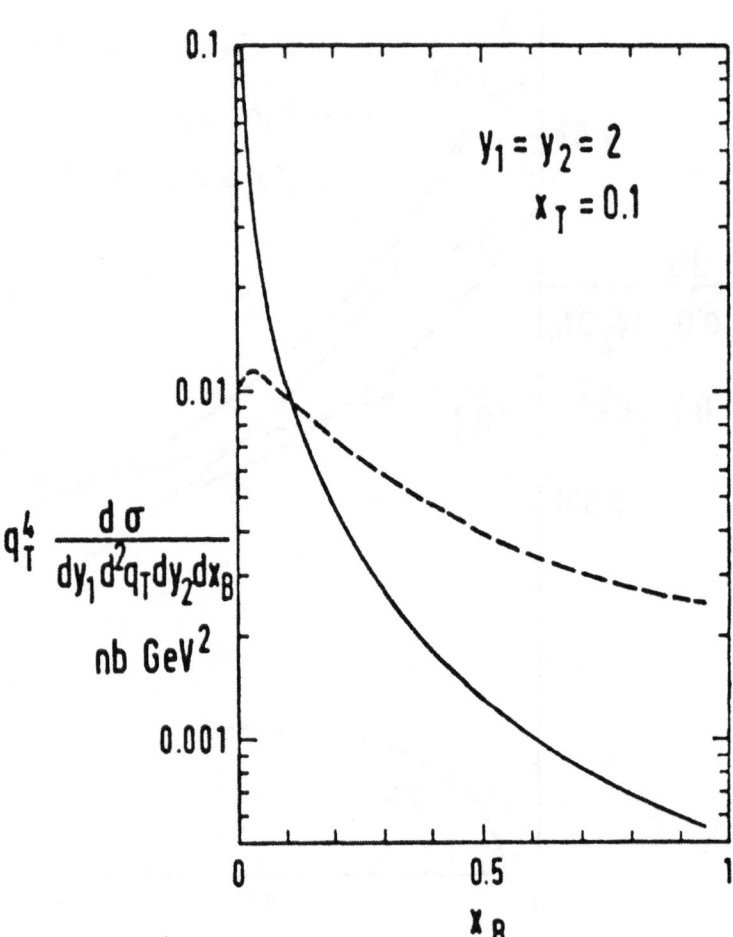

$y_1 = y_2 = 2$
$x_T = 0.1$

$q_T^4 \dfrac{d\sigma}{dy_1 \, d^2q_T \, dy_2 \, dx_B}$

$nb \; GeV^2$

x_B

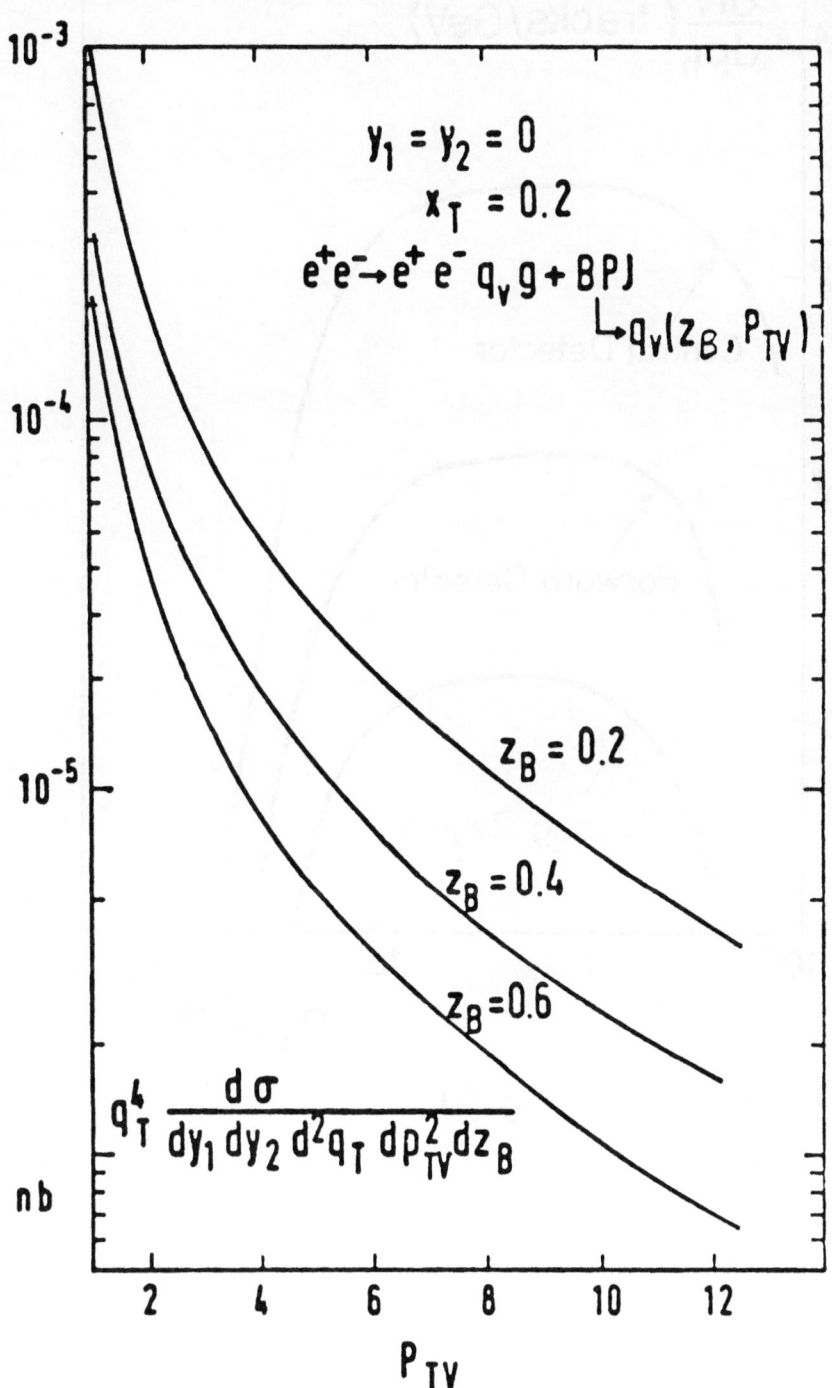

Fig. 20

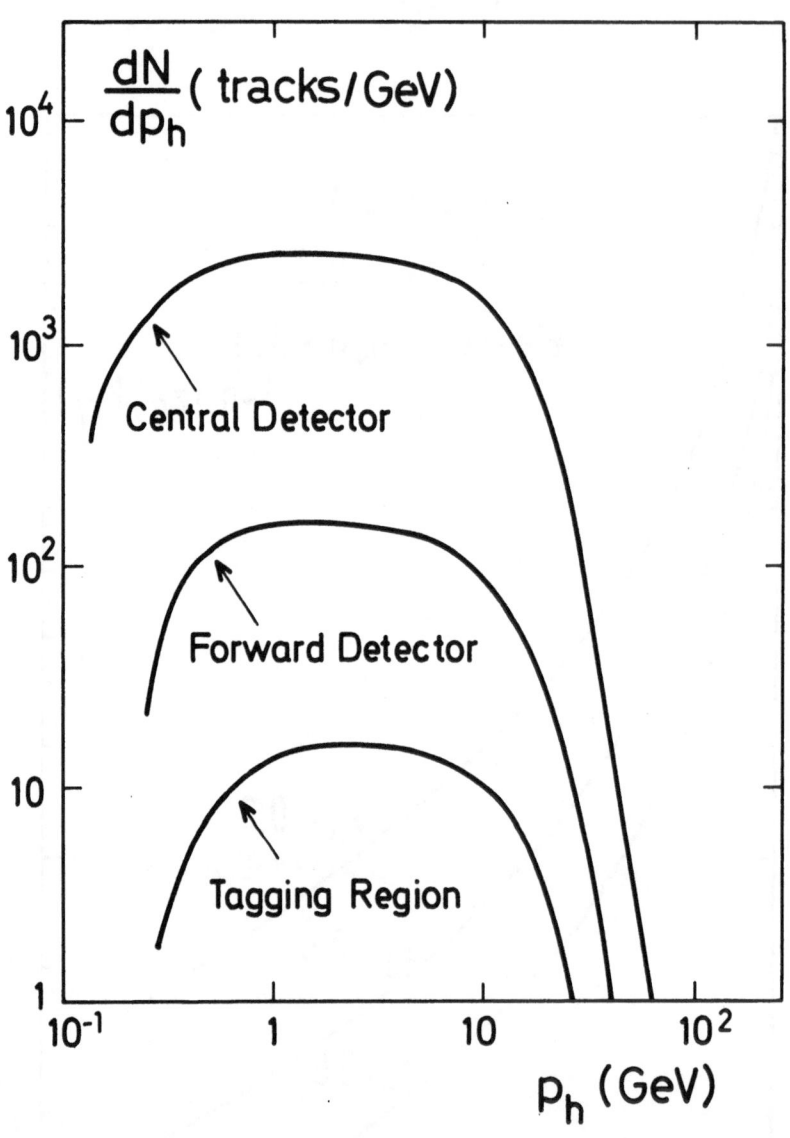

Fig. 21

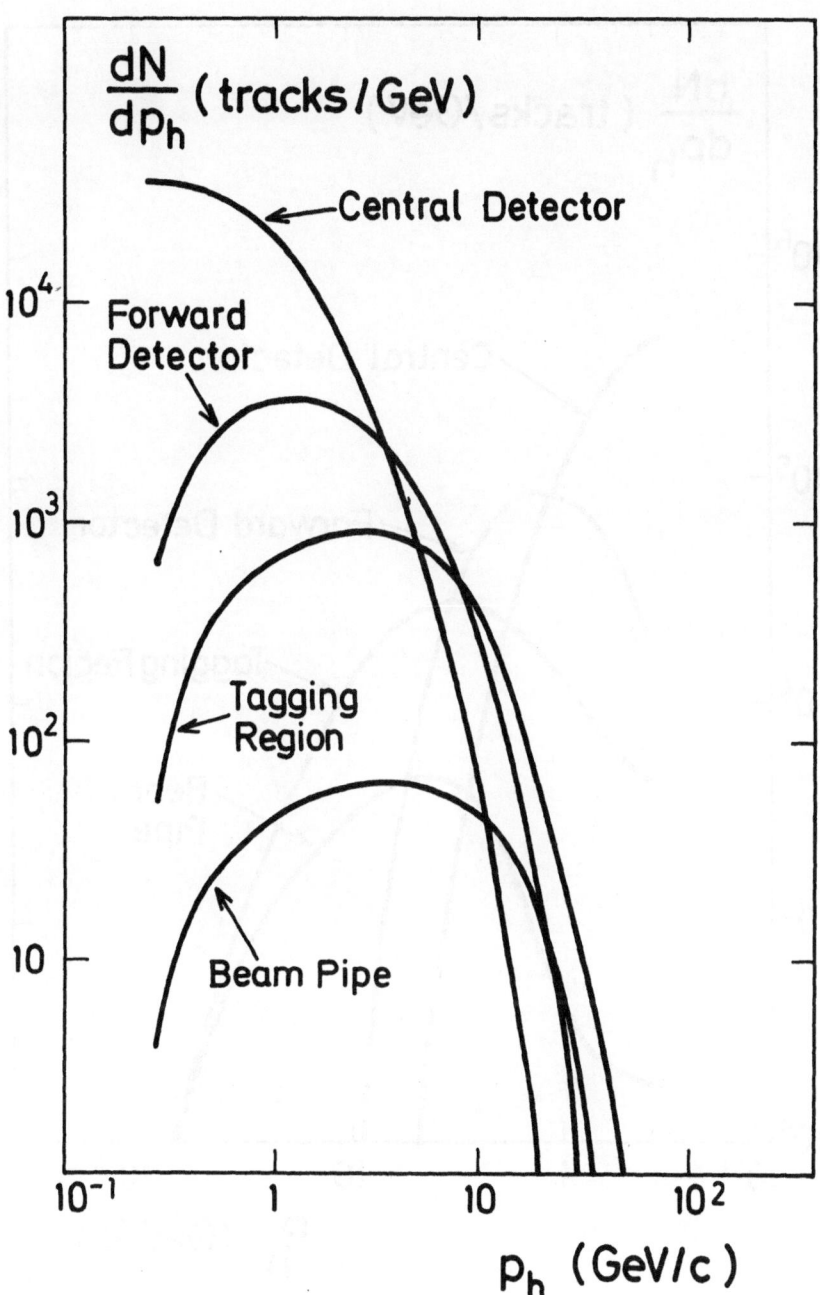

$\dfrac{dN}{dp_h}$ (tracks / GeV)

Central Detector

10^4

Forward
Detector

10^3

Tagging
Region

10^2

10

Beam Pipe

10^{-1} 1   10 10^2

p_h (GeV/c)

Fig. 22

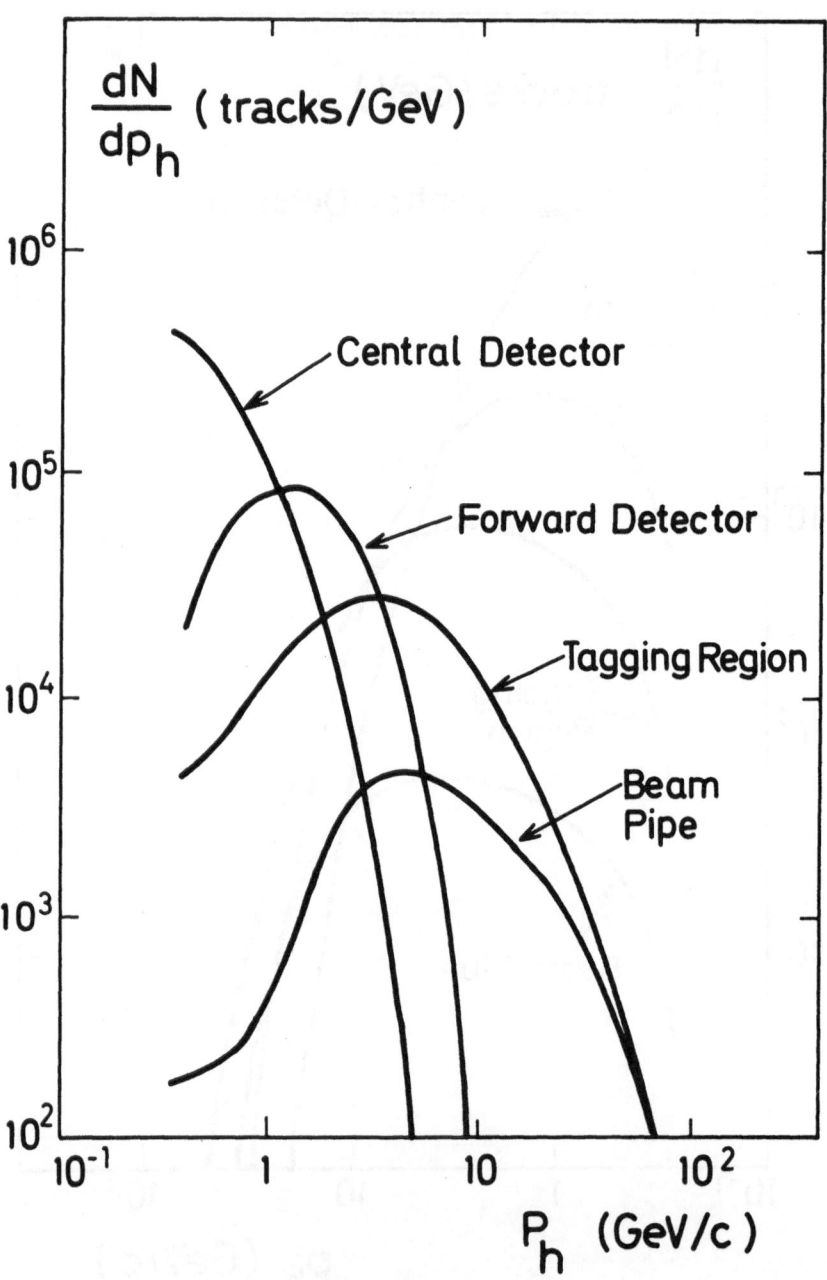

Fig. 23

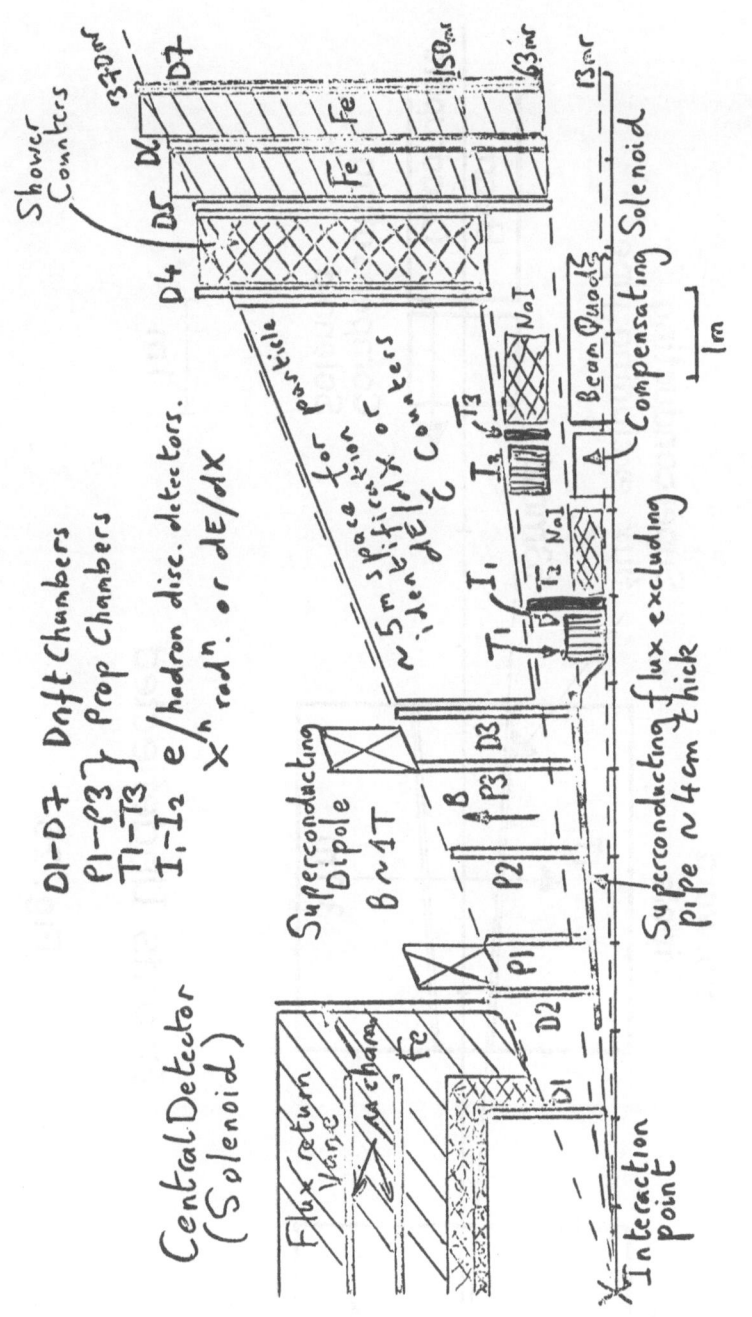

Fig. 24

294

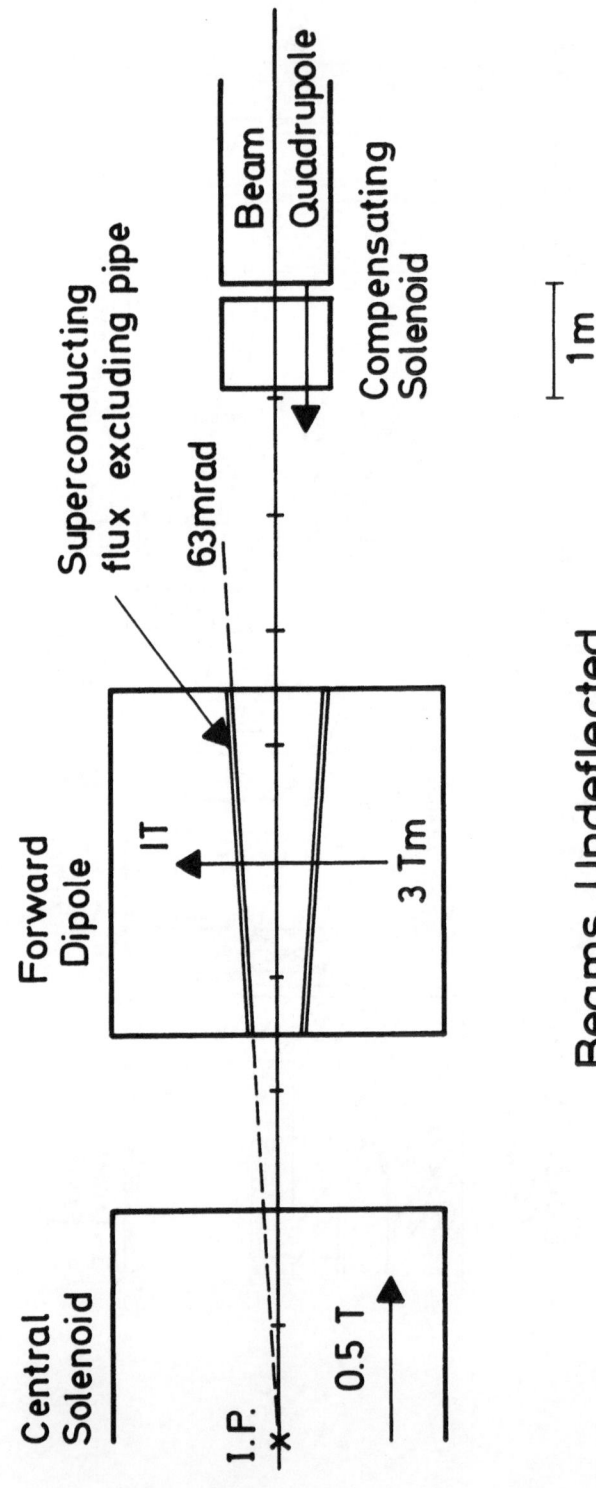

Top View

Fig. 25

Beams Undeflected

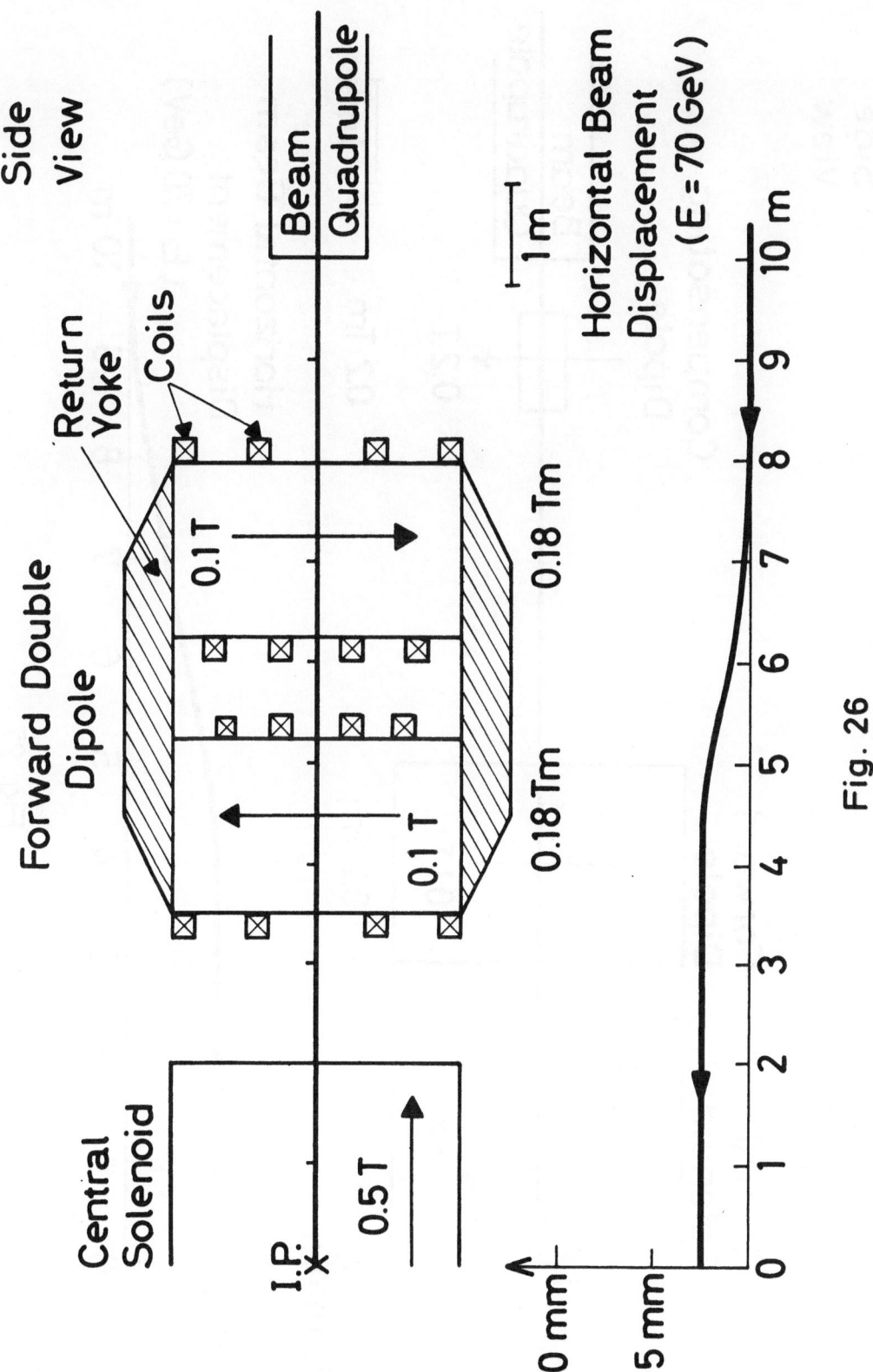

Fig. 26

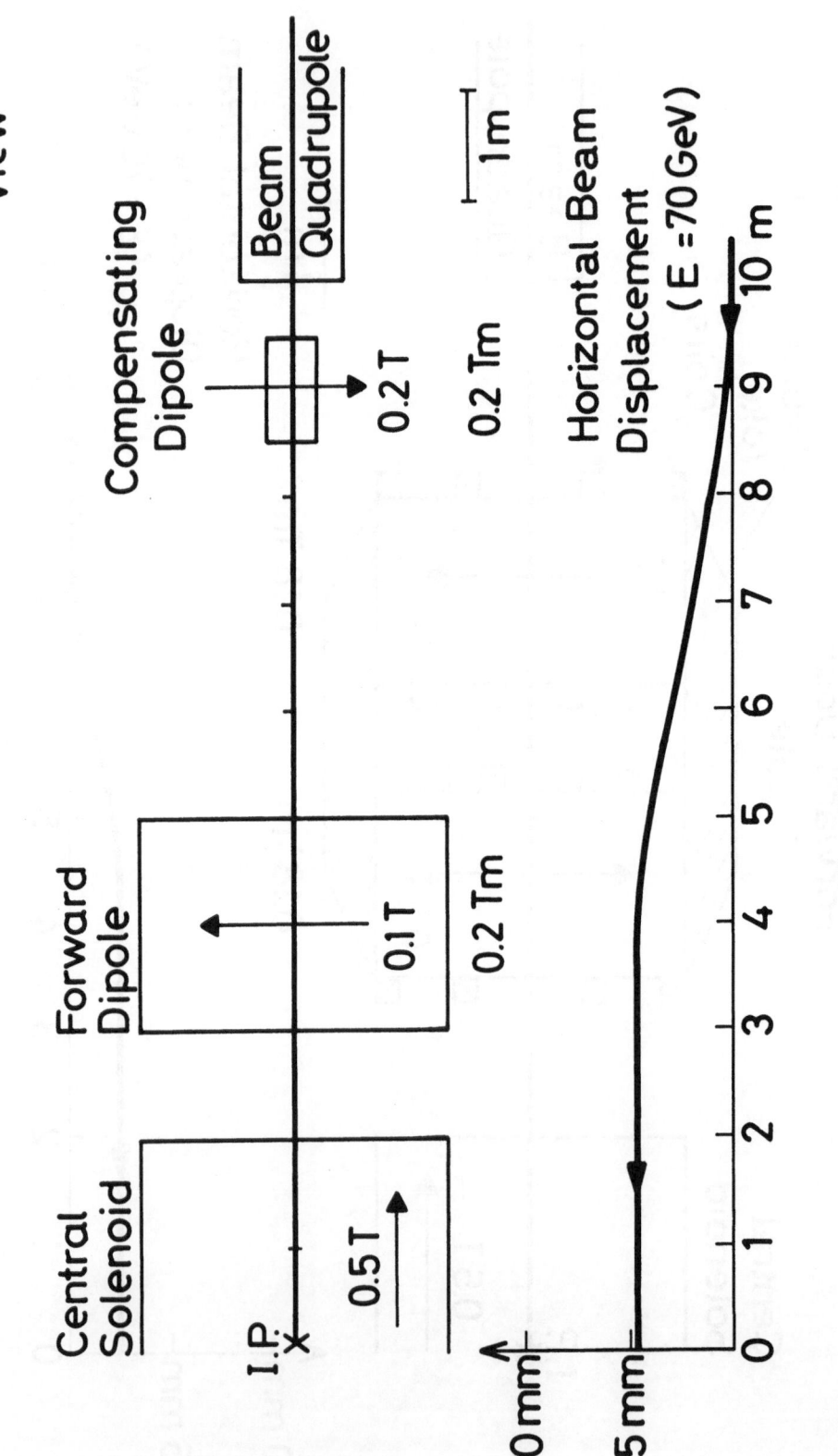

Fig. 27

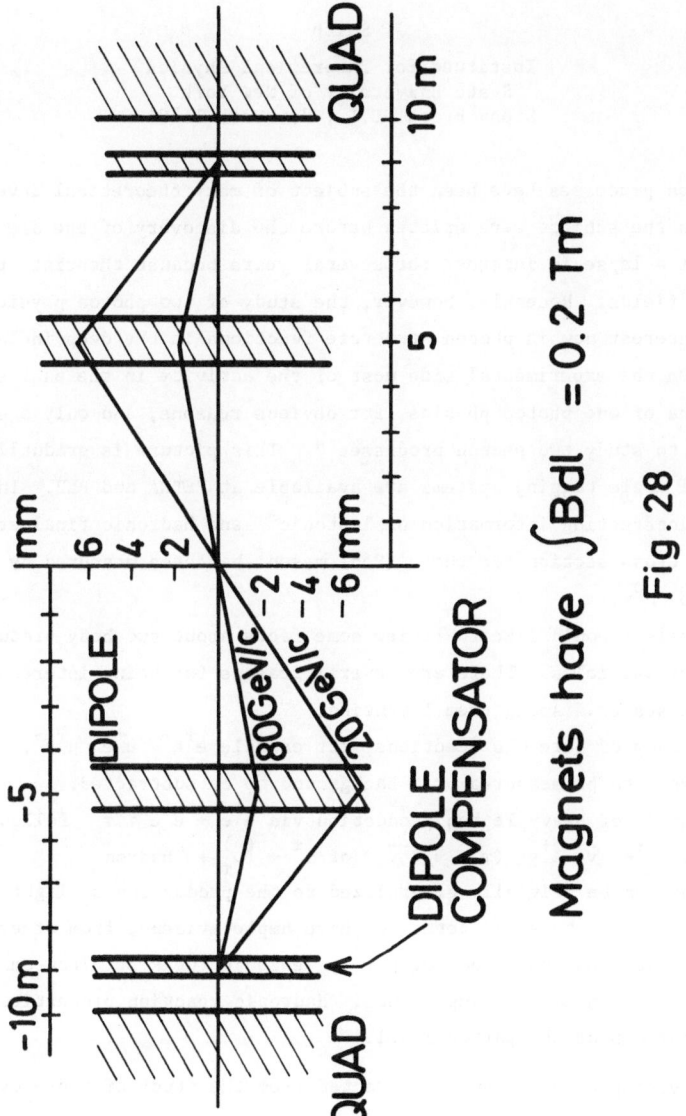

Fig. 28

PARTICLE PAIR PRODUCTION IN TWO-PHOTON REACTIONS

J. Smith

Institute for Theoretical Physics
State University of New York
Stony Brook, L.I., New York 11794

Two photon processes have been the subject of many theoretical investigations.[1] Most papers on the subject were written before the discovery of the J/ψ in 1974. Then there was a lapse in interest for several years because theorists turned to more exciting fields. Recently, however, the study of two photon physics has revived, with interest now in photon structure functions in the deep inelastic region[2] and jets.[3] On the experimental side most of the activity in the past six years was in the area of one photon physics, for obvious reasons, and only a small effort has been made to study two photon processes.[4] This picture is gradually changing, now that small angle tagging systems are available at PETRA and PEP. In fact, we already have interesting information on leptonic[5] and hadronic final states.[6] The $\gamma\gamma$ production cross section for the $\eta'(958)$ meson[7] has been measured by detecting its decay into $\rho^0\gamma$.

In this talk I would like to review some facts about two body production of light and heavy particles. There are several reasons for being interested in these specific final states. Among them I mention:

1. The study of pure QED reactions, for example $e^+e^- \to e^+e^-\ \mu^+\mu^-$, either as a signal to be measured or a background to be subtracted.

2. The study of heavy lepton production via $e^+e^- \to e^+e^-\tau^+\tau^-$ followed by the decays $\tau^+ \to \bar{\nu}_\tau\ \ell^+\nu_\ell$ $(\tau^- \to \nu_\tau\ell^-\bar{\nu}_\ell)$ or $\tau^\pm \to \overset{(-)}{\nu}_\tau + $ hadrons.

These reactions can be trivially generalized to the production of light u,d,s quarks and heavy c,b,t quarks to yield jets. We have ample evidence from other areas of high energy physics that the study of purely leptonic (weak or electromagnetic) interactions is of fundamental importance. Hadronic reaction properties can then be inferred from QCD or the parton model.

3. Interesting physics can be extracted from the study of two-body hadronic final states such as

$$e^+e^- \to e^+e^-\pi^+\pi^-\ , \qquad e^+e^- \to e^+e^-\rho^+\rho^- \ , \qquad e^+e^- \to e^+e^-D^+D^-.$$

The pion production reaction was originally suggested as a good way to measure $\pi\pi$ phase shifts, which cause a deviation from the point-like (no final state interaction) cross section. Rho meson production will probably be of a diffractive nature. Finally, the heavy mesons D (and F) can be pair produced in virtual $\gamma\gamma$ collisions, even though the cross section will fall rapidly just above the threshold due to form factor effects.

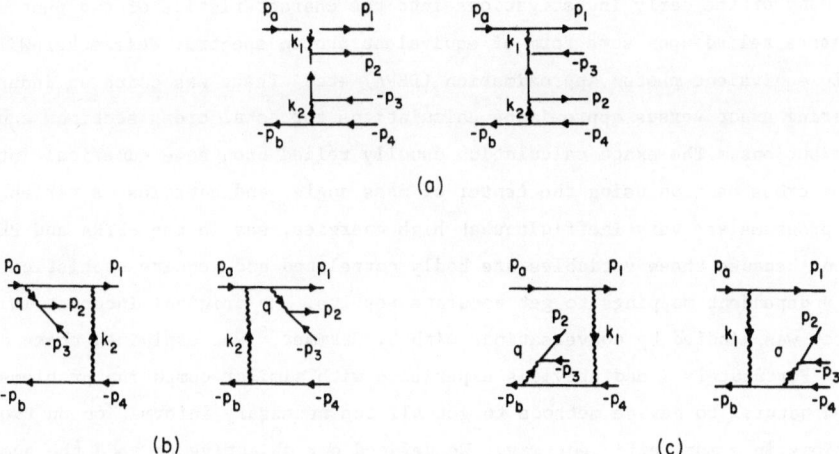

(a)

(b)   (c)

Fig. 1 The C even and C odd Feynman diagrams for the reaction $e^+e^- \to e^+e^-\mu^+\mu^-$

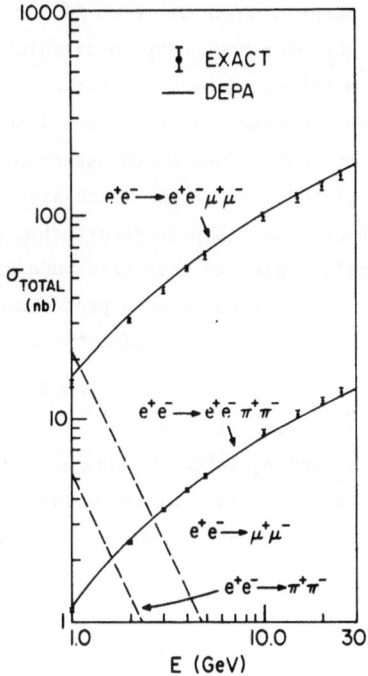

Fig.2 The cross sections for C even production of $\mu^+\mu^-$ and $\pi^+\pi^-$ pairs.

Many of the early investigations into the characteristics of two photon processes relied upon some form of equivalent photon spectra: Weiszacker-Williams (WW) double equivalent photon approximation (DEPA) etc. There was quite an industry of comparing exact versus approximate calculations for total cross sections and distributions.[1] The exact calculation usually relied upon some numerical integration of the cross section using the center of mass angles and energies as variables. Such programs are very inefficient at high energies, say in the PETRA and PEP energy region, because these variables are badly correlated and require sophisticated energy dependent mappings to get accurate results. My original interest in this subject was kindled by conversations with G. Grammer,[8] who explained these difficulties. Fortunately I had previous experience with similar computing problems, so it seemed natural to devise methods to get all the necessary information on two photon reactions in a more efficient way. We defined our objective to redo the numerical programs so that everything could be handled numerically, without resorting to any approximations at high energies. The natural variables in the process are of multipheripheral type, using invariant masses and momentum transfers.

This led to a series of papers on two-body reactions involving several collaborators.[9,10,11,12] The program never seems to finish because, as experimental techniques are improved, it becomes necessary to calculate even more Feynman diagrams, such as interferences between C even two photon reactions and C odd bremsstrahlung graphs. There the interference traces get impossibly long, so Vermaseren[13] has recently written a set of subroutines which use complex matrix methods to calculate the individual amplitudes of graphs which are then added directly. The real and imaginary parts of the final amplitude are then squared to get the cross section. Bjorken[14] originally suggested this trick and Brodsky and Ting[15] used it to calculate Bethe-Heitler reactions in muon proton scattering (which is the laboratory frame equivalent of the center-of-mass two photon reaction).

The first paper in reference 9 contains the results of a calculation of the C even and C odd cross sections for $e^+e^- \rightarrow e^+e^-\mu^+\mu^-$. The relevant Feynman diagrams are shown in Fig. 1, where p_2 and p_3 label the muon momenta. We compared the results for the total cross section and pair mass distribution in the C even case with those expected from the DEPA approximation, which uses the equivalent photon energy spectrum

$$N(\omega) = \frac{\alpha}{\pi} \left[\frac{E^2 + E'^2}{E^2} \left(\ell n \frac{E}{m_e} - \frac{1}{2} \right) + \frac{(E - E')^2}{2E^2} \left(\ell n \frac{2E'}{E-E'} + 1 \right) + \frac{(E+E')^2}{2E^2} \ell n \left(\frac{2E'}{E+E'} \right) \right]$$

where $\omega = E - E'$. The C odd cross section was compared with the WW approximation given by Baier and Fadin.[1] For the C even reactions with pions we assumed point-like coupling. Figs. 2 and 3 show the resulting cross sections, and the error bars

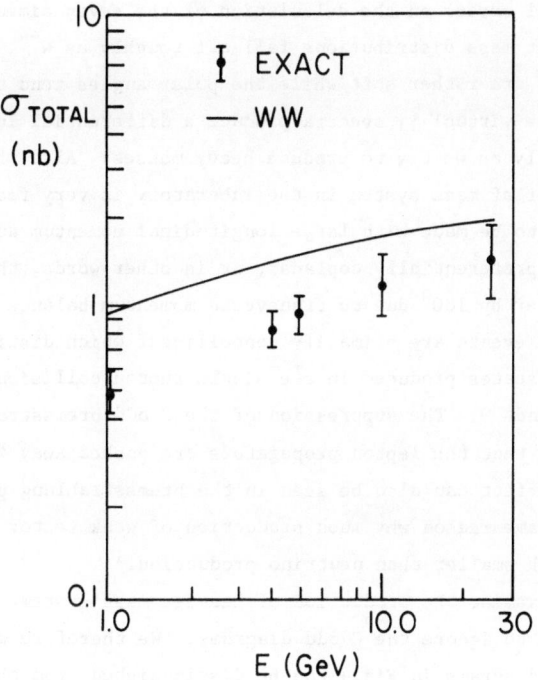

Fig. 3 The cross section for the C odd reaction $e^+e^- \rightarrow e^+e^-\mu^+\mu^-$

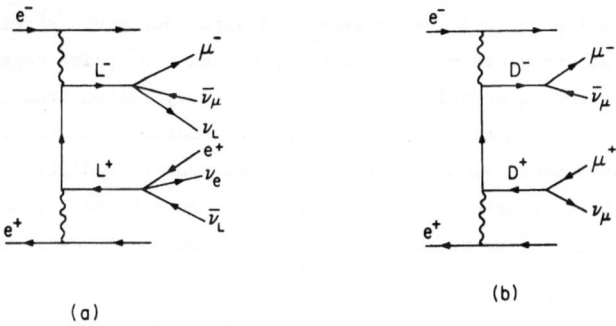

Fig.4 The Feynman diagrams for heavy lepton and heavy meson production.

are theoretical errors on the calculation of the seven dimensional integrals. The
dimuon invariant mass distributions fall off roughly as W^{-4}. The energies of the
muons and pions are rather soft while the polar angles tend to be very small. As
we all know, the virtual $\gamma\gamma$ spectra produce a differential luminosity which falls
extremely rapidly as we try to produce heavy masses. Also the rapidity distribution
of the $\gamma\gamma$ center of mass system in the laboratory is very flat.[16] Hence the produced
particles tend to be made with large longitudinal momentum and emerge at small angles.
Also, they are preferentially coplanar, or in other words, their azimuthal angles
ϕ_i tend to differ by 180° due to transverse momentum balance perpendicular to the
beam line. The events are primarily noncollinear which distinguishes them from the
two body final states produced in e^+e^- single photon collisions. Distributions are
given in reference 9. The suppression of the C odd bremsstrahlung cross section is
due to the fact that the lepton propagators are pushed away from their mass shell
values. This effect can also be seen in the bremsstrahlung production of ρ mesons[12]
and it is the same reason why muon production of weak vector bosons in a Coulomb
field is so much smaller than neutrino production.[17]

If we now examine the production of heavier mass systems such as τ or D mesons
then it is safe to ignore the C odd diagrams. We therefore want to find out whether
the two photon diagrams in Fig.4 can be distinguished from the normal one photon
production mechanisms. The C even cross sections are still large, even though the
large masses seriously reduce the cross section near threshold. A plot of the C
even two photon and one photon cross sections for the production of heavy leptons
and mesons (using only point-like coupling) are shown in Fig.5, where one can see
that the cross sections for one photon and two photon production of $\tau^+\tau^-$ pairs
intersect at 16 GeV beam energy. In the graphs the heavy lepton is labelled by L.

The heavy leptons and mesons decay instantaneously via three body leptonic and
semileptonic decays yielding $\mu^+\mu^-$, μ^+e^-, μ^-e^+ and e^+e^- pairs together with undetect-
able neutrinos and detectable hadrons. These heavy states tend to be produced with
smaller longitudinal momentum (at threshold of course they are at rest). Hence they
can emit decay leptons with sizable transverse momenta and perhaps contaminate experimental data on single photon production of heavy lepton or hadron pairs. To
answer this question, we therefore allowed the leptons to decay via $\tau^- \to \nu_\tau \mu^- \bar{\nu}_\mu$
and the hadrons via $D^- \to \mu^- \bar{\nu}_\mu$. Of course the latter decay rate is completely
negligible compared to the three body decays $D^- \to K^0 \mu^- \bar{\nu}_\mu$, but this decay gives
similar distribution to τ decay, so we used the $D^- \to \mu^- \bar{\nu}_\mu$ decay to simply examine
the difference in spectra between three body and two body decays. Realistically one
should treat the D particles as charmed quarks and add fragmentation functions, etc.
These refinement lead to lower rates due to fractional charges and to softer spectra
from the lower Q value in the decay. So our curves should only be taken as an indication of the relevant distributions.

The polar angle distributions of the heavy leptons and mesons are shown in

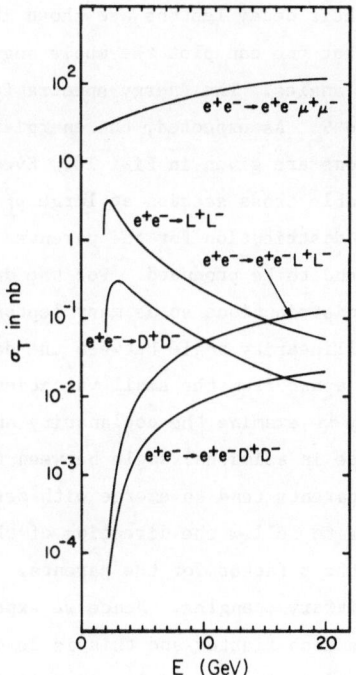

Fig.5 The cross sections for heavy lepton and heavy meson production.

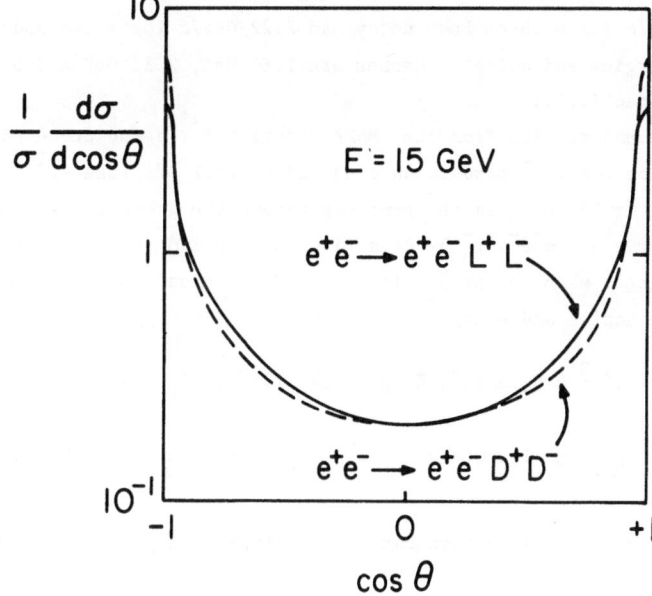

Fig. 6   The angular distribution for the parent heavy particles.

Fig.6, while those of their decay leptons are shown in Fig. 7. These distributions are sufficiently flat that one can plot the whole angular range and see how many leptons emerge at large angles. The energy spectra for the parents and daughters are given in Figs. 8 and 9. As expected, the energies tend to be rather soft. The daughter $p_\perp^2$ distributions are given in Fig. 10. Even though this distribution drops rapidly, there is a sizable cross section at large $p_\perp^2$ (say 1 or 2 GeV/c). We also show the invariant mass distribution for the parents and daughters in Figs. 11 and 12. Large mass pairs tend to be produced. For the daughters this follows from the roughly isotopic decay distributions so as many leptons are emitted backwards as forwards. Hence the collinearity angle between the daughters, which is shown in Fig.13, is rather flat, apart from the small variation due to helicity effects in the V-A decays. Finally we examine the coplanarity angle between the $e^+\ell^+$ and $e^-\ell^-$ planes, or the difference in azimuthal angle between the emerging daughter ℓ^+ and ℓ^-. Remember that the parents tend to emerge with zero acoplanarity angle and the decay leptons prefer to follow the direction of the parents. But these statements depend upon the β factor for the parents. Low energy heavy leptons will emit daughters with arbitrary ϕ angles. Hence we expect the daughter coplanarity angle distribution to be much flatter and this is demonstrated by the results in Fig. 14. Both decays lead to an essentially identical distribution.

For completness we give some average numbers for beam energies of 15 GeV on 15 GeV. The average coplanarity angle is given by $\cos\theta = 0.86$ and the average collinearity angle is $\cos\theta = 0.06$ for daughters from $\tau^+\tau$ decays. The average $P_\perp$ is 1.12 GeV/c for a three body decay and 2.22 GeV/c for a two body decay while the average energies and dilepton masses are 1.60 GeV, 2.15 GeV and 5.0 GeV/c^2, 8.5 GeV/c^2, respectively.

The recent results from the MARK J detector[18] demonstrate that the one photon cross section for $\tau^+\tau^-$ production does indeed fall off like s^{-1} and is approximately 100 pb at E=15 GeV. In the same experiment the group measured the QED cross sections for $e^+e^- \to e^+e^-\mu^+\mu^-$ by detecting $\mu\mu$, μe and $\mu\mu e$ final states. These cross sections agreed with the predictions from Vermaseren's Monte Carlo with the appropriate cuts on the angles and energies

$$168^0 \geqslant \theta_e \geqslant 12^0, \ E_e \geqslant 2 \text{ GeV}; \ 135^0 \geqslant \theta_{\mu_1} \geqslant 45^0 \ ,$$

$$P_{\perp,\mu_1} \geqslant 1.5 \text{ GeV and } 147^0 \geqslant \theta_{\mu_2} \geqslant 33^0, \ P_{\perp,\mu_2} \geqslant 0.8 \text{ GeV.}$$

The actual cross sections turn out to be $\sigma(\mu\mu) \simeq 50$ pb, $\sigma(\mu e) = 10$ pb and

$\sigma(e\mu\mu) \simeq 10$ pb. Adding in the branching ratios for τ decays, we expect $\sigma(e^+e^- \to \tau^+\tau^-) \times B^2 = 2.5$ pb with complete acceptance. Hence it is obvious that the $\mu^+\mu^-$ signal from the regular two photon reaction $e^+e^- \to e^+e^-\mu^+\mu^-$ will dominate this channel. As

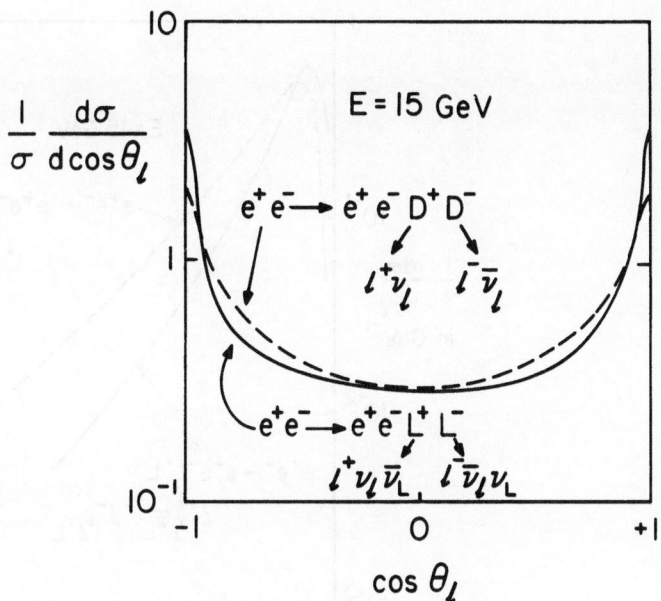

Fig.7 The angular distributions for the daughter leptons.

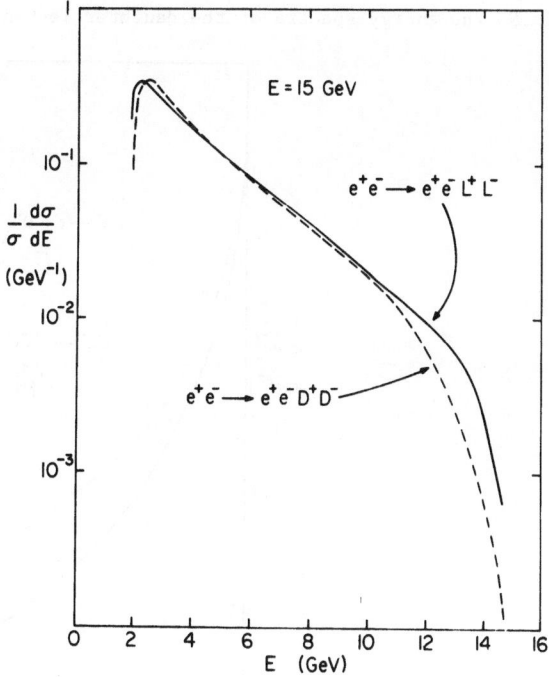

Fig.8   The energy spectra of the parent heavy particles.

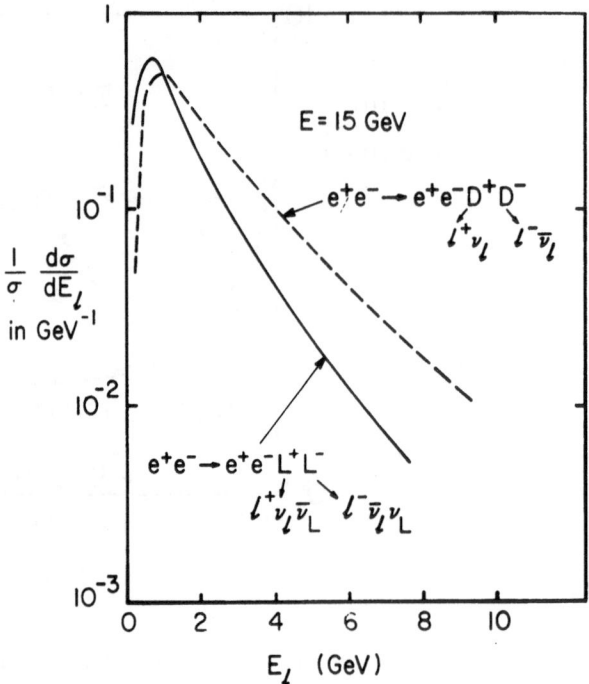

Fig.9 The energy spectra of the daughter leptons.

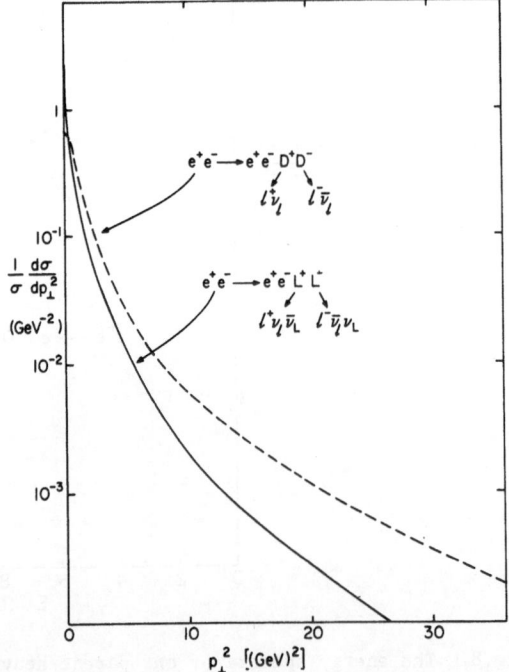

Fig.10 The distributions in p_T^2 for the daughter leptons.

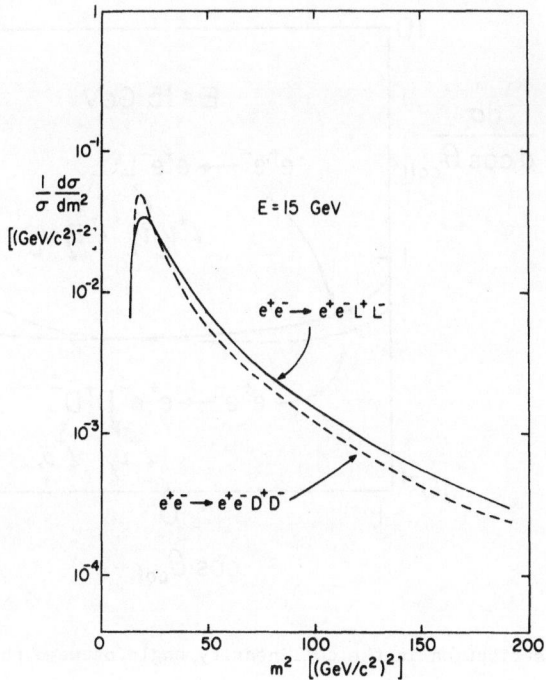

Fig.11 The distributions in m^2 for the parent heavy particles.

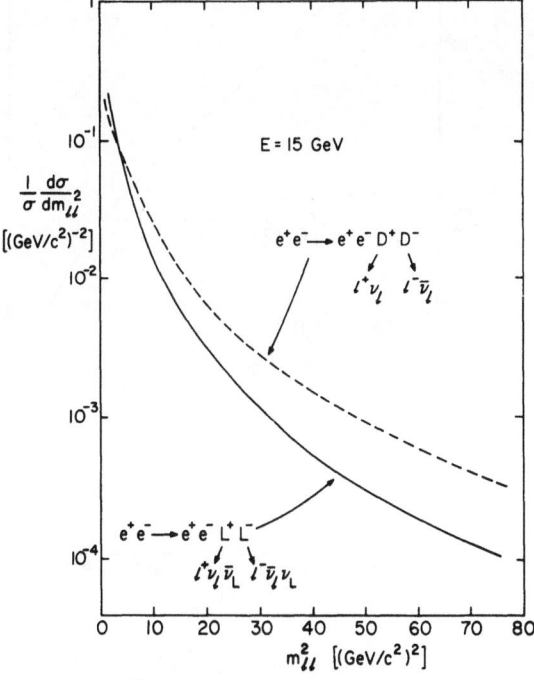

Fig.12 The distributions in m^2 for the daughter leptons.

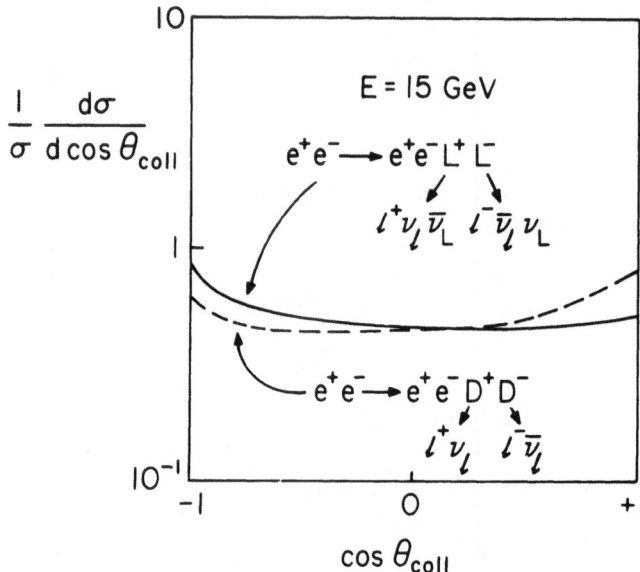

Fig. 13   The distribution in the collinearity angle between the daughters.

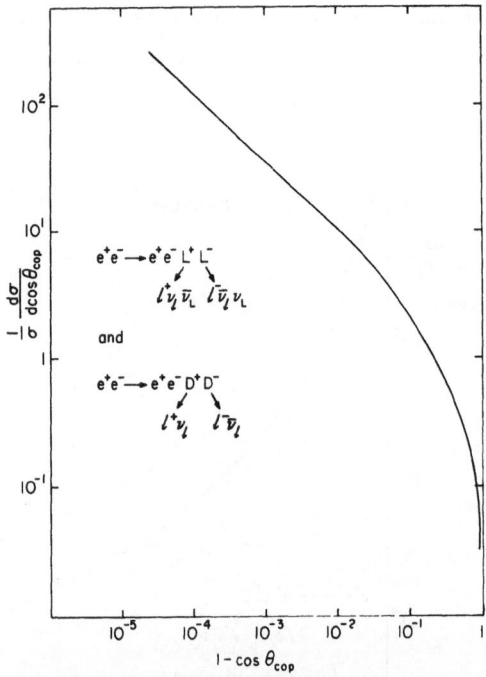

Fig.14   The distribution in the coplanarity angle between the daughters.

regards μe the two photon background can be eliminated by a large angular cut of $150^\circ \geqslant \theta_e \geqslant 30^\circ$ (it is reduced to $\sim 10^{-3}$ pb). Thus the $e^+e^- \to \tau^+\tau^-$ cross section was monitored by the μe and μ hadron signals which are reasonably collinear and co-planar. To make an estimate of whether the two photon reaction $e^+e^- \to e^+e^-\tau^+\tau^-$ could also be seen I made a simple check on the decay $e\mu$ signal assuming that the $p_\perp$ for both detected leptons is larger than 1.5 GeV. At a beam energy of 15 GeV this yields a signal of $B^2 \times 3$ pb. The corresponding result for the daughter lep-tons from the one photon reaction $e^+e^- \to \tau^+\tau^-$ is $B^2 \times 58$ pb. Hence it is not clear that the MARK J group would have seen the two photon signal.

There is a better chance to see this signal when one of the forward going leptons is tagged. For instance, the PLUTO collaboration have already distinguished two photon jets. A $\tau^+\tau^-$ signal would have similar characteristics but lower multipli-city. One should therefore keep a look out for non-collinear but coplanar μe, μ hadron, or hadron hadron events, which will indicate the onset of heavy lepton and/or heavy quark two photon physics.

REFERENCES

1. The canonical review articles are

   H. Terazawa, Rev. Mod. Phys. $\underline{45}$, 615 (1973);

   V. M. Budnev, I. F. Ginzburg, G. V. Meledin and V. G. Serbo, Phys. Reports $\underline{15}$, 181 (1975);

   See also the talks presented at the Colloquium on "Two Photon Physics in e^+e^- Colliding Beams", Journal de Physique, Colloque C2, Supplement No. 3 (1974). I apologize for not giving references to individual papers.

2. E. Whitten, Nucl. Phys. $\underline{B120}$, 189 (1977);

   C. H. Llewellyn Smith, Phys. Letts. $\underline{79B}$, 83 (1978);

   See the talk by T. Walsh in these proceedings.

3. S. J. Brodsky, T. De Grand, J. Gunion and J. Weis, Phys. Rev. $\underline{D19}$, 1418 (1979);

   K. Kajantie, Physica Scripta $\underline{29}$, 230 (1979); HUTFT 79-5, 79-7;

   K. Kajantie and R. Raitio, Phys.Letts.$\underline{87B}$, 133 (1979); Nucl.Phys. $\underline{B159}$,528 (1979);

   M. Abbud, R. Gatto and C. A. Savoy, Phys.Rev. $\underline{D20}$, 2224 (1979).

4. The talk by G. Barbiellini in these proceedings reviews the early experiments

5. D. P. Barber et al. Phys. Rev. Letters $\underline{43}$, 1915 (1979).

6. We refer to the talks by C. Berger and E. Hilger, who presented data from the PLUTO and TASSO collaborations at PETRA.

7. G. S. Abrams et al., Phys. Rev. Letters $\underline{43}$, 447 (1979).

8. See G. Grammer and T. Kinoshita, Nucl. Phys. $\underline{B80}$, 461 (1974).

9. R. Bhattacharya, J. Smith and G. Grammer, Phys. Rev. $\underline{D15}$, 3267 (1977).

   Similar calculations were done by F. Gutbrod and Z. J. Rek, Zeit. für Physik $\underline{1}$, 171 (1979).

10. J. Smith, J.A.M. Vermaseren and G. Grammer, Phys. Rev. $\underline{D15}$, 3280 (1977).

11. J. A. M. Vermaseren, J. Smith and G. Grammer, Phys. Rev. $\underline{D19}$, 137 (1979).

12. R. M. Godbole and J. Smith, Nucl. Phys. $\underline{B158}$ 234 (1979).

13. The talk by J.A.M. Vermaseren contains more details of these calculations.

14. J. D. Bjorken and M. C. Chen, Phys. Rev. $\underline{154}$, 1335 (1967).

15. S. J. Brodsky and S.C.C. Ting, Phys. Rev. $\underline{145}$, 1018 (1966).

    G. Reading Henry, Phys. Rev. $\underline{154}$, 1534 (1967).

16. We refer the reader to the talk by J. Field in these proceedings for a thorough discussion of luminosity, rapidity and other variables.

17. R. W. Brown and J. Smith, Phys. Rev. $\underline{D3}$, 207 (1971).

18. The talk by J. Bürger in these proceedings reviews the MARKJ results.

γγ → HADRONS AT LOW ENERGIES

M. Greco
INFN, Laboratori Nazionali di Frascati
00044 Frascati, Italy

1. - INTRODUCTION.

Ten years after the nice papers by Low[1] and Calogero and Zemach[2], the importance of the photon processes in e^+e^- colliding beam experiments was empha sized by three independent groups[3, 4, 5], who carried out a quite detailed analysis of the most relevant leptonic and hadronic reactions. Together with the first obser- vations[6] of γγ processes at Novosibirsk, Frascati and CEA those works stimula- ted a great deal of theoretical interest. Much of the work done in this field, particu larly γγ → hadrons, dates back in fact to the early seventies[7].

It is only after ten more years that we finally have detailed and quite precise measurements of the photon-photon cross sections $\sigma(\gamma\gamma \to$ hadrons). Exiting data have been reported to this Conference from various experimental groups operating at SPEAR and PETRA storage rings[8, 9].

In this talk I will review the theoretical predictions for γγ → hadrons at low c.m. energies, with special emphasis on those results which are in closer connec- tion with the data. Because, as already said, much of the work done in this field is not very recent and has been excellently reviewed earlier[7], I will not give very ma ny details, which can be found in the original papers.

The plan of the talk is the following. After a brief kinematical introduction, I will discuss in Sect. 3 the case of pseudoscalars production, which is by now rather well settled, on the light of the recent measurements of $\Gamma(\eta' \to \gamma\gamma)$[10, 11]. Scalar and tensor resonance production is considered in Sect. 4. Estimates of γγ total cross sections are finally discussed in Sect. 5, together with the question of the validity of simple resonance ⟷ Regge duality in presence of non-Regge terms in the absorp- tive part of the γγ scattering amplitude.

The production of pion pairs, as well as the applications of current algebra and soft pion theorems to the processes $\gamma\gamma \to (2n)\pi$, $(2n+1)\pi$, which have been alrea dy discussed[7] many times in great detail, will not be included in the written version of the talk, which is centered on various aspects of resonance production and $\gamma\gamma$ total cross sections. In the same spirit we also leave out a discussion of a duali ty sum rule between resonances and the quark box diagram in virtual $\gamma\gamma$ scattering. The interested reader can refer to the original work[12].

2. - KINEMATICS AND NOTATION.

The absorptive part of the forward $(q_1 = q_3, \; q_2 = q_4)$ current x current scattering is defined as

$$W^{\mu\nu\lambda\sigma}(q_1, q_2) = \frac{1}{2} \sum_n (2\pi)^4 \delta^4(q_1 + q_2 - P_n) \, T_n^{\mu\nu}(q_1, q_2) \, T_n^{\lambda\sigma*}(q_1, q_2) =$$

$$= \frac{1}{2} \int d^4x \, d^4y \, d^4z \, \exp\left[-\frac{i}{2} \left\{ (q_2 - q_1)(x - y) + (q_1 + q_2)z \right\} \right] \cdot \tag{1}$$

$$\cdot \left\langle 0 \left| \bar{T} \left[j^\lambda(\tfrac{x}{2}) \, j^\sigma(-\tfrac{x}{2}) \right] T \left[j^\mu(\tfrac{y}{2} + z) \, j^\nu(-\tfrac{y}{2} + z) \right] \right| 0 \right\rangle ,$$

where the vertex function $T_n^{\mu\nu}(q_1, q_2)$ is given by

$$T_n^{\mu\nu}(q_1, q_2) = i \int d^4x \, \exp(iQx) \left\langle n \left| T \left[j^\mu(\tfrac{x}{2}) \, j^\nu(-\tfrac{x}{2}) \right] \right| 0 \right\rangle . \tag{2}$$

We use the following notations

$$P = (q_1 + q_2), \quad Q = \frac{1}{2}(q_2 - q_1), \quad s = (q_1 + q_2)^2, \quad t = (q_1 - q_3)^2, \quad u = (q_1 - q_4)^2 .$$

The decomposition of $W^{\mu\nu\lambda\sigma}$ for virtual photons in terms of the eight helicity amplitudes may be found in ref. (13). When $q_1^2 = q_2^2 = 0$ only five helicity amplitudes survive and one can define the amplitudes $F_i^S(s, t, u)$ $(i = 1, \ldots, 5)$:

$$F_1^S \equiv f_{++,\,++}^S + f_{++,\,--}^S \;\underset{\text{even, } P=+1}{\sim}\; \sum_{J=0} F_{1J} ,$$

$$F_2^S \equiv f_{+-,\,++}^S + f_{+-,\,--}^S \;\underset{\text{even, } P=+1}{\sim}\; \sum_{J=2} F_{2J} ,$$

$$F_3^S \equiv f_{+-,\,+-}^S + f_{+-,\,-+}^S \;\underset{J=2}{\overset{P=+1}{\sim}}\; \sum F_{3J} , \tag{3}$$

$$F_4^S \equiv f_{++,\,++}^S - f_{++,\,--}^S \;\underset{\text{even, } P=-1}{\sim}\; \sum_{J=0} F_{4J} ,$$

$$F_5^S \equiv f_{+-,\,+-}^S - f_{+-,\,-+}^S \;\underset{J=2}{\overset{P=+1}{\sim}}\; \sum F_{5J} ,$$

where $f^s_{\lambda_1\lambda_2,\lambda_3\lambda_4}$ stand for the s-channel helicity amplitudes $(\lambda_i = \overset{+}{_-} 1)$ and the spin parity of the intermediate states which contribute in the s-channel are indicated in the r.h.s. of eq.(3).

By using the crossing properties of the amplitudes (3) and their partial wave expansion a new set of amplitudes can be found[12, 13] which are free of kinematical singularities and have appropriate Regge expansion.

The contribution of a given resonances of mass m_R, width Γ_R and spin J_R to the total cross section for a given helicity amplitude is

$$\sigma_{\lambda_1\lambda_2}(s) = \frac{1}{s} \, \mathrm{Im} \, f^s_{\lambda_1\lambda_2,\lambda_1\lambda_2}(s, t=0) \simeq 16\pi^2 (2J_R + 1) \frac{\Gamma^R_{\lambda_1\lambda_2}}{m_R} \, \delta(s - m^2_R) , \tag{4}$$

which leads, in the equivalent photon approximation, to

$$\sigma(ee \rightarrow eeR) \simeq 8\alpha^2 (2J_R + 1) \frac{\Gamma(R \rightarrow \gamma\gamma)}{m^3_R} \, \ln^2 \left(\frac{E}{m}\right) f\left(\frac{m^2_R}{4E^2}\right) , \tag{5}$$

with $f(y) = -(2+y)^2 \ln y - 2(1-y)(3+y)$.

3. - PSEUDOSCALAR MESON PRODUCTION.

The simplest example of hadron production by e-e collisions is the production of a single π^o, η, η'. The $P\gamma\gamma$ vertex function $T^{\mu\nu}_P$ of eq.(2) can be written in terms of one form factor as

$$T^{\mu\nu}_P = \epsilon^{\mu\nu\rho\sigma} q_{1\rho} \, q_{2\sigma} \, F_P(q^2_1, q^2_2) . \tag{6}$$

The normalization of F_{π^o} is given by the PCAC triangle anomaly[14]

$$F_{\pi^o}(q^2_1 = q^2_2 = 0) \Big|_{P^2 = 0} = -\frac{S_\pi}{2\pi^2 f_\pi} \simeq g_{\pi^o\gamma\gamma} , \tag{7}$$

where $S_\pi = \frac{1}{2}$ for fractionally charged quarks, $f_\pi \simeq 95$ MeV and the approximate equality refers to the extrapolation from $P^2 = 0$ to $P^2 = m^2_\pi$. As well known, eq.(7) predicts

$$\Gamma(\pi^o \rightarrow \gamma\gamma) = \frac{e^4}{64\pi} \, g^2_{\pi^o\gamma\gamma} \, m^3_\pi \simeq 7.3 \text{ eV} , \tag{8}$$

in excellent agreement with the experimental value $\Gamma(\pi^o \rightarrow \gamma\gamma) = (7.95 \overset{+}{_-} 0.55)$ eV.

According to the Bjorken-Johnson-Low theorem[15], the asymptotic limit of

$F_\pi(q_1^2, q_2^2)$, when $q_1^2 \to \infty$ with $q_1^2/q_2^2 \to 1$ is determined by the commutator of the currents and is given by

$$F_\pi(q_1^2, q_2^2) \longrightarrow \frac{2 f_\pi}{q_1^2} \,. \tag{9}$$

It would be very interesting to check this type of predictions, particularly for the case $P = \eta, \eta'$ discussed below, which seems more easily experimentally accessible.

In the simplest quark model, expressing the quark content of the photon as the usual mixture of $\varrho - \omega - \varphi$ mesons one obtains the following relations

$$g_{\eta\gamma\gamma} = \frac{1}{\sqrt{3}} g_{\pi^0\gamma\gamma}(\cos\theta_p - 2\sqrt{2}\sin\theta_p) \,, \quad g_{\eta'\gamma\gamma} = -\frac{1}{\sqrt{3}} g_{\pi^0\gamma\gamma}(\sin\theta_p + 2\sqrt{2}\cos\theta_p) \tag{10}$$

where θ_p is the usual $\eta - \eta'$ mixing angle. This is also equivalent to the statement that the $\eta - \eta'$ couplings are also determined by the quark charges, with the additional assumption of pure nonet symmetry

$$f_{\eta_1} = f_{\eta_8} = f_\pi \,. \tag{11}$$

Eqs. (10), using $\theta_p = -11^\circ$ from the naive mass formula, predict[16] $\Gamma(\eta \to \gamma\gamma) = 0.39$ keV, $\Gamma(\eta' \to \gamma\gamma) = 6.3$ keV in very good agreement with the experimental values $\Gamma(\eta \to \gamma\gamma) = (0.32 \pm 0.05)$ keV and $\Gamma(\eta' \to \gamma\gamma) = (5.9 \pm 1.6)$ keV[10]. These results definitively favour the fractionally-charged quark model with respect the integral-charged quark model, which predicts for $\eta' \to \gamma\gamma$ a decay rate four times larger. A possible violation of the condition (11), as recently discussed by Chanowitz[17], can however be checked by studying the large q^2 dependence of the $\eta - \eta'$ form factors, in analogy to eq. (9). Furthermore the value of the mixing angle $\theta_p = -11^\circ$ is also in agreement with all other $SU(3)$ radiative decays involving the $\eta - \eta'$ mesons[18].

Finally, for later purposes, let us observe that the sum

$$\sum_{\pi^0, \eta, \eta'} \frac{\Gamma(P_i \to \gamma\gamma)}{m_i^3} = \frac{4}{3} \frac{\Gamma(\pi^0 \to \gamma\gamma)}{m_{\pi^0}^3} = \frac{e^4}{16\pi} g_{\pi^0\gamma\gamma}^2 \tag{12}$$

is independent both of the pseudoscalar masses and the mixing angle θ_p.

4. - SCALAR AND TENSOR MESONS PRODUCTION.

The physics of $\gamma\gamma \to \pi\pi$ has been discussed in many places in the literature[7]. We shall only study here the strong interaction modifications due to the production of scalar and tensor resonances, also in order to estimate the resonant contribution to the total $\gamma\gamma$ cross section.

Let us consider first the $\sigma\gamma\gamma$ vertex, where σ indicates a scalar SU(3) singlet. Defining

$$T_\sigma^{\mu\nu}(q_1, q_2) = iA^{\mu\nu} F_\sigma(q_1^2, q_2^2) + i A'^{\mu\nu} F_\sigma'(q_1^2, q_2^2) \, , \tag{13}$$

with

$$A_{\mu\nu} = Q^2 P_\mu P_\nu + P^2 Q_\mu Q_\nu - (P \cdot Q)(P_\mu Q_\nu + P_\nu Q_\mu) + \left[(P \cdot Q)^2 - Q^2 P^2\right] g_{\mu\nu} \, ,$$

$$A'_{\mu\nu} = -\frac{1}{4} P_\mu P_\nu + Q_\mu Q_\nu + \frac{1}{2}(P_\mu Q_\nu - P_\nu Q_\mu) - (Q^2 - \frac{1}{4} P^2) g_{\mu\nu} \, , \tag{14}$$

the two form factors $F_\sigma(q_1^2, q_2^2)$ and $F_\sigma'(q_1^2, q_2^2)$ are related to the two independent helicity amplitudes as

$$T_\sigma^{++} = i(\nu^2 - m^2 Q^2)F_\sigma - i(Q^2 - \frac{1}{4} m_\sigma^2)F_\sigma' \, , \quad T_\sigma^{00} = i\sqrt{(Q^2 + \frac{1}{4} m_\sigma^2)^2 - \nu^2} \; F_\sigma' \, , \tag{15}$$

with $\nu = P \cdot Q$. The coupling constant $g_{\sigma\gamma\gamma}$ for the $\sigma \to \gamma\gamma$ decay is given by

$$g_{\sigma\gamma\gamma} = \frac{1}{4}\left[m_\sigma^2 F_\sigma(0,0) + 2 F_\sigma'(0,0)\right] \, , \tag{16}$$

with

$$\Gamma(\sigma \to \gamma\gamma) = \frac{e^4}{16\pi} g_{\sigma\gamma\gamma}^2 m_\sigma^3 \, . \tag{17}$$

By use of the canonical trace anomaly of the energy momentum tensor Crewther and Chanowitz and Ellis[19] have obtained the low energy theorem

$$F_\sigma'(0,0) = \frac{R}{6\pi^2 f_\sigma} \, , \tag{18}$$

with the usual definition of $R \equiv \sigma(e\bar{e} \to \text{hadrons})/\sigma(e\bar{e} \to \mu\bar{\mu})$. In deriving eq. (18) it has been assumed that the scalar meson σ dominates the trace of the energy momentum tensor, which define f_σ as

$$\langle\sigma(P)|\,\theta_{\mu\nu}(0)\,|0\rangle = \frac{i}{3} f_\sigma (m_\sigma^2 g_{\mu\nu} - P_\mu P_\nu) \, . \tag{19}$$

Assuming a smooth extrapolation to $q^2 = 0$ of the asymptotic relation $m_\sigma^2 F_\sigma = F_\sigma'$, obtained[20] for $q_1^2 = q_2^2 \to \infty$ one finally has

$$g_{\sigma\gamma\gamma} = \frac{R}{8\pi^2 f_\sigma} \,, \tag{20}$$

in exact analogy to the π^0 case (eq. (7)). The estimates for f_σ are in the range $f_\sigma \sim 150$ MeV, for $\Gamma(\epsilon = \sigma \to \pi\pi) \sim 400$ MeV, and $f_\sigma \sim f_\pi \sim 100$ MeV, for $\Gamma(\epsilon \to \pi\pi) \sim \sim 700$ MeV. Using $R \simeq 2.5$, $f_\sigma \sim 100$ MeV and, for comparison with the FESR estimates given below, $m_\sigma \sim 700$ MeV, one gets $\Gamma(\sigma \to \gamma\gamma) \sim 6$ keV. It is obvious that this result strongly depends on m_σ. However the ratio $\Gamma(\sigma \to \gamma\gamma)/m_\sigma^3$, on which we will come back later, should be rather stable, within a factor of two.

Previous estimates of $\Gamma(\sigma \to \gamma\gamma)$ have been obtained by using finite energy sum rules (FESR) and duality. More in detail the product $g_{\sigma\gamma\gamma} g_{\sigma\pi\pi}$ of coupling constants has been estimated from various authors[21], the results depending however on the way the sum rules are saturated. Then using $m_\sigma \sim 700$ MeV, $\Gamma(\sigma \to \pi\pi) \sim \sim 200\text{-}400$ MeV one finds $\Gamma(\sigma \to \gamma\gamma) \sim 6\text{-}22$ keV.

More recently non-relativistic quark models have also been applied[22] to light mesons decaying into two photons. The results however are rather questionable for the strong assumption of the non-relativistic treatment of bound states of n, d quarks. Values $\Gamma(\sigma \to \gamma\gamma) \sim 8.4\, G$ keV and $\Gamma(S^* \to \gamma\gamma) \sim 13\, B$ keV are found[2, 3].

The experimental verification of the above predictions through the direct measurement of the $\pi\pi$ production cross section seems hardly feasible since the ϵ is a very broad resonance. A similar conclusion applies also to the S^*-resonance in the reaction $\gamma\gamma \to K\bar{K}$, because the S^* is too much close to the $K\bar{K}$ threshold.

The situation is much neater in the case of tensor mesons. Nice evidence for f production has been in fact reported to this Conference by the PLUTO and TASSO Collaborations[9] at PETRA and Mark II[8] at SPEAR. Various theoretical estimates for $f \to \gamma\gamma$ are also quite consistent each other.

Assuming tensor meson dominance for the energy momentum tensor, together with vector meson dominance for the electromagnetic current, Renner[24] has estimated $\Gamma(f \to \gamma\gamma) \simeq 8$ keV.

A similar value, $\Gamma(f \to \gamma\gamma) \sim 6$ keV has been found by Schrempp-Otto, Schremmp and Walsh[21] by use of FESR. They also predict $|T_{+-}|^2 \gg |T_{++}|^2$, namely the f couples mainly to two photons of opposite helicity. This can be simply tested by measuring the angular distribution of the two pions produced in the f decay, which is expected $\propto \sin^4\theta$. The same conclusion has been drawn by Grassberger and Kögerler[25] on the basis of various sum rules of the type discussed below. A value $\Gamma(f \rightarrow \gamma\gamma) \sim 12$ keV, with $|T_{+-}|^2 : |T_{++}| = 6 : 1$, has been also more recently predicted in refs. (26).

The following superconvergence relation for real photons has been discussed extensively in the literature[27]:

$$\int_0^\infty \frac{dv}{v} \left[\sigma_{++}(v) - \sigma_{+-}(v) \right] = 0 . \tag{21}$$

Furthermore it has been checked[12] that the box diagram, considered as a possible source of non-Regge terms, as discussed in detail in the next section, does not affect the validity of (21). Then resonance saturation of the sum rule with low lying pseudoscalar, scalar and tensor mesons leads to

$$\sum_P \frac{\Gamma(P_i \rightarrow \gamma\gamma)}{m_i^3} + \sum_S \frac{\Gamma(S_i \rightarrow \gamma\gamma)}{m_i^3} + 5 \sum_T \frac{\Gamma(T_i \rightarrow ++)}{m_i^3} \simeq 5 \sum_T \frac{\Gamma(T_i \rightarrow +-)}{m_i^3} . \tag{22}$$

Then eqs. (12), (17), (20) and the naive quark model lead to the values $\Gamma(f \rightarrow \gamma\gamma) \simeq$ $\simeq 9$ keV, for $|T_{++}|^2 \ll |T_{+-}|^2$. Notice that the pseudoscalar and scalar contributions are in the ratio $4 : 6$. Therefore if only pseudoscalars are taken into account in eq. (22) one gets the lower bound $\Gamma(f \rightarrow \gamma\gamma) \gtrsim 3.6$ keV.

In the light of the above argument the value $\Gamma(f \rightarrow \gamma\gamma) = (2.3 \pm 0.5)$ keV, obtained by the PLUTO collaboration[9], seems rather low, and should be therefore confirmed by the other experiments at PETRA and SPEAR before definite conclusions can be drawn. It should be pointed out however that a value $\Gamma(f \rightarrow \gamma\gamma) \sim 2.6$ keV is predicted[22] by the non-relativistic quark models mentioned above, in connection with the ϵ resonance. The same kind of criticism of course applies also here.

5. - TOTAL CROSS SECTION.

Very exciting data from PLUTO and TASSO collaborations have been presented to this Conference. They show a rather large hadronic production at low c.m. energies which fall rapidly down to a constant value of about 0.25 μb, in excellent agreement with the prediction[3, 4, 5] of factorization of the cross section at high energy (universal Pomeron coupling)

$$\sigma_{\gamma\gamma}(s) \rightarrow \sigma_0 = \frac{\left[\sigma_T(\gamma N)\right]^2}{\left[\sigma_T(NN)\right]} \sim 0.24 \ \mu b . \tag{23}$$

Early predictions of the energy dependent component of $\sigma_{\gamma\gamma}(s)$ are also based on the usual tools of hadronic physics, specifically, resonance-Regge duality and factorization. This leads[28] to

$$\sigma_{\gamma\gamma}(s) = \sigma_o + \sigma_1(s) \simeq 0.24 \ \mu b + \frac{0.27}{\sqrt{s}} \ \mu b \ GeV \ . \tag{24}$$

However if one compares the integrated yield of $\sigma_1(s)$ in the low energy domain with the resonance estimates of the preceding sections, in particular the scalar and tensor contributions to $\sigma_{\gamma\gamma}^{Res}$, one finds[12] a large discrepancy by more than a factor of 3 :

$$\int_{\sim m_\varrho^2}^{\sim 3 \ GeV^2} \frac{ds}{s} \ \sigma_1(s) \lesssim \frac{1}{3} \sum_{0^+,2^+} \int \frac{ds}{s} \ \sigma_{\gamma\gamma}^{Res}(s) \ . \tag{25}$$

This result suggest that simple resonance $\longleftrightarrow$ Regge duality might not be valid in $\gamma\gamma$ scattering. A good reason for that is the plausible presence of non-Regge terms (fixed poles, Knönecker delta singularities, ...) in the absorptive part of the current current elastic amplitude, in contrast to the case of current scattering off a hadron where only the real part of the amplitude is affected by these non-Regge terms.

The box diagram (one quark loop) has been suggested[12] as a simple model for non-Regge terms in $\gamma\gamma$ scattering, on the basis of an explicit calculation of forward scattering of charged SU(2) currents where it has been found to consistently satisfy the appropriate Ward identities.

Then writing

$$\sigma_{\gamma\gamma}^{tot}(s) \simeq \sigma_{\gamma\gamma}^{Regge}(s) + \sigma_{\gamma\gamma}^{box}(s) \tag{26}$$

with $\sigma_{\gamma\gamma}^{Regge}(s)$ given by eq. (24) and

$$\sigma_{\gamma\gamma}^{box}(s) \sim \frac{4\pi\alpha^2}{s} \sum_i Q_i^4 \ \ln(\frac{s}{m_{q_i}^2}) \ , \qquad (s \gg m_{q_i}^2) \tag{27}$$

one obtains a rough estimate for $\sigma_{\gamma\gamma}^{tot}$, which however approximates quite well the experimental results. This is shown in Fig. 1, where the data from PLUTO are compared with eqs. (26-27) with $\sum Q_i^4 = 2/3$ and $m_q \simeq 300$ MeV.

Notice that, in contrast to what happens in e^+e^- annihilation in the one photon channel, the charm contribution to $\sigma_{\gamma\gamma}^{tot}(s)$ through eq. (27) is strongly depressed relatively to the flat background, because of the $(1/s)$ factor.

A careful theoretical reanalysis of the resonant-Regge terms in eq. (24), as

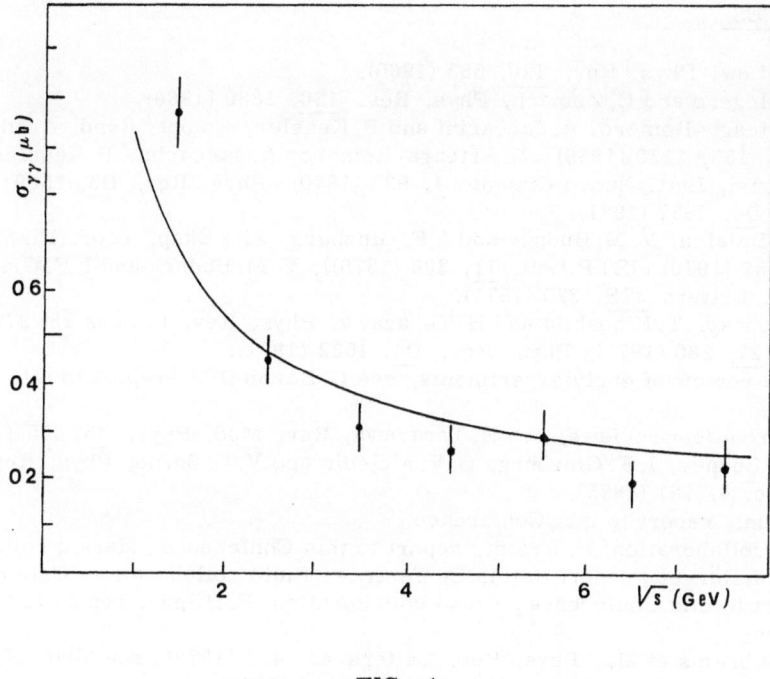

FIG. 1

well as a more detailed study of the energy dependence of the measured $\sigma_{\gamma\gamma}^{tot}$ are highly desirable in order to draw more quantitative conclusions on the relative importance of the various components of the total cross section. Nevertheless, we can conclude that the agreement obtained so far between theory and experiments is very encouraging and hope that this new generation of data will stimulate fresh theoretical ideas in $\gamma\gamma$ physics.

REFERENCES.

(1) - F.E.Low, Phys. Rev. $\underline{120}$, 582 (1960).

(2) - F. Calogero and C. Zemach, Phys. Rev. $\underline{120}$, 1860 (1960).

(3) - N. Arteaga-Romero, A. Jaccarini and P. Kessler, Compt. Rend. Acad. Sci. $\underline{269B}$, 153, 1129 (1969); N. Arteaga-Romero, A. Jaccarini, P. Kessler and J. Parisi, Lett. Nuovo Cimento $\underline{4}$, 933 (1970); Phys. Rev. $\underline{D3}$, 1569; Phys. Rev. $\underline{D4}$, 2927 (1971).

(4) - V.E. Balakin, V.M. Budnev and I.F. Ginzburg, Zh. Eksp. Teor. Fiz. Pis'ma $\underline{11}$, 559 (1970) (JETP Lett. $\underline{11}$, 388 (1970)); V.M. Budnev and I.F. Ginzburg, Phys. Letters $\underline{37B}$, 320 (1971).

(5) - S. Brodsky, T. Kinoshita and H. Terazawa, Phys. Rev. Letters $\underline{25}$, 972 (1970); ibid. $\underline{27}$, 280 (1971); Phys. Rev. $\underline{D4}$, 1532 (1971).

(6) - For a review of early experiments, see G. Barbiellini, report to this Conference.

(7) - For excellent reviews, see H. Terazawa, Rev. Mod. Phys. $\underline{45}$, 615 (1973); V.M. Budnev, I.F. Ginzburg, G.V. Meledin and V.G. Serbo, Phys. Reports $\underline{15}$, no. 4, 181 (1975).

(8) - P. Jenni, report to this Conference.

(9) - Jade collaboration: H. Wriedt, report to this Conference; Mark J collaboration: J. Bürger, report to this Conference; Pluto collaboration: C. Berger, report to this Conference; Tasso collaboration: E. Hilger, report to this Conference.

(10) - G.S. Abrams et al., Phys. Rev. Letters $\underline{43}$, 477 (1979); see also ref.(8).

(11) - D.M. Binnie et al., Phys. Letters $\underline{83B}$, 141 (1979).

(12) - M. Greco and Y. Srivastava, Nuovo Cimento $\underline{43A}$, 88 (1978).

(13) - V. Budnev, V. Chernyak and I. Ginzburg, Nuclear Phys. $\underline{34B}$, 470 (1971).

(14) - S.L. Adler, Phys. Rev. $\underline{177}$, 2426 (1969); J.S. Bell and R. Jackiev, Nuovo Cimento $\underline{60A}$, 47 (1969).

(15) - J.D. Bjorken, Phys. Rev. $\underline{148}$, 1467 (1966); K. Johnson and F.E. Low, Progr. Theor. Phys. Suppl. $\underline{37-38}$, 74 (1966).

(16) - A. Bramon and M. Greco, Phys. Letters $\underline{48B}$, 137 (1974).

(17) - M.S. Chanowitz, Phys. Rev. Letters $\underline{44}$, 59 (1980).

(18) - For a review, see M. Greco, Proceedings of the XII Rencontre de Moriond, ed. by Tran Than Van, Flaine (1977).

(19) - J. Crewther, Phys. Rev. Letters $\underline{28}$, 1421 (1972); M.S. Chanowitz and J. Ellis, Phys. Rev. $\underline{D7}$, 2490 (1973).

(20) - M. Greco and H. Inagaki, Phys. Letters $\underline{65B}$, 267 (1976).

(21) - A.Q. Sarker, Phys. Rev. Letters $\underline{28}$, 1421 (1970); A. Bramon and M. Greco, Lett. Nuovo Cimento $\underline{2}$, 522 (1971); B. Schrempp-Otto, F. Schrempp and T.F. Walsh, Phys. Letters $\underline{36B}$, 463 (1971).

(22) - V.M. Budnev and A.E. Kaloshin, Phys. Letters $\underline{86B}$, 351 (1979).

(23) - The quantities C and B are defined as $C = (\sin\theta_s - 2^{-1/2}\cos\theta_s)^2$ and $B = \sin\theta_s - 2^{-3/2}\cos\theta_s)^3$ where θ_s is the $\epsilon - S^*$ mixing angle. It has been also assumed $m_\epsilon = 1300$ MeV.

(24) - B. Renner, Phys. Letters $\underline{33B}$, 599 (1970).

(25) - P. Grassberger and R. Kögerler, Nuclear Phys. $\underline{106B}$, 451 (1976).

(26) - M. Greco, Nuovo Cimento $\underline{42A}$, 315 (1977); W.N. Cottingham and I.H. Dunbar, J. Phys. $\underline{G5}$, L155 (1979).

(27) - P. Roy, Phys. Rev. $\underline{D9}$, 2631 (1974); V. Budnev, I. Ginzburg and V. Serbo, Lett. Nuovo Cimento $\underline{7}$, 13 (1974); S. Gerasimov and J. Moulin, Nuclear Phys. $\underline{98B}$, 349 (1975); P. Grassberg and R. Kögerler, ref.(24); M. Greco and Y. Srivastava, ref.(12).

(28) - J. Rosner, BNL preprint CRISP 71-26 (1971), (unpublished; S. Brodsky, Proceedings of the Intern. Coll. on Photon-Photon Collisions, Paris 1973, J. Physique, C-2, Suppl. $\underline{3}$, 69 (1974).

TWO-PHOTON COLLISIONS AND QCD[*]

J. F. Gunion
University of California at Davis
Physics Department, Davis, California 95616

and

Stanford Linear Accelerator Center
Stanford University, Stanford, California 94305

Abstract

A critical review of the applications of QCD to low- and high-p_T interactions of two photons is presented. The advantages of the two-photon high-p_T tests over corresponding hadronic beam and/or target tests of QCD are given particular emphasis.

Results for the two-photon interactions are now becoming available from experiments at PETRA and two-photon experiments at PEP will soon begin. Thus, it seems an opportune time to review and assess the implications of current theoretical ideas, especially of quantum chromodynamics, for photon-photon collisions. Particular emphasis will be placed upon the unique characteristics of a photon target, photon beam combination as compared to the hadron target situation. A brief outline of the talk is:

I)   Remarks on low-p_T physics in $\gamma\gamma$ collisions.

II) Review of the salient features of high-p_T hadron target collisions -- problems and ambiguities.

III) High-p_T 2γ physics.

    1. Inclusive jet and single particle production

        (a) 2-jet topology.

        (b) 3-jet topology.

        (c) 4-jet topology.

        (d) Higher twist and vector dominance backgrounds, and the importance of single particle spectra.

    2. Exclusive reactions.

IV) Conclusions.

I have attempted to keep the number of equations minimal and emphasize physical tests and conclusions. This talk does not cover the deep inelastic photon target situation, reviewed by T. Walsh at this conference.

---

[*] Work supported in part by the Department of Energy under contract DE-AC03-76SF00515 and the A. P. Sloan Foundation.

I. Low-p_T Physics in 2γ Collisions

The fundamental observable in 2γ collisions at low-p_T is the total cross section, $\sigma_{\gamma\gamma}$. Various guesses as to the form of this cross section have appeared in the literature over the years.[1] I will present a point of view which differs slightly from those given earlier.

Most workers agree that $\sigma_{\gamma\gamma}$ cannot be obtained entirely from vector dominance. The photons have additional point-like couplings which presumably yield an addition to $\sigma_{\gamma\gamma}$ above the vector dominance estimate. These point-like couplings can, in particular, lead to fixed-pole ($J = 0$) contributions to the large s limit of $\sigma_{\gamma\gamma}$. A typical vector dominance estimate of $\sigma_{\gamma\gamma}$ would include only the diagrams of Fig. 1. There we employ the Low-Nussinov[2] gluon exchange model for the Pomeron and represent Regge exchanges as due to quark exchange (as in the duality approach) with all possible gluon radiative corrections.

The point-like coupling diagrams of Fig. 2 are not included in the VMD diagrams of Fig. 1. Of course, interference terms between vector-dominated (V) type couplings and point-like (PL) couplings are also possible. An estimate for the Pomeron contribution to $\sigma_{\gamma\gamma}$, which includes both V and PL couplings can be obtained from factorization,

$$\lim_{s \to \infty} \sigma_{\gamma\gamma} = \lim_{s \to \infty} \frac{\sigma_{\gamma p}^2}{\sigma_{pp}} \sim 240 \text{ nb} \tag{1.1}$$

since the Pomeron contribution to the γp cross section should contain both V and PL γ couplings. Certainly in the gluon-exchange Pomeron model the above result is explicitly correct. An estimate for the Regge contribution[1] from VMD is

$$\sigma_{VV}^{\text{Non-Pomeron}} \sim \frac{270 \text{ nb}}{\sqrt{s(\text{GeV}^2)}} \tag{1.2}$$

Experimentally it is clear that such a Regge contribution is too small to describe the data[3] which prefers a fit with

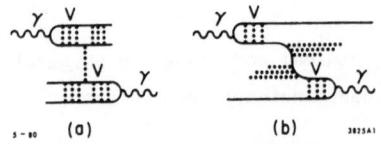

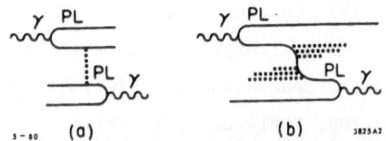

Fig. 1. Vector dominance diagrams for $\sigma_{\gamma\gamma}$: (a) Pomeron or gluon exchange; and (b) Regge exchange, represented by radiatively modified quark exchange.

Fig. 2. Purely point-like photon couplings to: (a) the Pomeron; and (b) a typical Reggeon.

$$\sigma_{\gamma\gamma}^{\text{Non-Pomeron}} \sim \frac{840 \text{ nb}}{\sqrt{s(\text{GeV}^2)}} \qquad (1.3)$$

or else a smaller Regge term with some $J = 0$ fixed-pole contribution

$$\sigma_{\gamma\gamma}^{\text{Non-Pomeron}} \sim \frac{A}{\sqrt{s}} + \frac{B}{s} \qquad (1.4)$$

Theoretically

$$\sigma_{\gamma\gamma}^{\text{Non-Pomeron}} = \sigma_{VV}^{\text{Non-Pomeron}} + 2\sigma_{V \, PL}^{\text{Non-Pomeron}} + \sigma_{PL \, PL}^{\text{Non-Pomeron}} \qquad (1.5)$$

will clearly be larger than the vector-dominance estimate, but cannot be explicitly calculated. It has, however, been suggested that estimates using finite energy sum rules and duality may be possible. One such scheme[1] is to make the fixed-pole plus vector-dominated Regge contributions to $\sigma_{\gamma\gamma}^{\text{Non-Pomeron}}$ dual to the low energy resonances; this implicitly assumes that the point-like photon components couple to resonances but not to Regge terms. It seems more reasonable to me that the point-like photon resonance component is dual to a point-like Regge component. This is illustrated in Fig. 3(a). Corresponding statements apply to mixed vector-dominated and point-like couplings; an example is given in Fig. 3(b). Thus one would maintain a standard form for the finite energy sum rule

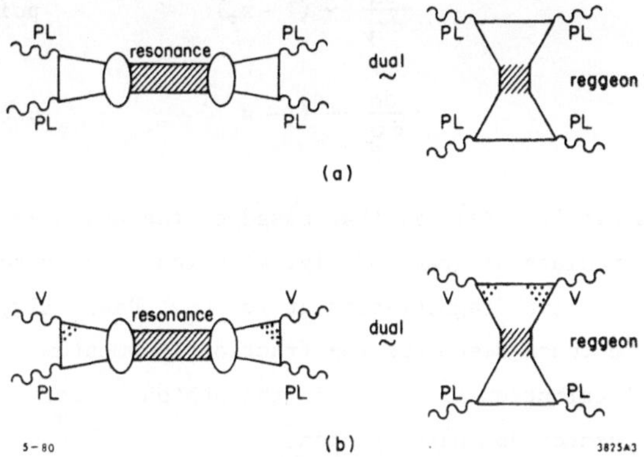

(a)

(b)

Fig. 3. Duality for $\sigma_{\gamma\gamma}$ -- each $\gamma\gamma \rightarrow$ Resonance $\rightarrow \gamma\gamma$ diagram has its corresponding Reggeon diagram: (a) for all point-like (PL) couplings; and (b) for a sample mixed PL, vector-dominated (V) coupling diagram.

$$\int_{s_0}^{s_1} \sigma_{\gamma\gamma}^{\text{Resonance}} \frac{ds}{s} = \int_{s_0}^{s_1} \sigma_{\gamma\gamma}^{\text{Regge}} \frac{ds}{s} \qquad (1.6)$$

and the fact that the experimentally determined left-hand, resonance side of (1.6) is larger than the contribution to the right-hand side coming from pure vector dominance, $\sigma_{VV}^{\text{Regge}}$ (Eq. (1.2)), is easily explained by the additional $\sigma_{V \, PL}^{\text{Regge}}$ terms of (1.5) which should be included in (1.6). The Regge fit, (1.3), is such that (1.6) is approximately satisfied. Ideally one would like to calculate the left-hand side of (1.6) with high precision (perhaps possible as more resonance data becomes available), determine $\sigma_{\gamma\gamma}^{\text{Regge}}$ from (1.6) (i.e., determine A in Eq. (1.4)) and then use the experimental measurement of $\sigma_{\gamma\gamma}^{\text{Non-Pomeron}}$ to decide if a fixed-pole contribution (B/s in Eq. (1.4)) is present. Certainly one cannot rule out a small fixed-pole term at the moment. However, there seems to be no obvious justification for identifying the coefficient, B, of the fixed-pole 1/s behavior with the coefficient of 1/s obtained from the simple bare-fermion-loop box diagram. One must consider simultaneously all diagrams with gluon and/or fermion bubble corrections to the bare fermion loop. The $s \to \infty$ limit of the sum of all such diagrams should then be separated into Regge and fixed-pole terms; this calculation is not possible without first essentially solving the confinement problem.

An interesting question is whether there is any direct way of exposing the point-like component of the photon using low p_T observations. In fact most workers[4,5] who have examined the fragmentation $\gamma \to$ fast meson, M agree that a point-like photon component will result in a softer-than-expected spectrum at high x_F $\left(x_F = (E_M + P_M^z)/ (E_M + P_M^z)_{max} \right)$:

$$\frac{dN}{dx_F} \sim (1 - x_F)^0 \qquad \text{point-like photon}$$

$$(1.7)$$

$$\frac{dN}{dx_F} \sim (1 - x_F)^1 \qquad \text{vector-dominated photon}$$

A sample model[4] is that based on the QCD brem-strahlung diagrams of Fig. 4, which illustrate $\gamma \to$ meson fragmentation at low p_T. Basically, the meson remembers the fractional momentum distribution of quarks in the photon.[6] For a vector-dominated photon[7]

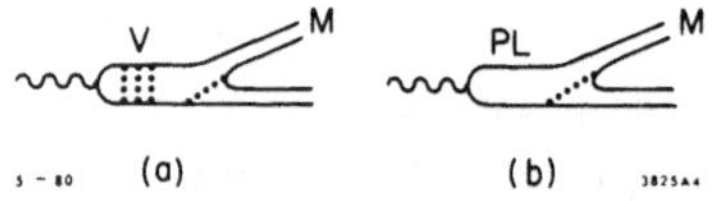

Fig. 4. Bremsstrahlung diagrams for $\gamma \to$ meson, M: (a) for the vector-dominated photon; and (b) for the point-like photon.

$$G_{q/\gamma_V}(x) \sim (1-x)^1 \qquad (1.8a)$$

whereas for a point-like photon

$$G_{q/\gamma_{PL}}(x) \sim (1-x)^0 \qquad (1.8b)$$

The weaker suppression predicted for a point-like photon is apparently present[9] in $\gamma p \to$ fast meson, M, where $M = \pi^+$ or π^- is observed in the photon fragmentation region. The same photon fragmentation spectrum should be observed in $\gamma\gamma$ collisions.

In summary, even at low p_T, we already have and can expect to find additional evidence in $\gamma\gamma$ collisions for a non-vector-dominated component of the photon. Such evidence is, however, less direct than that which can be obtained from $\gamma\gamma$ collisions at high transverse momentum.

II. Problems and Ambiguities in High-p_T Hadron Collisions

First let us recall the salient features of high-p_T collisions. The typical structure of a high-p_T inclusive jet cross section is illustrated in Fig. 5. It has been proven[10] that, in leading log, the correct procedure is to compute the jet cross section by convoluting p_T^2 dependent distribution functions for the secondaries, a and b coming from A and B respectively, which participate directly in the hard scattering subprocess $a+b \to c+d$, with the cross section for that subprocess: either c or d can form the observed jet. Thus one has for c = jet

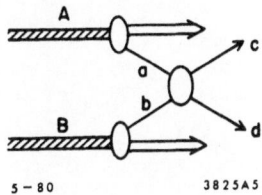

5 - 80   3825A5

Fig. 5. High-p_T jet production.

$$E_c \frac{d\sigma}{d^3 p_c} = \frac{1}{\pi} \int dx_a\, dx_b\, G_{a/A}(x_a, p_T^2)\, G_{b/B}(x_b, p_T^2) \frac{d\sigma^{a+b \to c+d}}{dt'}\, s'\delta(s'+t'+u')$$

$$(2.1)$$

where $s' = x_a x_b s$, $t' = x_a t$, $u' = x_b u$, $p_T^2 = \frac{u't'}{s'} = \frac{ut}{s}$, and s, t and u are the Mandlestam invariants appropriate to A+B $\to$ c+X. The distribution functions have the standard scale-breaking behavior and can presumably be measured via deep inelastic and/or massive μ-pair production.

The subprocess cross sections $\frac{d\sigma^{a+b \to c+d}}{dt'}$ are computable in the appropriate large s', fixed angle limit in QCD even when some of the participating particles are non-elementary.[11] The general form of the result is,[†] up to possible anomalous dimension powers of log p_T^2/Λ^2,

$$\left. \frac{d\sigma^{a+b \to c+d}}{dt'} \right|_{\substack{\text{fixed} \\ \text{angle}}} \sim \frac{1}{s'^{N-2}} \left[\alpha_s(p_T^2) \right]^{N-2} f_{abcd}(\theta'_{c.m.})$$

$$(2.2)$$

where in (2.1) s' is of order $4p_T^2$ and $N = n_a + n_b + n_c + n_d$ is the total number of elementary constituents participating in the subprocess. The strong coupling constant has the standard form

$$\alpha_s(p_T^2) = \frac{4\pi}{\left(11 - \frac{2}{3}n_f\right) \log p_T^2/\Lambda^2}$$

$$(2.3)$$

Scaling laws of this type are directly testable in exclusive scattering. For instance, ignoring the α_s variation, one predicts and observes

---

[†] Here we use a somewhat simplified expression. In actuality a given QCD diagram has α_s's evaluated at various fractions of p_T^2 (a given internal gluon typically transfers only part of the overall momentum transfer). Strictly speaking, though, such corrections are part of the next-to-leading log correction to the leading log result (2.2).

$$\frac{d\sigma}{dt}\bigg|_{\substack{\text{fixed}\\\text{angle}}} \underset{\sim}{\overset{s\to\infty}{\longrightarrow}} \begin{cases} \dfrac{1}{s^{10}} & ; \; pp \to pp \\[2ex] \dfrac{1}{s^{8}} & ; \; \pi p \to \pi p \\[2ex] \dfrac{1}{s^{7}} & ; \; \gamma p \to \pi p \\[2ex] \dfrac{1}{s^{6}} & ; \; \gamma p \to \gamma p \end{cases} \tag{2.4}$$

In addition, the constituent interchange[12] (CIM) class of QCD diagrams (examples appear in Fig. 6) describe the angular dependence and crossing properties of such exclusive reactions extremely well. The normalizations in most cases have not been computed from first principles yet because of the large number of contributing graphs but the simpler diagrams for the pion and proton form factors have been computed and normalized. As an example consider $G_M^{proton}(Q^2)$. One obtains[13]

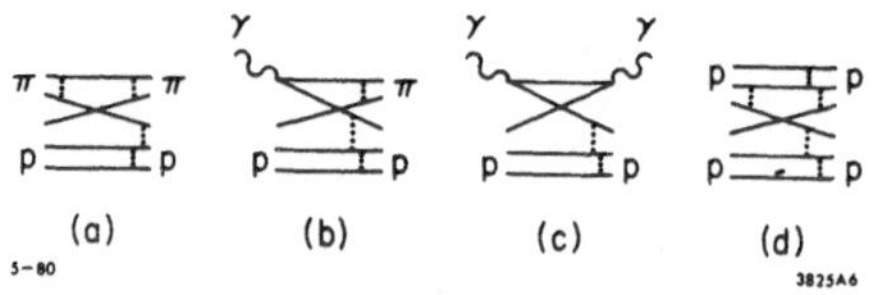

Fig. 6. Examples of CIM type QCD diagrams for: (a) $\pi p \to \pi p$; (b) $\gamma p \to \pi p$; (c) $\gamma p \to \gamma p$; and (d) $pp \to pp$.

$$G_M^{proton}(Q^2) \sim \frac{32\pi^2}{9Q^4} \alpha_s^2(Q^2) \sum_{n,m=0}^{\infty} b_{n,m}\left[\log Q^2/\Lambda^2\right]^{-\gamma_n-\gamma_m} \tag{2.5}$$

where the b_n are computable given certain proton wave function information and γ_n is the standard anomalous dimension appearing in deep inelastic scattering. This expression agrees very well with data provided $\Lambda^2 < .01$ GeV^2 (i.e., well below the "standard" $\Lambda^2 = .25$ GeV^2 value). This is illustrated in Fig. 7, taken from Ref. 11.

Focusing for a moment on the reactions involving photons in (2.4), it should be remarked that the experimentally observed decrease in the inverse s power from $1/s^8$ to $1/s^7$ and $1/s^6$ as one proceeds from a reaction involving no photons to ones with one and two photons respectively is direct evidence for the point-like photon component. If the photon were purely vector-dominated $\pi p \to \pi p$, $\gamma p \to \pi p$ and $\gamma p \to \gamma p$ should all have the same fixed-angle s-dependence. Of course, the observed simple power laws should, theoretically, be modified by appropriate powers of $\alpha_s(p_T^2)$, unless α_s is slowly varying as when Λ is small. The requirement of small Λ becomes even more crucial for the $1/s^{10}$ prediction for $pp \to pp$ which agrees well with data but is accompanied by ten powers of $\alpha_s(p_T^2)$. Unless Λ is very small the extra variation with p_T^2 introduced in this way destroys this agreement.

While predictions for exclusive reactions in QCD appear to be successful, ignoring worries about the size of Λ as reflected in the variation of α_s, it is clear that inclusive reactions have the potential for probing more directly the most elementary QCD reactions such as $qq \to qq$, $qg \to qg$, etc. which with $N = 4$ yield, naively, $1/p_T^4$ behavior in inclusive high p_T scattering (s' in (2.2) converts roughly to $4p_T^2$ in the inclusive cross section (2.1)). Unfortunately, while the subprocesses important in inclusive reactions are simpler the theoretical ambiguities are more numerous. In comparing theory to experiment in the inclusive situation the following problems arise.

(1) The question of the exact subprocess cross section behavior, i.e., the uncertain size of Λ. This is the exact analogue of the exclusive reaction uncertainty.

(2) Corrections from next-to-leading-log terms, not incorporated in the leading order result ((2.1) + (2.2)). These corrections are in part to the distribution functions in (2.1) and in part to the subprocess

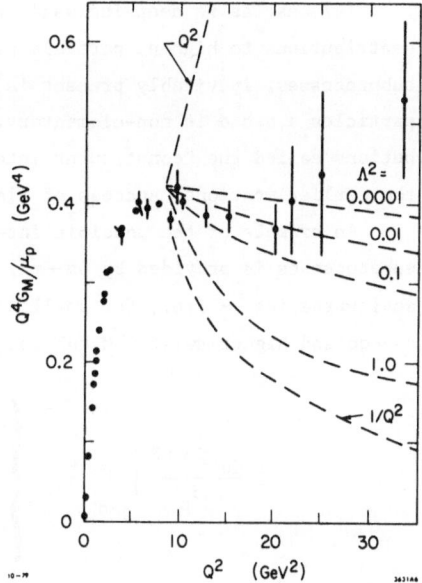

Fig. 7. Prediction for $Q^4 G_M^p(Q^2)$ for various QCD scale parameters Λ^2 (in GeV^2).

cross sections (2.2). The combined next order correction to inclusive jet production via the subprocess $qq \to qq$ has been found to be of order 50%-100%.[14] Next order corrections to the fixed angle predictions (2.4) also occur but have not yet been calculated. The size of all such non-leading-log terms decreases for smaller Λ.

(3) The distribution functions $G_{a/A}$ and $G_{b/B}$ are not so easily determined. Even for a and b being quarks the connection between deep inelastic data and the distribution function is not necessarily straight forward. In particular there are many indications and computations[15] which suggest that higher twist terms are an important component of the observed deep inelastic scale-breaking. This means that the distribution functions extracted from the deep inelastic data could have much less scale breaking than might be naively anticipated when assuming a moderate value of the scale breaking parameter Λ in (2.3). Substantial higher twist contributions lead to a smaller value of Λ as extracted from deep inelastic data and effect the exact shape and normalization of the distribution functions so obtained. Since the higher twist contributions decrease with Q^2 more rapidly than the leading scale-breaking terms an analysis at high Q^2 would not suffer from these ambiguities. Unfortunately this requires large Q^2 values over a large range in $\log Q^2$. This is not available; a recent analysis[16] which employs existing neutrino data, with a moderately high low-Q^2 cutoff does, in fact, obtain (with large errors) a lower Λ value than previous analyses, namely $\Lambda = 100 \pm 100$ MeV.

(4) Just as in deep inelastic scattering there may be substantial higher twist contributions to high-p_T particle production.[17] These arise from hard scattering subprocesses, inevitably present in QCD, in which one or more of the participating particles a,b,c,d is non-elementary. A specific model of such higher twist contributions called the "constituent interchange model" has been proposed[17] because of the earlier mentioned success of closely related diagrams for exclusive reactions.

An example of the possible interplay between higher twist and elementary QCD subprocesses is provided by $pp \to \pi X$. There one predicts (in the absence of substantial scale-breaking -- i.e., for small Λ) on the basis of elementary subprocesses such as $qq \to qq$ and higher twist CIM subprocesses such as $qM \to qM$ (M = meson) (Fig. 8)

$$E \frac{d\sigma^{pp \to \pi X}}{d^3 p}\bigg|_{90^\circ} \sim \begin{cases} \dfrac{1}{p_T^4} f(x_T) & qq \to qq \quad \text{etc.} \\[2em] \dfrac{1}{p_T^8} f'(x_T) & qM \to qM \end{cases} \qquad (2.6)$$

where $x_T = 2p_T/\sqrt{s}$ and p_T is the transverse π momentum. The higher twist $qM \to qM$ contribution is only significant because it is enhanced by the so-called trigger bias effect;[18] the particular subprocess $qM \to q\pi$ produces the π meson directly whereas it must appear as a q or g fragment for the more elementary contributions. The effect is a suppression of the elementary-QCD

$qq \to qq$ $gq \to gq$ $gg \to gg$
(a)

$M \quad M$ / $q \quad q$
(b)

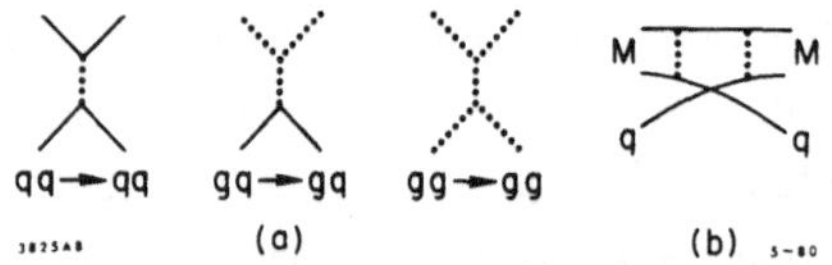

Fig. 8. Examples of QCD subprocess diagrams for: (a) elementary participants--quarks and/or gluons; and (b) $qM \to qM$ where M = meson.

subprocesses relative to the CIM subprocesses by a factor of 10 compared to their relative importance in producing jets.

We will return to jets in a moment but let us first look at n_{eff} defined by

$$E \frac{d\sigma^{pp \to \pi X}}{d^3 p}\bigg|_{90^\circ} \sim \frac{1}{p_T^{n_{eff}}} g(x_T) \qquad (2.7)$$

as extracted from data. A graph[19] obtained from ISR data is shown in Fig. 9. There appears to be a transition from higher n_{eff} powers near 8 to a lower n_{eff} power. If one looks at these $\sqrt{s} = 53$ GeV/$\sqrt{s} = 62$ GeV extractions this lower power is probably between 4 and 5. Clearly one might be tempted to say that there is, indeed, a mixture of CIM and elementary-QCD subprocesses. A superposition of the expressions in (2.6) does indeed describe this transition. The scenario is slightly different, however, if Λ is not small but of order $\Lambda = .5$ GeV. There is then scale breaking in the quark and gluon distribution functions and non-negligible α_s variation in the elementary-

QCD subprocesses. Both effects tend to increase n_{eff} so that in the experimental energy and x_T range n_{eff} should not have fallen below about 5.5.

In any case, one might be inclined to believe that CIM processes are at least responsible for the $n_{eff} \simeq 8$ region at lower x_T. This unfortunately, is also subject to debate. Feynman and Field[20] and others[21] claim that one should use on-shell quark-quark scattering but incorporate the effects of "smearing" over the intrinsic transverse momentum, k_T, of the quarks in the proton. This procedure requires parameters and cutoffs but probes the $t' \to 0$ singularities of the elementary-QCD $\frac{d\sigma}{dt'}$'s in such a way as to yield high n_{eff} values for these elementary subprocess contributions at moderate p_T. In a Feynman diagram sense, however, the initial quarks or gluons

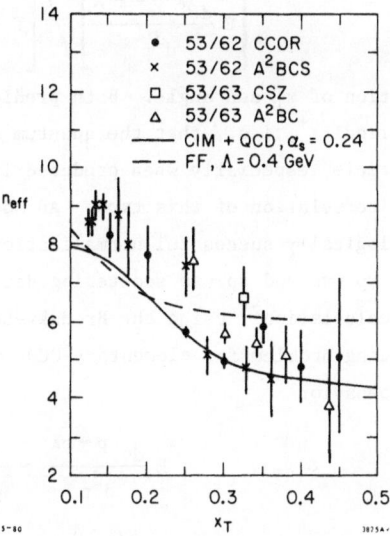

Fig. 9. Effective power n_{eff} dependence on x_T for π^0 production at the ISR. Only the highest energy $\sqrt{s} = 53$ GeV/$\sqrt{s} = 63$ GeV extraction is shown; it probes the highest p_T values. Lower energy extractions, e.g., $\sqrt{s} = 31$ GeV/$\sqrt{s} = 53$ GeV, will always yield higher n_{eff} values since they probe lower p_T values where higher twist terms are more important. Two sample n_{eff} predictions are shown. One uses a superposition of the two terms in (2.6) with no scale breaking. The other is from the Feynman-Field (FF) model with "k_T smearing" and the QCD scale breaking parameter $\Lambda = .4$ GeV.

entering the subprocess are necessary off-shell. This off-shellness would shield[22] the above $t' \to 0$ singularity and make it impossible to obtain high n_{eff} values from $qq \to qq$, $qg \to qg$, etc.; higher twist contributions would then be required to explain the lower p_T region of the data. Simultaneously triggering on two high-p_T particles on opposite sides of the beam axis can be shown[23] to eliminate any possible k_T "smearing" effects. As data becomes available regarding such symmetric triggers the question of higher-twist versus smearing as an explanation of moderate p_T data should be resolved. Correlations between the quantum numbers of the two symmetric particle triggers will also serve[23] to discriminate between the possible contributing subprocesses; elementary-QCD subprocesses lead to little correlation while higher twist-CIM subprocesses give rise to substantial correlations.

(5) Which of the many alternatives scenarios for n_{eff} considered above actually holds, if any, is not clear. In addition, there are other difficulties. Both elementary-QCD and CIM subprocesses, especially the latter, have difficulty[24] in describing the ratio

$$\left[E \, \frac{d\sigma^{\pi^- p \to \pi^- X}}{d^3 p} \right] \bigg/ \left[E \, \frac{d\sigma^{\pi^- p \to \pi^+ X}}{d^3 p} \right] \qquad (2.8)$$

as a function of x_T and angle. Both predict that the final π^- is more easily made than the final π^+, i.e., that the quantum numbers of the beam are transmitted to the final particle, especially when produced in the forward direction. The data shows[24] almost no correlation of this type. An additional problem for the CIM is that the phenomenologically successful normalization of the $qM \to qM$ subprocess, obtained from exclusive $\gamma p \to \pi p$ and $\pi p \to \pi p$ scattering data,[17] is much larger than that obtained in a recent calculation[25] using the Brodsky-Lepage techniques.[11] Balancing this is the long standing problem for elementary-QCD subprocesses of their failure to predict the observed behavior

$$E \, \frac{d\sigma^{pp \to pX}}{d^3 p} \sim \frac{1}{p_T^{12}} \, f(x_T) \qquad (2.9)$$

which is a natural result[17] in the CIM due to the existence of the subprocess $qp \to qp$ with fixed angle $1/s^6$ behavior (which translates into the $1/p_T^{12}$ of Eq. (2.9)). The behavior (2.9) suggests that the phenomenologically determined CIM subprocess normalizations are not unreasonable.

(6) As if all the above did not provide sufficient reason for failing to draw firm conclusions from existing high-p_T single particle data a recent "jet" experiment[26] fails to see jets. Only the elementary-QCD subprocesses should contribute significantly to jet cross sections. (The CIM subprocesses are no longer enhanced by the trigger bias effect which results when a single high-p_T particle is triggered on.) Asymptotically as $p_T^{jet} \to \infty$, one should see planar events with four jets, two at high-p_T on opposite sides of the beam axis and two along either beam axis. While the results are preliminary this planar structure is not currently observed when the experiment triggers on a total large transverse energy E_T on one side of the beam. Instead the total E_T is composed of many low-p_T particles in a variety of different azimuthal and rapidity locations. In addition the cross section is more than a factor of 50 above that predicted by the elementary-QCD subprocesses. One guesses that the typical event is not a single high-p_T hard scattering but rather a multiple scattering event, each of the multiple collisions being at low-p_T. Because the trigger is not forcing all the high-p_T momentum to be carried on a single particle (as in the single π trigger), the background has become overwhelming. Or perhaps the theory is wrong. At best, far larger values of E_T are required before the power-law-behaved single-hard-scattering cross sections will stand out above the background.

Even though hadron-hadron collisions producing a high-p_T hadron are still not unambiguously interpretable, it might be that photon+hadron $\to$ hadron or hadron+hadron $\to$ photon collisions might be beset by fewer problems. To some extent this is the case and it is at least possible to learn from these reactions some facts which will be relevant when we come to photon-photon collisions. Due to lack of space I will discuss only hadron-hadron collisions with production of a single high-p_T unaccompained photon. There has been an increasing amount of experimental data[27] in this area.

To keep the discussion simple consider the two principle competing processes in pp collisions; an elementary-QCD process $gq \to \gamma q$ yielding $1/p_T^4$ behavior for the inclusive cross section $E_\gamma(d\sigma/d^3p_\gamma)$ and a higher twist (CIM) process $Mq \to \gamma q$ yielding $1/p_T^6$ behavior at fixed x_T. If we compute the γ/π^o ratio it might be anticipated that some of the scale-breaking and other Λ dependent effects might cancel between the γ and π^o predictions. Naively,[28] i.e.; for small Λ one obtains as $p_T \to \infty$, $x_T = 2p_T/\sqrt{s}$ fixed

$$\frac{E_\gamma \dfrac{d\sigma^{pp \to \gamma X}}{d^3p_\gamma}}{E_\pi \dfrac{d\sigma^{pp \to \pi X}}{d^3p_\pi}} \propto \begin{cases} \dfrac{1}{(1-x_T)^{1 \text{ or } 2}} & \text{elementary-QCD} \\[4mm] p_T^2 & \text{higher twist-CIM} \end{cases} \qquad . \quad (2.10)$$

If elementary-QCD processes dominate both γ and π^o production the γ/π^o ratio should be p_T independent at fixed x_T and a decreasing function of $\sqrt{s}$ at fixed p_T. If higher twist (CIM) diagrams dominate both reactions the γ/π^o ratio should behave as p_T^2 (compare $1/p_T^6$ for $Mq \to \gamma q$ to $1/p_T^8$ for $Mq \to \pi^o q$) with little or no x_T (i.e., $\sqrt{s}$) dependence at fixed p_T. Including other less important subprocesses does not essentially alter the above comparison.

Experimentally the latest ISR data,[27] which compares $\sqrt{s} = 31$, 45 and 63 GeV, indicates a result much nearer the naive CIM expectation, see Fig. 10. The normalization of the γ/π^o ratio is roughly a factor of 2 below that originally predicted in Ref. 28, i.e., within the anticipated phenomenological normalization uncertainties. However, to a small extent "scale breaking"[29] and to a much larger extent "low k_T smearing"[30] do not entirely cancel out of the γ/π^o ratio. In addition Brodsky and I[31] have completed a preliminary computation of the $Mq \to \gamma q$ subprocess normalization in terms of the PCAC constant, f_π, following the techniques of Ref. 11 and find a result consistent with the $qM \to qM$ results of Ref. 25 -- namely a much smaller normalization than obtained by the earlier phenomenological techniques.[17,28] Thus, the interpretation of hadron-hadron production of high-p_T γ's even though relatively simpler than production of hadrons at high-p_T remains uncertain at this point in time. There is strong evidence for the point-like photon component but the precise subprocess mechanism by which it enters is not yet clear. Where, then, do we turn in order to obtain direct experimental verification of the existence of the most elementary QCD reactions, such as

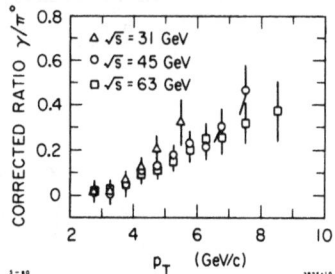

Fig. 10. Comparison of $\sqrt{s} = $ 31, 45 and 63 GeV data for γ/π^o at high-p_T. Data is from M. Diakonov et al., Ref. 27.

quark-quark scattering, and direct experimental information on the normalization of higher twist subprocesses? It appears that photon-photon collisions at high-p_T provide a much less ambiguous probe of both types of subprocess.

III. High-p_T Production in Photon-Photon Reactions[32-36]

III.1. Inclusive Jet and Single Particle Production

We now turn to a discussion of photon-photon collisions producing a high-p_T hadron or jet. Such reactions provide a probe of the same subprocesses as discussed in Section II, elementary-QCD and higher twist-CIM, but will be much less ambiguous in their interpretation. For the calculable point-like photon component: (a) there are many reactions which are predicted to be Λ independent in leading order; (b) there are no hadron wave function, smearing, etc. ambiguities; (c) all subprocess and wave function normalizations are explicitly calculable; (d) the contributions coming from the vector-dominated component of the photon can be shown to be small in most situations; and (e) in addition, different types of subprocesses can often be distinguished through their final state topology. We will divide up the high p_T 2γ subprocesses according to final state topology.

A. 2-Jet Processes[32-36]

In two-photon collisions the simplest subprocess is that illustrated in Fig. 11, $\gamma\gamma \rightarrow q\bar{q}$. The final state is exactly the same as that produced via annihilation (aside from the e^+ and e^- spectators) except that the q and $\bar{q}$ jets do not carry the full energy of the incoming e^+ and e^-. The same type of diagram can, of course, be drawn for $e^+e^- \rightarrow e^+e^-\mu^+\mu^-$. If the intermediate quark propagator in $\gamma\gamma \rightarrow q\bar{q}$ behaves just like the elementary muon propagator in $\gamma\gamma \rightarrow \mu^+\mu^-$ then the $\gamma\gamma \rightarrow q\bar{q}$ process yields (using the equivalent photon approximation)

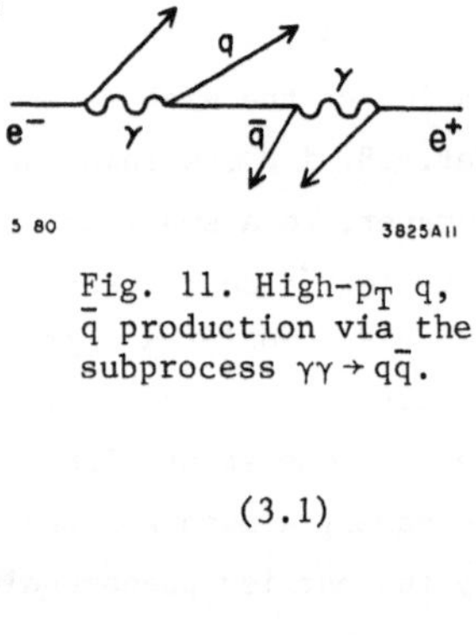

Fig. 11. High-p_T q, $\bar{q}$ production via the subprocess $\gamma\gamma \rightarrow q\bar{q}$.

$$E_{Jet} \frac{d\sigma^{ee \rightarrow eeJet}}{d^3 P_{Jet}} \sim \left(\frac{\alpha}{2\pi} \ln \eta\right)^2 \frac{\alpha^2}{P_T^4} f\left(x_R^{Jet}, \theta_{c.m.}^{Jet}\right) \tag{3.1}$$

where $x_R = E_{Jet}/E_e$ and

$$\eta = \begin{cases} \dfrac{s}{4m_e^2} & \text{no electron tag} \\[3ex] \dfrac{\theta_{max}^2}{\theta_{min}^2} & \text{tagged electron} \end{cases} \tag{3.2}$$

As in annihilation it is convenient to compare directly to the analogous μ–pair cross section. We obtain

$$\frac{d\sigma(e^+e^- \rightarrow e^+e^-q\bar{q})}{d\sigma(e^+e^- \rightarrow e^+e^-\mu^+\mu^-)} = R_{\gamma\gamma} \qquad (3.3)$$

with

$$R_{\gamma\gamma} = 3\sum_q e_q^4\left(1 + \mathcal{O}(\alpha_s/\pi)\right) \qquad (3.4)$$

For standard fractional charges and two flavor generations $R_{\gamma\gamma} = 34/27$ in leading order. In a model where the photon has a color singlet <u>and</u> a color octet component but experiments are done below color threshold, it is still possible[37] that the octet components of the initial photons could mix together to yield a final state color singlet and yield a much higher value, namely $R_{\gamma\gamma} = 10/3$. In such a model the quark charges are integral above color threshold.

A useful characterization of the magnitude of the two-jet cross section is its contribution to the standard R value of e^+e^- annihilation. Figure 12 from Ref. 34 shows that the contribution ΔR (in the fractional charge model) from all $e^+e^- \rightarrow e^+e^-q\bar{q}$ events with $p_T^{quark} > p_T^{min}$ can be very substantial, especially at LEP energies, even for quite large values of p_T^{min}. In the PEP–PETRA range, at $\sqrt{s} = 30$ GeV, one finds

$$(\Delta R)_{p_T^{min}} = \frac{5}{\left[p_T^{min}(\text{GeV/c})\right]^2} \qquad (3.5)$$

i.e., such processes occur at a rate which is 30% of that for $e^+e^- \rightarrow \mu^+\mu^-$ when $p_T^{min} = 4$ GeV.

As for the usual annihilation R, $R_{\gamma\gamma}$ has $\mathcal{O}(\alpha_s/\pi)$ corrections coming from virtual and real gluon radiation. One anticipates that these will be of the same order as found for R ($\approx 10\%$). A calculation[38] of the virtual part of the corrections appears to confirm this expectation.

The following important points should be noted.

(1) Production of two jets (and only two jets) is significant only if the point-like γ component is present. The cross section for two vector meson dominated photons to collide producing two and only two jets is small. More often residues in the beam directions remain as discussed in Section II. A more precise vector-dominance

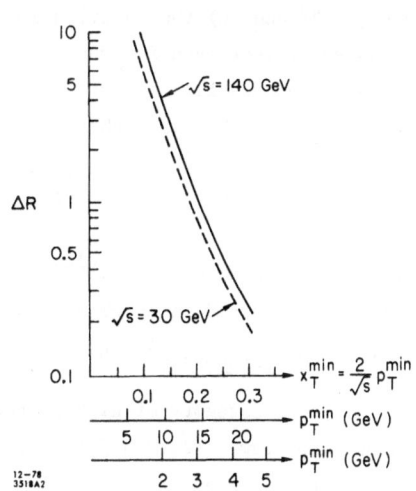

Fig. 12. The contribution to R from γγ → q$\bar{q}$ two-jet processes at $\sqrt{s}$ = 30 and 140 GeV (from Ref. 34).

background estimate appears later, it is basically a higher twist contribution with strong inverse p_T damping.

(2) A failure to see these purely two-jet events implies that perturbative ideas cannot be applied to short distance physics. This would represent a dramatic breakdown of the general approach to short distance reactions motivated by asymptotic freedom.

In fact, two of the PETRA groups, PLUTO and TASSO, have now seen[39] a few two-jet events with one tagged electron. Such tagging eliminates a possibly important background, discussed in Ref. 34, from annihilation events with two bremsstrahlung photons. As the statistics improve we can clearly expect to either confirm or disprove the perturbative predictions based on Fig. 11.

There is a second process, $\gamma\gamma \rightarrow gg$, Fig. 13, which produces events with exactly two gluon jets. The diagram involves a fermion loop and occurs at order $(\alpha_s/\pi)^2$. It might be anticipated that it would be negligible; however, explicit calculations[40,41] show that

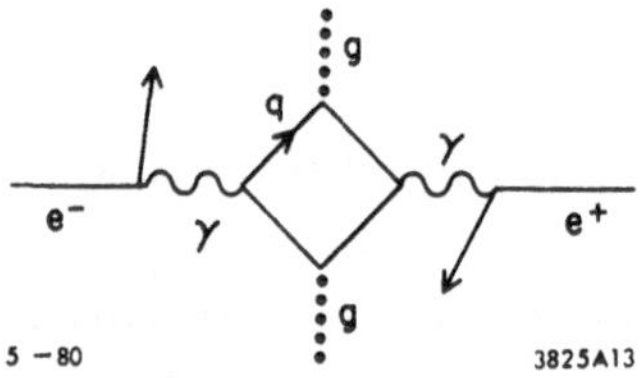

$$\frac{E\left(d\sigma^{ee \rightarrow ee \text{ gluon jet}}/d^3p\right)}{E\left(d\sigma^{ee \rightarrow ee \text{ q or } \bar{q} \text{ jet}}/d^3p\right)} \approx .1 \qquad (3.6)$$

Fig. 13. Production of two high-p_T gluon jets via $\gamma\gamma \rightarrow gg$.

essentially independent of x_T and angle. Because of their similar angular dependence, kinematics cannot be used to separate gluon jets from quark jets. The only physically distinguishable features of a gluon jet in this case will be (a) the anticipated higher average multiplicity of gluon jets;[42] (b) the broader transverse spread of a gluon jet;[43] and (c) the polarization of the gluons.[44] With regard to the latter, Ref. 44 shows that even for unpolarized initial photons the gluons are preferably produced with polarization parallel ($\parallel$) to the scattering plane as opposed to perpendicular ($\perp$). They find that

$$P(\theta) \equiv \frac{d\sigma_\perp - d\sigma_\parallel}{d\sigma_\perp + d\sigma_\parallel} \qquad (3.7)$$

becomes as large as -0.3 at $\theta_{c.m.}^{gluon} = 90°$. This polarization is, however, only indirectly reflected in an oblateness of the gluon jet.[45]

B. 3-Jet Processes[32-35]

This next category of high p_T reactions is distinguished by having two high p_T jets and one beam direction jet in the final state. Elementary QCD subprocesses which produce such a configuration are shown in Fig. 14; they are (a) $\gamma q \rightarrow gq$ and (b) $\gamma g \rightarrow q\bar{q}$. In two photon collisions these subprocesses yield one and only one beam jet. The beam jet comes from the remnants of the photon which provides the quark or gluon initiating the subprocess. The generic form of the cross section for production of a high-p_T jet is, in the two cases

$$E \frac{d\sigma^{Jet}}{d^3p} \sim \begin{cases} \frac{1}{\pi}\int dx_\gamma dx_q \ G_{\gamma/e}(x_\gamma) \ G_{q/e}(x_q, p_T^2) \ \frac{d\sigma^{\gamma q \to gq}}{dt'} \ s'\delta(s'+t'+u') & \text{(3.8a)} \\\\ \frac{1}{\pi}\int dx_\gamma dx_g \ G_{\gamma/e}(x_\gamma) \ G_{g/e}(x_g, p_T^2) \ \frac{d\sigma^{\gamma g \to q\bar{q}}}{dt'} \ s'\delta(s'+t'+u') & \text{(3.8b)} \end{cases}$$

where we have used the leading log result (2.1) and (2.2). The difference between this situation and the hadron target/beam case is that all the above distribution functions are explicitly calculable. The form of $G_{\gamma/e}$ is obtained[46] from QED and the triggering conditions of the final electron. The distributions $G_{q,g/e}$ are obtained by convolution from

$$G_{q,g/e}(x, p_T^2) = \int_x^1 \frac{dz}{z} \ G_{q,g/\gamma}\left(\frac{x}{z}, p_T^2\right) G_{\gamma/e}(z)$$

(3.9)

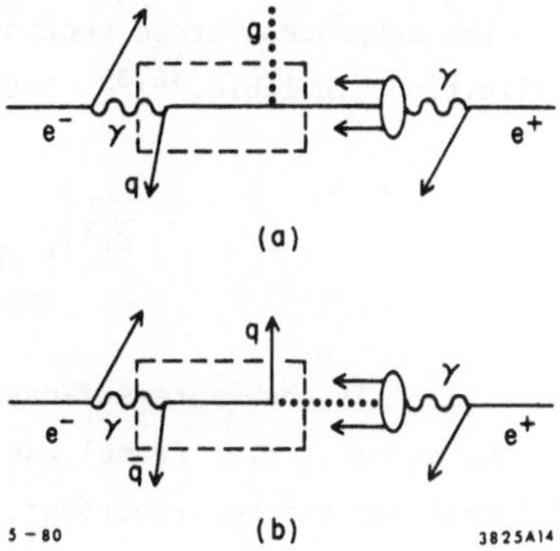

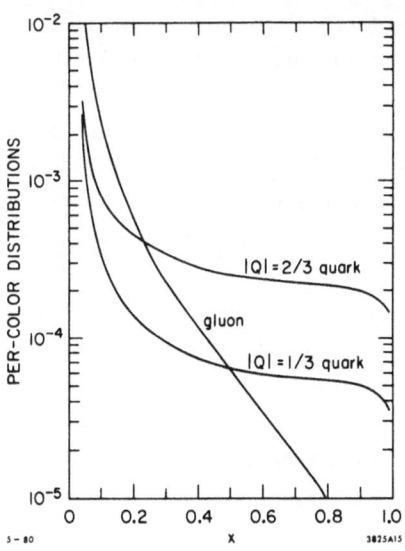

(a)

(b)

5 – 80   3825A14

Fig. 14. Elementary-QCD processes with 3-jet final state topology. The --- box encloses the high-p_T subprocess.

Here the non-trivial $G_{q,g/\gamma}$ distribution, which will develop a p_T^2 dependence, is completely calculable, for the point-like photon component. The leading log results were obtained originally by Witten[47] and were rederived diagrammatically in Refs. 6, 33, and 48. More recently the next order corrections have been obtained[49] and do not greatly modify the leading log results in the moderate x range. The form of the leading log result is shown in Fig. 15. (For a more detailed discussion, see T. Walsh's talk at this conference or W. Frazer's and W. Bardeen's talks at the Lake Tahoe Conference.[50]) Both these distributions take the leading log form

$$G_{\begin{bmatrix} q \\ g \end{bmatrix}/\gamma}(x, "Q^2") = \frac{\alpha}{\alpha_s("Q^2")} \begin{bmatrix} q(x) \\ g(x) \end{bmatrix}$$

(3.10)

where "Q^2" represents the momentum scale of the short distance probe; in our case "Q^2"

Fig. 15. Plotted are leading log quark and gluon distributions $G_{q/\gamma}(x, "Q^2")$ and $G_{g/\gamma}(x, "Q^2")$ divided by log "Q^2"/Λ^2.

is of order p_T^2. (Written in terms of $1/\alpha_s$, this form also incorporates $\ell n \, \ell n \, "Q^2"/\ell n \, "Q^2"$ corrections to α_s.) The important feature is the factorization of the $"Q^2"$ and x dependences; both increase as $\log \, "Q^2"/\Lambda^2$. In addition g(x) and especially q(x) have much weaker fall off as $x \to 1$ than comparable distributions for the vector-dominated component of the photon (see Ref. 6, for example). Thus the high-p_T situation (in which $"Q^2"$ is of order p_T^2 and x is of order x_T) is dominated by the point-like component of the photon distribution functions.

The subprocess cross sections for $\gamma q \to gq$ and $\gamma g \to q\bar{q}$ are, of course, also explicitly calculable.[32,34] Both take the form

$$\left. \frac{d\sigma}{dt'} \right|_{\substack{fixed \\ angle}} = \frac{\alpha_s(p_T^2)}{4 \, p_T^2} \, f(\theta'_{c.m.}) \tag{3.11}$$

where $\theta'_{c.m.}$ is the center-of-mass scattering angle of the final quark or gluon jet (in the γq or γg c.m. frame) and p_T its transverse momentum; $f(\theta'_{c.m.})$ is, of course, different for the two reactions. Combining Eqs. (3.8)–(3.11) we see that $\alpha_s(p_T^2)$ cancels between $d\sigma/dt'$ and $G_{\begin{bmatrix} q \\ g \end{bmatrix}/\gamma}$ and that the jet cross section is completely scale invariant in leading log:

$$E \, \frac{d\sigma^{Jet}}{d^3 p} \sim \frac{1}{(p_T^{Jet})^4} \, F\!\left(x_R^{Jet}, \theta_{c.m.}^{Jet}\right) \tag{3.12}$$

where $x_R^{Jet} = E_{Jet}/E_e$ and $\theta_{c.m.}^{Jet}$ is the overall center-of-mass jet scattering angle. The function F is completely determined by the convolution integrals in Eq. (3.8). The $\gamma q \to gq$ contribution is generally larger than the $\gamma g \to q\bar{q}$ contribution because g(x) < q(x) (Fig. 15) over much of the x range.

In Fig. 16 we compare, at $\theta_{c.m.}^{Jet} = 90°$, the cross section $E(d\sigma^{Jet}/d^3p)$ for jet production due to the 3-jet processes, $\gamma q \to gq$ and $\gamma\bar{q} \to g\bar{q}$, to that coming from the $\gamma\gamma \to q\bar{q}$ and $\gamma\gamma \to gg$ 2-jet processes. As expected $d\sigma(\text{3-jet}) < d\sigma(\text{2-jet})$. Clearly, in order to see a subprocess like $\gamma q \to gq$ over the $\gamma\gamma \to q\bar{q}$ subprocess, it is necessary to distinguish the 3-jet from the 2-jet topology.

The fact that there are only two 3-jet processes at the elementary-QCD level makes it possible to also imagine testing the associated predictions[51] for energy distribution in the beam jet and for angular distributions of the opposite side jet relative to the trigger jet. But the most crucial observation is that in leading log none of the Λ dependent complexities associated with hadron targets arise. These elementary-QCD subprocesses yield a α_s, and hence Λ, independent answer in leading order.[52] Next-to-leading log corrections need to be computed for these jet cross sections but are probably no larger than those found in hadron target scattering.

C. 4-Jet Processes[32-35]

Examples of subprocesses which contribute to the 4-jet topology (2 balancing high-p_T jets and 2 beam direction jets) at the elementary QCD level are shown in Fig. 17. Of course, many diagrams and processes are not shown. Note that since the point-like-photon quark distribution, $q(x)$, is so much larger (Fig. 15) than the gluon distribution, $g(x)$, the subprocesses $qq \to qq$, $q\bar{q} \to q\bar{q}$ and $\bar{q}\bar{q} \to \bar{q}\bar{q}$ dominate.

The generic form of the cross section from the $qq' \to qq'$ subprocess is

$$E \frac{d\sigma^{Jet}}{d^3p} \sim \frac{1}{\pi} \int dx_q \, dx_{q'} \, G_{q/e}(x_q, p_T^2)$$

$$\times G_{q'/e}(x_{q'}, p_T^2) \frac{d\sigma^{qq' \to qq'}}{dt'} s'\delta(s'+t'+u')$$

$$(3.13)$$

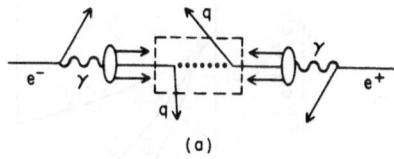

(a)

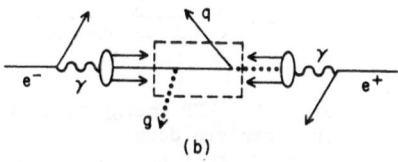

(b)

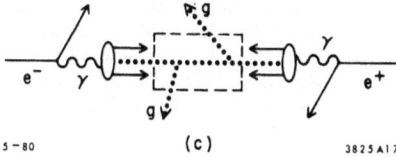

(c)

5-80   3825A17

Fig. 17. Examples of elementary-QCD subprocesses contributing to the 4-jet topology: (a) $qq \to qq$; (b) $qg \to qg$; and (c) $gg \to gg$. The order is one of decreasing importance as the gluon distribution for the point-like photon is generally smaller than the quark distribution.

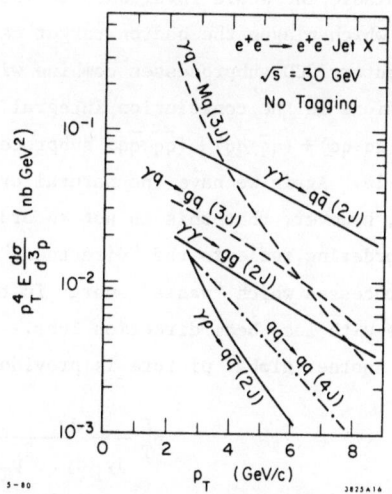

Fig. 16. A comparison of $E \frac{d\sigma^{Jet}}{d^3p}\Big|_{\theta_{c.m.}=90^0}$ coming from various 2-jet (——), 3-jet (---), and 4-jet (-·-·) processes. Except for the $\gamma p \to q\bar{q}$ curve all subprocess contributions are calculated using the point-like photon component only; in particular initial quarks are obtained from the p_T^2 dependent point-like photon distribution. In the curves labeled $\gamma q \to Mq$, $\gamma q \to gq$ and $qq \to qq$ we have also included the antiquark contributions $\gamma\bar{q} \to M\bar{q}$, $\gamma\bar{q} \to g\bar{q}$ and $q\bar{q} \to q\bar{q}/\bar{q}\bar{q} \to \bar{q}\bar{q}$, respectively.

with $G_{q/e}G_{q'/e} \propto \left[1/\alpha_s(p_T^2)\right]^2$, see Eq. (3.10), and, from (2.2),

$$\frac{d\sigma^{qq' \to qq'}}{dt'} \sim \frac{1}{p_T^4} \alpha_s^2(p_T^2) f^{qq' \to qq'}(\theta'_{c.m.}).$$

$$(3.14)$$

Thus the $\alpha_s(p_T^2)$'s cancel and one obtains

$$\left(E \frac{d\sigma^{Jet}}{d^3p}\right)_{qq' \to qq'} \sim \frac{1}{p_T^4} F^{qq' \to qq'}\left(x_R^{Jet}, \theta_{c.m.}^{Jet}\right)$$

$$(3.15)$$

As in the 3-jet case $F^{qq' \to qq'}$ is completely

calculable and scale invariant; in leading order the Λ dependence and other ambiguities which plague the hadron target case are absent. Of course, all the other
elementary-QCD subprocesses combine with their appropriate quark/gluon distribution
functions in the convolution integral so that this same $\alpha_s(p_T^2)$ cancellation occurs.
The $(qq \to qq) + (q\bar{q} \to q\bar{q}) + (\overline{qq} \to \overline{qq})$ subprocess contribution to $E(d\sigma^{Jet}/d^3p)$ is shown in
Fig. 16. Again we have the natural ordering $d\sigma(4\text{-jet}) < d\sigma(3\text{-jet}) < d\sigma(2\text{-jet})$.
Note, however, that this is not an ordering in α_s (the α_s's always cancel). Rather
the ordering reflects the "directness" with which the final jet is produced; those
subprocesses which "waste" energy in the beam direction are suppressed relative to
those with less beam direction loss.

Another global picture is provided by Fig. 18 from Ref. 34. There we plot

$$p_T^4 \frac{d\sigma}{dy_1 \, dy_2 \, d^2p_T} \bigg|_{y_1 = y_2 = 0, \ p_T^1 = p_T^2 = p_T}$$

where we imagine triggering on two jets, back-to-back
at $90°$ with equal p_T's. This plot incorporates all
subprocesses of the elementary-QCD type which contribute to the various jet topologies.

D. Higher Twist and Vector-Dominance Effects -- Single Particle Spectra

(a) Higher Twist and Single Particle Spectra

At this point the alert reader might ask if there
are not some higher twist diagrams which complicate
the high-p_T jet situation. The answer, fortunately,
is in general "no." Only one such diagram is likely
to be important. It contributes to the 3-jet case
and is based on the subprocess $\gamma q \to Mq$ (the reversal
of which was discussed in Section II). The center-
of-mass diagram, Fig. 19, is exactly analogous to
that drawn for $\gamma q \to gq$, Fig. 14, but with the gluon
replaced by a meson. As discussed in Section II,
the $\gamma q \to Mq$ subprocess yields $1/p_T^6$ behavior compared
to $1/p_T^4$ behavior
for the $\gamma q \to gq$
subprocess:

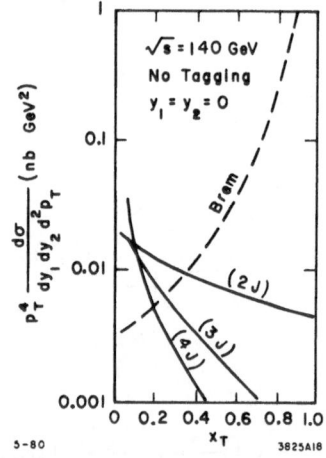

Fig. 18. LEP energy plot of
the complete double jet
trigger cross sections coming
from elementary-QCD processes
of each topology type. Also
shown is the double bremsstrahlung background which
can be eliminated by single
tagging, see Ref. 34.

$$\frac{d\sigma^{\gamma q \to Mq}}{dt'} \sim \frac{\alpha_s^2(p_T^2)}{p_T^6} \hat{f}(\theta'_{c.m.}) \qquad (3.16)$$

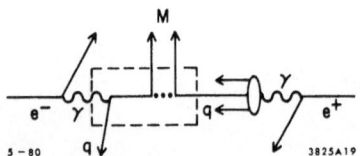

Fig. 19. Higher twist 3-jet
topology diagram based on the
subprocess $\gamma q \to Mq$.

with $\hat{f}$ completely calculable aside from its

normalization. Either the final meson or final quark can yield an observed "jet;" if the "jet" derives from the meson it will have limited multiplicity and should reconstruct to a resonance. The form of the cross section is

$$E \frac{d\sigma^{Jet}}{d^3p} \sim \frac{1}{\pi} \int dx_\gamma \, dx_q \, G_{\gamma/e}(x_\gamma) \, G_{q/e}(x_q, p_T^2) \frac{d\sigma^{\gamma q \to Mq}}{dt'} \, s'\delta(s'+t'+u')$$

$$\sim \frac{1}{p_T^6} \alpha_s(p_T^2) \, \hat{F}\left(x_R^{Jet}, \theta_{c.m.}^{Jet}\right) \tag{3.17}$$

The function $\hat{F}$ is completely determined by the above convolution integral and $\hat{f}$ of Eq. (3.16). Thus only its normalization is uncertain. This normalization is of great theoretical interest.

The normalization of $\hat{f}(\theta'_{c.m.})$, i.e., of $\gamma q \to Mq$, has been determined phenomenologically[17,32] in a variety of different ways. The crucial ingredient is the normalization of the meson wave function. This has been obtained from:

(a) $\pi p \to \pi p$ elastic scattering which in the CIM is normalized by the proton form factor and the π wave function.

(b) $Mq \to \pi q$ exclusive scattering, the normalization of which is determined by the size of the CIM contribution to $pp \to \pi X$ at high-p_T.

(c) $Mq \to \gamma q$, the normalization of which is determined by the size of the CIM contribution to $pp \to \gamma X$.

All three determinations are completely consistent with one another. For instance if $\pi p \to \pi p$ elastic scattering is used to determine the normalization of the meson wave function, then the higher twist CIM diagrams based on the subprocesses $Mq \to \pi q$ and $Mq \to \gamma q$ yield excellent fits to $pp \to \pi X$ and $pp \to \gamma X$ high-p_T data, respectively. It is also possible to calculate the meson wave function normalization using the techniques of Ref. 11.[31] A much smaller normalization is obtained; but it is difficult to ignore the phenomenological successes (a)-(c). Thus a clean determination of the normalization of $\hat{f}(\theta'_{c.m.})$ is highly desirable.

In Fig. 16 we compare the CIM 3-jet process contribution to jet production at $\theta_{c.m.}^{Jet} = 90°$ to those previously discussed -- using the normalization of (a)-(c) above. Only at lower p_T values does the CIM process clearly dominate the 3-jet topology contribution to jet production.

However, as mentioned in Section II, if we trigger on a single fast meson, e.g., a π^+, then the $\gamma q \to \pi^+ q$ subprocess can produce the π^+ directly whereas the $\gamma q \to gq$ and $\gamma g \to q\bar{q}$ subprocesses as well as the much more important $\gamma\gamma \to q\bar{q}$ subprocess must create the π^+ as a fragment of a quark or gluon. The jet cross section from the $\gamma q \to Mq$ subprocess is roughly a factor of 20 above the direct π^+ cross section (a factor 10 is due to the restriction to $M = \pi^+$ compared to arbitrary M and a factor of 2 from loss of the q jet triggering possibility). In comparison $\gamma\gamma \to q\bar{q}$ with q or $\bar{q}$ fragmenting to the π^+ is suppressed by roughly a factor of 100 relative to its jet cross section, mostly because of the trigger bias effect.[18] Thus there is a relative enhancement of the CIM subprocess by a factor of 5 in the single π^+ trigger situation. The three most important single pion spectrum contributions are given in Fig. 20 (at $\theta^{\pi^+}_{c.m.} = 90^\circ$) where it is seen that the CIM term dominates over the p_T range shown, if the normalization from (a)-(c) is employed.

Preliminary "single tag" spectra for $\left(dN^{ee \to ee\pi X}/dp_T^2\right)$ have been obtained by both TASSO and PLUTO.[39] Both exhibit a sharp break at $p_T^2 \approx 1$ (GeV/c)^2 and a large high-p_T tail. A detailed study of these spectra should clearly determine whether or not there is room for a large CIM contribution; triggering efficiencies, etc., must be unfolded before comparison to theory can be made.

(b) <u>Vector Dominance Backgrounds</u>

In Figs. 16 and 20 we have also plotted the largest vector dominance backgrounds, to jet and single π^+ production respectively. In the jet case we have compared the 2-jet contribution coming from the subprocess $\gamma\rho \to q\bar{q}$ (i.e., one of the initial two photons is vector dominated as shown in Fig. 21a) to the purely point-like diagram based on $\gamma\gamma \to q\bar{q}$. The process $\gamma\rho \to q\bar{q}$ is a higher-twist relative of $\gamma\gamma \to q\bar{q}$ yielding $1/p_T^6$ vs. $1/p_T^4$ behavior for the jet cross section. Not surprisingly it is very much suppressed. This is typical of vector dominance backgrounds. <u>For a given topology</u>

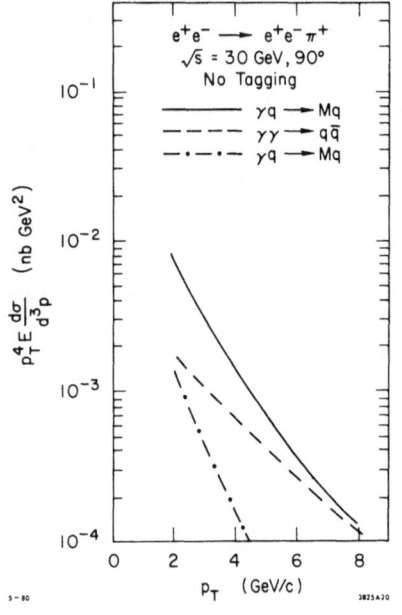

Fig. 20. Contributions to π^+ production at 90° from: (a) $\gamma q \to \pi^+ q$ where the q comes from the point-like photon component; (b) $\gamma\gamma \to q\bar{q}$ with either the q or $\bar{q}$ fragmenting to the observed π^+; (c) $\gamma q \to \pi^+ q$ where the initial q comes from the vector-dominated photon component.

(2, 3 or 4 jet) the vector dominance related backgrounds are suppressed relative to the analogous point-like photon contributions by at least a factor of 10.

Another example in the jet case (not plotted in Fig. 16) is the comparison of 4-jet contributions from the qq → qq scattering subprocess. If the initial quarks are both from ρ-dominated photons the contribution is a factor of 10^3 below the contribution where both of the initial quarks come from the point-like photon components, as in Fig. 17a. This reflects the dominance of point-like vector-dominated components of the quark distribution in a photon. Thus

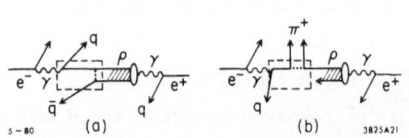

Fig. 21. Vector dominance backgrounds to (a) 2-jet production; (b) π^+ production.

even if the $\rho\rho \to$ 4-jet cross section is an unexpectedly small percentage of the calorimetric trigger rate, as appears to be the case in hadron-hadron collisions (see item (6) under Section II), it may still be that the simple point-like process $\gamma_{PL}\gamma_{PL} \to$ 4-jet will stand above the hadron-related background.

Our final vector dominance example is the contribution to single π^+ production coming from the subprocess $\gamma q \to \pi^+ q$ with the initial quark coming from a ρ-dominated photon, Fig. 21b. Figure 20 shows that this background is suppressed by a factor of ≥ 5 relative to the analogous contribution coming from $\gamma q \to \pi^+ q$ with the initial quark from a point-like photon distribution.

To summarize Section III.1 on inclusive high-p_T reactions in $\gamma\gamma$ collisions we first reemphasize that high-p_T physics is relatively clean in this situation compared to typical hadronic collisions. Possible problems are confined to:

* Non-leading-log corrections -- possibly as small as 10% for the 2-jet topology and probably about 50% for 3 and 4-jet topologies.

* Incorporating the non-zero $\langle q^2 \rangle$ of photons coming from an e^- or e^+ that is triggered on.

* Thresholds -- in leading log these effects enter through the number of photon quark components we consider. As p_T^2 increases one, for instance, passes new 2-jet $\gamma\gamma \to q\bar{q}$ thresholds when $W_{subprocess}^2 \approx 4p_T^2 \geq 4m_q^2$. Effects of new quark thresholds which enter through α_s tend to cancel since α_s's cancel in most cases.

III.2. <u>Exclusive High p_T Reactions in $\gamma\gamma$ Collisions</u>[53]

Two photon collisions provide tests of QCD which are analogous in simplicity to, but more versatile than, those provided by form factor measurements. I will discuss briefly only two examples:

(a) $\underline{\gamma^*(Q^2)\gamma \to \pi^0}$

If we define the invariant Feynman amplitude as

$$\mathcal{M} = ie^2 F_{\pi\gamma}(Q^2)\, \varepsilon_{\mu\nu\rho\sigma}\, p_\pi^\nu\, q^\rho\, \varepsilon^\sigma \tag{3.18}$$

then the exact prediction is equivalent to exposing one gluon in the π^0 wave function as in Fig. 22,

$$F_{\pi\gamma}(Q^2) \;\overset{Q^2 \to \infty}{\sim}\; \frac{2f_\pi}{Q^2} \tag{3.19}$$

with no α_s or Λ dependence in this leading order. (The PCAC constant as defined here is $f_\pi = 0.093$ GeV.) This is the same prediction as obtained many years earlier using PCAC and the Bjorken, Johnson, Low

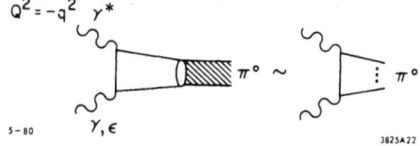

Fig. 22. Dominant diagram for $\gamma^*(Q^2)\gamma \to \pi^0$.

limit.[1] It is a fundamental and simple test of the perturbative approach to short distance physics.

(b) $\gamma\gamma \to M\bar{M}$

Here we consider only real photons. The "Born" diagrams which contribute in the large Q^2 fixed angle limit are shown in Fig. 23. After including all leading log corrections the prediction takes the form

$$
\mathcal{M}_{\gamma\gamma \to M\bar{M}} = \frac{e^2 \, 16\pi \, \alpha_s(Q^2)}{3} \left[\epsilon_1 \cdot \epsilon_2 \, \mathcal{M}_1\left(z, \log Q^2/\Lambda^2\right) \right.
$$

$$
\left. + \frac{4\epsilon_1 \cdot k \epsilon_2 \cdot k}{Q^2} \, \mathcal{M}_2\left(z, \log Q^2/\Lambda^2\right) \right] \tag{3.20}
$$

where $z = \cos\theta_{c.m.}$ of the final M and

$$
\mathcal{M}_i = \sum_{n,m} b^i_{n,m}(z) \left(\log Q^2/\Lambda^2\right)^{-\gamma_n - \gamma_m} \tag{3.21}
$$

The γ_n and γ_m are the anomalous dimensions which appear in deep inelastic scattering and the form factor prediction, Eq. (2.5), and the leading terms $b^i_{00}(z)$ are calculable in terms of the PCAC f_π. Note that the $b^i_{n,m}$'s also depend on scattering

Fig. 23. Diagrams for $\gamma\gamma \to M\bar{M}$.

angle z so that the z dependence of the cross section can in principle be used to separate the different b's from one another.

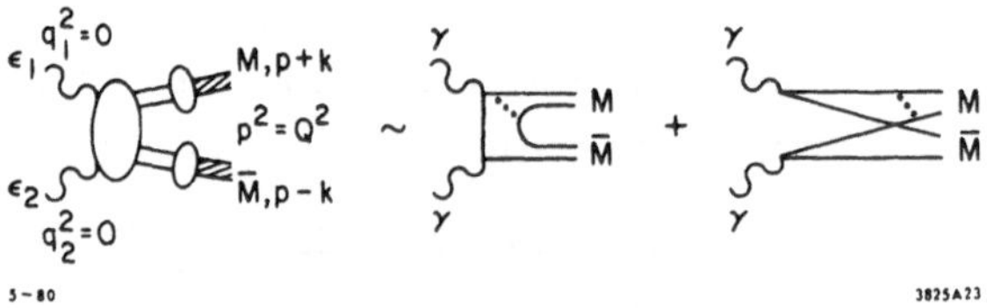

Again the above are fundamental yet simple predictions of QCD. These exclusive reactions may be simpler to analyze (despite their relatively small cross section) than the inclusive reactions. However, only in the inclusive reactions is it possible to directly probe the simple QCD processes $qq \to qq$, $\gamma\gamma \to q\bar{q}$, etc. Clearly both types of experiment deserve long-term attention and analysis.

IV. Conclusions

We summarize the main points of this talk.

(a) The precise low energy $\sigma_{\gamma\gamma}$ form needs more theoretical attention. It is abundantly clear that estimates of the non-Pomeron contributions to $\sigma_{\gamma\gamma}$ using vector dominance will yield only part of the answer. Additional Regge and/or fixed pole contributions, related to the point-like component of the photons, will be present.

(b) We are fortunate in the high-p_T domain to have a point-like photon component. Already experiments have seen the fundamental 2-jet signal predicted on the basis of the perturbative process $\gamma\gamma \to q\bar{q}$. In comparison hadronic collisions present a confusing picture, even refusing (according to a recent experiment) to yield a clear jet trigger signal without first biasing the event by requiring a single fast high-p_T particle.

(c) Most two-photon high-p_T predictions exhibit exact scaling and are independent of the QCD Λ parameter in leading order. All normalizations are computable. The tests of the underlying elementary-QCD cross sections are correspondingly clean.

(d) Higher twist and vector dominance related backgrounds are generally negligible except for the interesting $\gamma\gamma \to \pi^+ X$ high-p_T single particle cross section which provides a simple and fundamental measure of the normalization of an important higher-twist subprocess, $\gamma q \to \pi q$. Experimental results on $dN^{\gamma\gamma \to \pi}/dp_T^2$ will shortly provide meaningful constraints.

(e) Fundamental, but simple, QCD predictions have been obtained for several basic exclusive channels in photon-photon collisions.

Acknowledgements

I would like to thank P. Kessler and G. Parisi for their hospitality at the Amiens Two-Photon Workshop (1980) for which this talk was prepared. I would also like to thank S. J. Brodsky for communicating his recent exclusive results to me.

References

1. See M. Greco's talk at the "Workshop on Two Photon Collisions," Amiens, France (1980) and references therein.

2. F. E. Low, Phys. Rev. D12 (1975) 163; S. Nussinov, Phys. Rev. Lett. 34 (1975) 1286; J. F. Gunion and D. E. Soper, Phys. Rev. D15 (1977) 2617.

3. See the talks of Ch. Berger (PLUTO) and E. Hilger (TASSO) at the "Workshop on Two Photon Collisions," Amiens, France (1980).

4. J. F. Gunion, Phys. Lett. 88B (1979) 150.

5. V. Chang and R. Hwa, Phys. Lett. 85B (1979) 285.

6. See for example, W. R. Frazer and J. F. Gunion, Phys. Rev. D20 (1979) 147.

7. The actual prediction incorporating spin is $(1-x)^2$ + constant, $(1-x)$ is an adequate approximation over the moderate x domain. See G. Farrar and D. R. Jackson, Phys. Rev. Lett. 35 (1975) 1416.

8. The actual prediction is $\dfrac{(1-x)^0}{a+b \ln (1/1-x)}$ indicating a logarithmic suppression as $x \to 1$. See Ref. 6.

9. A. M. Boyarski et al., Phys. Rev. D14 (1976) 1733. This is 18 GeV $\gamma p \to \pi^+$, π^- data. Higher energy data from the tagged photon beam at FNAL will hopefully confirm this lower energy result.

10. See for example, W. Frazer and J. F. Gunion, Phys. Rev. D19 (1978) 2447 and references therein.

11. For a summary of recent progress in this area see S. J. Brodsky, SLAC-PUB-2447, presented at the Summer Institute on Particle Physics, SLAC (1979).

12. R. Blankenbecler, S. J. Brodsky and J. F. Gunion, Phys. Rev. D8 (1973) 4117; Phys. Lett. 39B (1972) 649; P. V. Landshoff and J. C. Polkinghorne, Phys. Rev. D10 (1974) 891 and references therein; M. K. Chase and W. J. Stirling, Nucl. Phys. B133 (1978) 157. For a general review see D. Sivers, R. Blankenbecler and S. J. Brodsky, Phys. Rep. 23C (1976) 1 and Ref. 11.

13. See Ref. 11 and G. P. Lepage and S. J. Brodsky, Phys. Rev. Lett. 87B (1979) 359.

14. R. K. Ellis, M. Furman, H. Haber and I. Hinchliffe, LBL-10304 (1979).

15. See L. F. Abbott and R. M. Barnett, SLAC-PUB-2227, submitted to Annals of Phys.; R. Blankenbecler and I. Schmidt, Phys. Rev. D16 (1979) 1318; W. R. Frazer and I. Schmidt, Phys. Rev. D16 (1979) 1318; W. R. Frazer and J. F. Gunion, SLAC-PUB-2489 (1980). The general form of higher twist contributions was first discussed in R. Blankenbecler, S. J. Brodsky and J. F. Gunion, Phys. Rev. D12 (1975) 3469.

16. R. M. Barnett, private communication.

17. See R. Blankenbecler, S. J. Brodsky and J. F. Gunion, Phys. Rev. D18 (1978) 900; D. Jones and J. F. Gunion, Phys. Rev. D20 (1979) 232.

18. S. D. Ellis, P. V. Landshoff and M. Jacob, Nucl. Phys. B108 (1978) 93.

19. Experimental points are from R. Stronowski, SLAC Summer Institute on Particle Physics (1979). Experimental references can be obtained there. The $\sqrt{s} = 31/\sqrt{s} = 53$ n_{eff} extractions are removed from Stronowski's graph. A different range of p_T is being probed compared to the $\sqrt{s} = 53/\sqrt{s} = 62$ extractions (at $x_T = .4$, $p_T = 6$ to 10 vs. $p_T = 10$ to 12). More higher twist–CIM component is expected (and seen) in the lower $\sqrt{s}$ extraction; correcting for this yields approximate agreement between the two extractions. I have also added to Stronowski's graph recent n_{eff} values obtained by the Athens, Athens, Brookhaven, CERN collaboration (A^2BC), C. Kourkoumelis et al., CERN-EP/80-07.

20. See R. P. Feynman, R. D. Field and G. C. Fox, Nucl. Phys. B128 (1979)   ; Phys. Rev. D18 (1978) 3320. A variety of others have also pursued such studies. For references see Refs. 11 and 19. See also R. Field, Proceedings of the Tokyo International Conference on High Energy Physics (1978).

21. See Ref. 19 for a discussion and further references.

22. W. E. Caswell, R. Horgan and S. J. Brodsky, Phys. Rev. D18 (1978) 2415; R. Horgan and P. Scharbach, Phys. Lett. 81B (1979) 215.

23. J. F. Gunion and B. Peterson, U.C. Davis preprint UCD-79-5, Phys. Rev. to be published. Elementary QCD processes were considered in more detail by R. Baier, J. Cleymans and B. Petersson, Phys. Rev. D17 (1978) 2310.

24. H. Frisch et al., Phys. Rev. Lett. 44 (1980) 511.

25. G. R. Farrar and G. C. Fox, Rutgers preprint, RU-79-170 (1980).

26. See P. Seyboth's contribution at the Rencontre de Moriond, Les Arcs, France (1980).

27. R. M. Baltrusaitis et al., FNAL-PUB-79/38 Exp (1979); E. Amaldi et al., Nucl. Phys. B150 (1979) 326; M. Diakonov et al., CERN-EP/80-02 and 80-03.

28. R. Ruckl, S. J. Brodsky and J. F. Gunion, Phys. Rev. D18 (1978) 2469.

29. A. P. Contogouris, S. Papadopoulos and M. Hongoh, Phys. Rev. D16 (1979) 2607.

30. L. Cormell and J. F. Owens, FSU-HEP-800307 (1980).

31. S. J. Brodsky and J. F. Gunion, in progress.

32. S. J. Brodsky, T. DeGrand, J. F. Gunion and J. Weis, Phys. Rev. Lett. 41 (1978) 672; and Phys. Rev. D19 (1979) 1418.

33. C. H. Llewellyn Smith, Phys. Lett. 79B (1978) 83.

34. K. Kajantie, Phys. Scripta 29 (1979) 230; K. Kajantie and R. Raitio, Nucl. Phys. B159 (1979) 528.

35. M. Abud, R. Gatto and C. A. Savoy, Phys. Rev. D20 (1979) 2224; and Phys. Lett. 84B (1979) 229.

36. S. Berman, J. Bjorken and J. Kogut, Phys. Rev. D4 (1971) 3388.

37. See also M. Chanowitz, Proceedings of the XIIth Rencontre de Moriond (1977), edited by Tran Thanh Van; P. V. Landshoff, LEP Summer Study, 1-13 October 1978; and S. J. Brodsky and J. M. Weis, Memorial Symposium on Strong Interactions (1978), University of Washington. H. Lipkin, Nucl. Phys. B155 (1979) 104, feels that color fluctuations invalidate this result; however the contribution is basically of a "Z-graph" type (refering to time ordered perturbation theory) which would not be affected by color fluctuations. The issue is not yet settled.

38. F. Berends, Z. Kunszt and R. Gastmans, DESY preprint, DESY-80/08 (1980).

39. See the talks by Ch. Berger (PLUTO) and E. Hilger (TASSO) at the Amiens "Two-Photon Workshop (1980).

40. R. Cahn and J. F. Gunion, Phys. Rev. D20 (1979) 2253.

41. K. Kajantie and R. Raitio, Phys. Lett. 87B (1979) 133.

42. See for example, S. J. Brodsky and J. F. Gunion, Phys. Rev. Lett. 37 (1976) 402; K. Konishi, A. Ukawa and G. Veneziano, Phys. Lett. 78B (1978) 243.

43. See G. Veneziano, XIXth International Conference on High Energy Physics, Tokyo (1978).

44. A. Devoto, J. Pumplin, W. Repko and G. L. Kane, Michigan State University preprint (1979).

45. S. J. Brodsky, T. A. DeGrand and R. F. Schwitters, Phys. Lett. 79B (1978) 244.

46. See the talks by P. Kessler, G. Parisi and J. Field at the Amiens "Two-Photon Workshop" (1980) for a critical review of various approximations to $G_{\gamma/e}$. In the results quoted here I always use the equivalent photon spectrum.

47. E. Witten, Nucl. Phys. B120 (1977) 189.

48. Y. Dokshitser, D. Dyakonov and S. Troyan, SLAC-TRANS-183; R. J. DeWitt, L. Jones, J. Sullivan, D. Nillen and H. Wyld, Phys. Rev. D19 (1979) 2046.

49. W. Bardeen and A. Buras, Phys. Rev. D20 (1979) 166; D. Duke and J. Owens, Florida State University, FSU-HEP-802 401 (1980). These latter authors now agree with Bardeen and Buras.

50. T. Walsh, Amiens "Two-Photon Workshop" (1980); W. Bardeen, Lake Tahoe Two-Photon Conference (1979); W. Frazer, Lake Tahoe Two-Photon Conference (1979).

51. J. Field, E. Pietarinen and K. Kajantie, DESY preprint 79/85 (1979).

52. Actually (3.10) and (3.11) are still correct when the important $\ln \ln p_T^2 / \ln p_T^2$ corrections to $\alpha_s(p_T^2)$ are included and thus the scale invariance of (3.12) is only corrected by terms down by a full power of $\ln p_T^2$.

53. This section is based on work due to S. J. Brodsky and G. P. Lepage, Ref. 11 (and references therein) and work in progress.

Photon-Photon Physics in the Deep Inelastic Region

T.F. Walsh
DESY
Hamburg

Workshop on $\gamma\gamma$ Interactions

Amiens, France

April 8-12, 1980

Abstract

The topics discussed are

1. The real photon structure functions
2. Theoretical issues in $\gamma\gamma$ reactions with one or both photons off mass shell
3. Experimental questions

At this meeting we heard a report by Berger on the observation of inelastic electron photon scattering at PETRA.[1] In the near future we can expect measurements of F_i^γ at Q^2 around 2-5 GeV^2. For a given detector, the accessible Q^2 range increases roughly as E_B (the beam energy). So in the distant future, the F_i^γ could be measured at LEP for $Q^2 \approx$ 15-20 GeV^2. (This time scale is not unnatural; changes in this field occur typically on a scale $\sim$ 5 years.) So it is a good time to recapitulate. This is a brief discussion of what we know theoretically about real photon structure functions and why it is interesting to measure them.

The photon structure functions are a very clean laboratory for theorists. QCD is an elegant theory, but not a trivial one. There is a lot which is only partially understood. Deep inelastic reactions are at present certainly under the best theoretical control. We need experiments on something besides the proton and neutron. Now the photon target is slowly becoming available.

I. REAL PHOTON STRUCTURE FUNCTIONS

The cross section for scattering of an e^+ or e^- on a real photon target in $e^+e^- \longrightarrow e^+e^-$ + hadrons (Fig. 1)[2]

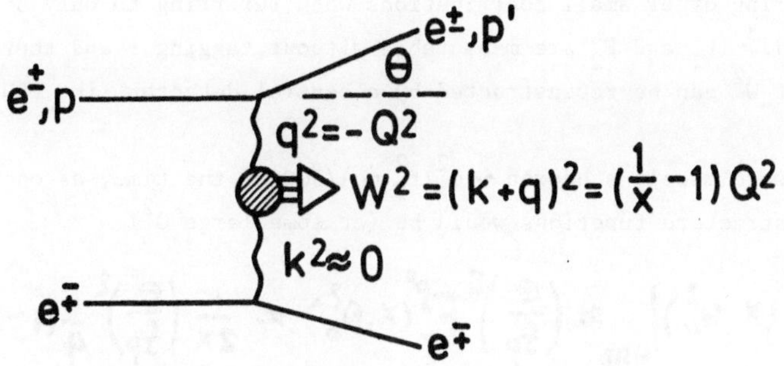

Fig. 1

is

$$\frac{d\sigma}{dx\,dy\,d\phi/2\pi} = \frac{4\pi\alpha^2\,p\cdot k}{Q^4}\left[1 + (1-y)^2\right] *$$

(1)

$$* \left\{ 2x\,F_T^\gamma + \epsilon(y)\,F_L^\gamma + \epsilon(y)\,\epsilon\left(\frac{E_\gamma}{E_B}\right)F_X^\gamma\,\cos 2\phi \right\}$$

where x and y are neutrinolike variables

$$x = Q^2/2p\cdot k, \qquad y = q\cdot k/p\cdot k$$

and

$$\epsilon(y) = \frac{2(1-y)}{1+(1-y)^2}$$

Defined similarly to the conventional structure functions,

$$F_1^\gamma(x,Q^2) = F_T^\gamma(x,Q^2)$$

$$F_2^\gamma = 2x\,F_T^\gamma + F_L^\gamma$$

(2)

$$F_3^\gamma = F_X^\gamma$$

F_3 is actually the structure function for a transversely polarized photon (analogous to the cross section difference $\sigma_\parallel - \sigma_\perp$ for parallel and perpendicular $\gamma(k)$, $\gamma(q)$ transverse polarizations). It is only measurable if the small angle electron in Fig. 1 is tagged at some non-zero angle. ϕ is then the angle between the planes of the scattered large and small angle $e^\pm$. Actually, for non-zero tagging angle, $k^2 \neq 0$.

So we are ignoring other small contributions when referring to only 3 structure functions in (1).[3] (F_1 and F_2 are measurable without tagging – and therefore for tiny $k^2 \sim 0(-m_e^2)$ – if W^2 can be reconstructed by a central detector like PLUTO.)

If the photon were merely a hadron $\sim e^2/f_\rho^2 \sim 1/300$ of the time, as once seemed to be the case, its structure functions would be (at some large Q_o^2)

$$F_T^\gamma(x,Q_o^2)\bigg|_{HAD} \approx \left(\frac{e}{f_\rho}\right)^2 F_T^\rho(x,Q_o^2) \approx \frac{1}{2x}\left(\frac{e}{f_\rho}\right)^2 \frac{1}{4}(1-x)$$

$$F_L^\gamma(x,Q_o^2)\bigg|_{HAD} \approx 0\left(\frac{\langle p_\perp^2 \rangle}{Q_o^2}\right) + 0\left(\alpha_s(Q_o^2)\right) \tag{3}$$

$$F_X^\gamma(x,Q_o^2)\bigg|_{HAD} \approx 0\left(\frac{\langle p_\perp^2 \rangle}{Q_o^2}\right) + 0\left(\alpha_s(Q_o^2)\right)$$

(The expression for $F_T^{\rho^o} = F_T^{\pi^o}$ from the quark model and π structure function "data" in ref. (4).) The two terms in $F_{L,X}^\gamma$ are the parton model expressions $0(\langle p_\perp^2 \rangle/Q_o^2)$ and the QCD radiative corrections, which are first order in $\alpha_s(Q^2)$. Beyond Q_o^2, $F_T^{\rho^o}$ evolves slowly with Q^2, decreasing as an inverse power of $\ln Q^2$ at any fixed x. In any case, $F_T^{\rho^o}$ vanishes as $x \to 1$.

If (3) were all there was to photon structure, it would still be interesting. But (3) is not all there is. The real photon can disassociate at a point into a $q\bar{q}$ pair, one of which is hit by the virtual photon (Fig. 2). The contribution of the box diagram to this pointlike piece is [5]

Fig. 2

$$F_T^\gamma(x,Q^2)\Big|_{Box} = \frac{\alpha}{2\pi} \sum e_i^4 (x^2 + (1-x)^2) \ln W^2/\Lambda^2$$

$$\rightarrow \frac{\alpha}{2\pi} \sum e_i^4 (x^2 + (1-x)^2) \ln Q^2/\Lambda^2$$

$$F_L^\gamma(x,Q^2)\Big|_{Box} = \frac{4\alpha}{\pi} \sum e_i^4 x^2(1-x) \tag{4}$$

$$F_X^\gamma(x,Q^2)\Big|_{Box} = -\frac{\alpha}{\pi} \sum e_i^4 x^3$$

This (parton) contribution already shows the important physics: it dominates as $x \rightarrow 1$ and $F_2 \propto \ln Q^2$. (Also, F_L^γ and F_X^γ scale.)[5]

F_2^γ contains two terms. The one proportional to $(1-x)^2$ is a $\gamma(q) + \gamma(k) \rightarrow \gamma(q) + \gamma(k)$ helicity amplitude where the total spin projection along the collision axis is $J_z = \pm 2$ (helicities $+- \rightarrow +-$ or $-+ \rightarrow -+$). The term proportional to x^2 has $J_z = 0$ along the collision axis ($++ \rightarrow ++$ or $-- \rightarrow --$). In the limit $Q^2 = -q^2 \rightarrow 0$ $x \rightarrow 0$ and $J_z = \pm 2$ dominates. For $x \rightarrow 1$ (i.e. fixed W^2 as $Q^2 \rightarrow \infty$), $J_z = 0$ dominates. The argument of the log is $W^2 = (\frac{1}{x} - 1)Q^2$ because it arises from a phase space integral. This is bounded by the CM energy W.

An Aside on Resonances

It's instructive to look at the contribution of a single resonance to F_i^γ. The π^0 final state is a good example, because all others turn out to behave similarly.[6] F_L vanishes and

$$F_{\binom{T}{X}}^\gamma = \binom{+}{-}\frac{1}{4} \delta(W^2 - m_\pi^2)|T^{++}|^2 \tag{5}$$

where

$$T^{++} = \frac{1}{2} Q^2 F_\pi(q^2, k^2 = 0)$$

(The contribution of a normal or abnormal parity resonance to F_X is + or -). It turns out that in the Bjorken-Johnson-Low limit

$$\lim_{-q^2 - k^2 \rightarrow +\infty} T^{++} = \frac{-\sqrt{2}f_\pi}{3} \tag{6}$$

Moreover, one can write an integral representation for $F_\mu(q^2, k^2)$ in terms of a spectral weight. If the weight is concentrated around equal q^2 and k^2, T^{++} is smooth as $k^2/q^2 \longrightarrow 0$ and [6]

$$T^{++}(q^2, 0) \approx -\frac{\sqrt{2} f_\pi}{3} \tag{7}$$

This scaling behavior is independent of QCD radiative corrections (there are none in the BJL limit). The behavior (7) is what one would expect if, from simple duality ideas, the resonances modulated the $x \longrightarrow 1$ limit of F_T and F_X. We expect that both approach finite constants at fixed W^2 and $Q^2 \gg W^2$. Incidentally, this makes it clear that we should not expect $F_T \propto \ln Q^2$ independent of x as $x \longrightarrow 1$. $F_T \propto \ln W^2$ is more reasonable.

The helicity structure of the resonance contributions is interesting. At $Q^2 = 0$ the $f^0(1250)$ is produced in a $J_z = \pm 2$ state and with θ_π measured relative to the collision axis,

$$f^0(1250) \rightarrow \pi^+ \pi^- \quad : \quad \frac{d\sigma}{d\cos\theta_\pi} \propto \sin^4\theta_\pi \tag{8}$$

However, at <u>large</u> Q^2, the f^0 is produced in a $J_z = 0$ state[6] and

$$f^0(1250) \rightarrow \pi^+ \pi^- \quad : \quad \frac{d\sigma}{d\cos\theta_\pi} \propto (2 - 3\sin^2\theta_\pi)^2 \tag{9}$$

Of course, this says nothing about the rapidity of the change from (8) to (9).

QCD Radiative Corrections

The box diagram (or parton model) result for the pointlike term in F_T^γ is changed by QCD radiative corrections.[7] The box result, written as an integral over the quark $p_\perp^2$ is of the form[2]

$$F_T^\gamma \sim \alpha \int \frac{dp_\perp^2}{p_\perp^2} \sim \alpha \ln Q^2 \tag{10}$$

Radiation of N gluons will give a contribution of order (Fig. 3)

$$\alpha \int \frac{d\,p_\perp^2}{p_\perp^2}\ \alpha_s \int \frac{d\,p_\perp^2}{p_\perp^2}\ \cdots \quad \alpha\alpha ln Q^2\,(\alpha_s\,ln Q^2)^N \qquad (11)$$

(The $p_\perp^2$ integrals have nested limits, but as $Q^2 \to \infty$ this does not matter and (11) results). A sum of terms like (11) resembles a geometrical series which sums up to a constant times the lowest order result $\alpha\ ln Q^2$. Thus the $ln Q^2$ behavior survives, but we expect that the x-dependence will differ from (4). It will be softer because the quarks loose momentum by gluon radiation.

The QCD corrections to F_L differ. To lowest order in α_s, the integrand to F_L^γ

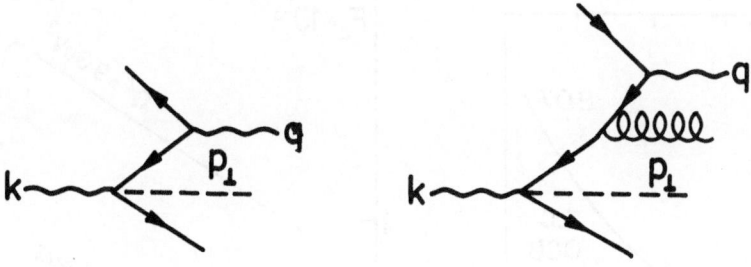

Fig. 3

vanishes at $p_\perp = 0$ for massless quarks. It can be written as $p_\perp^2/Q^2 + O(\alpha_s(Q^2)) +$ higher order terms,

$$F_L^\gamma \sim \alpha \int \frac{d p_\perp^2}{p_\perp^2}\ \frac{p_\perp^2}{Q^2} + \alpha \int \frac{d p_\perp^2}{p_\perp^2}\ \alpha_s(Q^2) + \cdots$$

$$(12)$$

$$\sim \alpha \cdot const + O(\alpha_s\ ln Q^2) + O(\alpha_s^2\ ln Q^2)$$

The first two terms give constants as $Q^2 \to \infty$; all others vanish as powers of $1/ln Q^2$. The numerical QCD corrections to the first (parton) term turn out to be very small. It is not likely that they will ever be measured (if QCD is correct).

Perhaps startlingly, there are no O(1) QCD corrections to F_x.[8] QCD effects are of order $\alpha_s(Q^2)$, and vanish as $Q^2 \to \infty$. This can be seen by a simple argument. F_x is not actually a cross section. It is the difference of cross sections for parallel and perpendicular polarized $\gamma(k), \gamma(q)$. By symmetry, the probability to find a q in $\gamma(k)$ must vanish at $p_\perp^2 = 0$ - no matter how many gluons have been radiated, provided their momenta and angles are integrated over. This is because $F_x^\gamma \propto \sigma_\parallel - \sigma_\perp$. The angular factor this introduces is $sin^2\theta_q \propto p_\perp^2/Q^2$. This means that F_x^γ takes the form

$$F_x^\gamma \sim \alpha \int \frac{dp_\perp^2}{p_\perp^2} \frac{p_\perp^2}{Q^2} + \alpha \int \frac{dp_\perp^2}{p_\perp^2} \alpha_s(Q^2) \frac{p_\perp^2}{Q^2} + \cdots \tag{13}$$

$$\sim \alpha \cdot \text{const} + O(\alpha_s(Q^2))$$

Thus the QCD corrections to the box diagram result vanish as $Q \rightarrow$

Fig. 4 shows the behavior of F_2

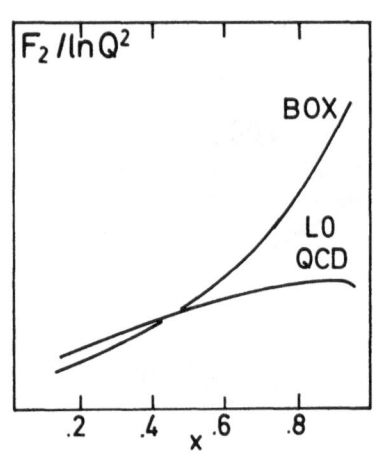
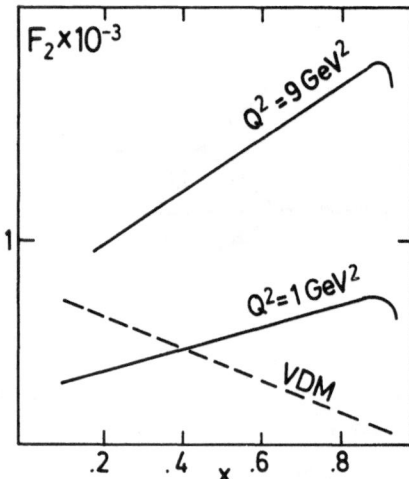

Fig. 4

II. ISSUES WITH ONE OR TWO PHOTONS OFF MASS-SHELL

1) F_T^γ, F_L^γ and F_x^γ probe the $p_\perp$ structure of QCD corrections in different ways. If it is at all possible, all three should be measured.

2) In inelastic electron-photon-scattering there is no target mass. So there are no target mass corrections as in eN scattering. For the pointlike term there are also no corrections from the non-zero radius of the target. We also expect $F_2 \propto \ln Q^2$, rising with Q^2, unlike ℓN scattering, where it decreases. It should prove difficult to explain the Q^2 dependence of F_2 solely as a hadronic higher-twist effect, as appears possible for lepton-nucleon scattering.

3) Higher order QCD corrections to F_2 have been calculated[9]. These can be important

at very large x. Much of the correction may be kinematic in nature. The argument of the log in F_2^γ is not Q^2, but W^2. For $x > 1/2$, $W^2 < Q^2$ and the kinematic limits in $\gamma\gamma \to q\bar{q} + q\bar{q}G + \ldots$ are set by W^2, not Q^2. (For example propagators can go off shell by an amount $W^2 < Q^2$). This suppresses F_2 at $x > 1/2$, compared to the leading $\ln Q^2$ result. At small x, where $W^2 \gg Q^2$, the hadronic structure of γ dominates and no clear statement can be made. This kinematic effect is clearly nonleading, as $\ln W^2 = \ln Q^2 + \ln(1/x - 1)$.

4) QCD predicts a simple relation between $e\,\gamma(k) \to e'X$ and $e^+e^- \to \gamma(q) \to \gamma(k) + X$. In the parton model this is trivial, and it is straightforward even including QCD corrections. Moreover, $e^+e^- \to \gamma(k) + X$ should be measurable at very large $Q^2 \sim 100 \text{ GeV}^2 - 1000 \text{ GeV}^2$ at CESR and PETRA or PEP.[10]

Futurism

F_T is well-defined at LEP, where one can define an off-shell extension

$$F_T^\gamma (Q^2, |k^2|) = \frac{1}{2}\left[W_{++\to++} + W_{+-\to+-} \right] \qquad (14)$$

In the limit $|k^2|$, Q^2, $W^2 \to \infty$ with fixed $x = 1 + W^2/Q^2$ and $Q^2/|k^2|$ very large there are no QCD radiative corrections and

$$F_T^\gamma (Q^2, |k^2|) \xrightarrow[|k^2| \ll Q^2]{} \frac{\alpha}{2\pi} \sum e_i^4 \left(x^2 + (1-x)^2\right) \ln \frac{Q^2}{|k^2|} \qquad (15)$$

the box diagram (or parton) result. By increasing $|k^2|$ from zero one "turns off" the QCD radiative corrections. Presumably the scale for this is set by the QCD dimensional parameter Λ.

It will clearly be interesting to measure (14) at LEP to see how fast the QCD radiative corrections disappear. (As yet there are no detailed theoretical calculations of this effect known to me.[11])

Another thing which may barely be measurable at LEP is the large $|k^2|$, $|q^2|$ behavior for fixed W^2. QCD radiative corrections do survive in this limit. The box result is modified.[12]

III. EXPERIMENT

There is clearly a background to eγ scattering in Fig. 1. It is the inelastic
compton process of Fig. 5.[13]

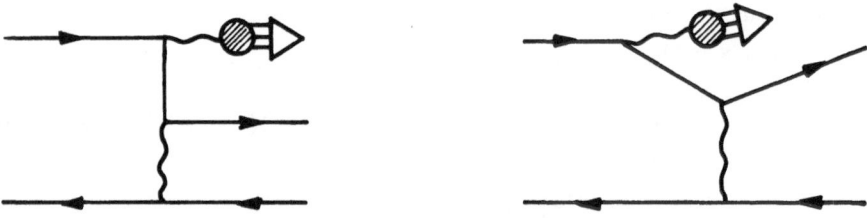

Fig. 5

For small angle scattering, where an e$^-$ is tagged at some Q^2 and not too large angle
to the e$^-$ beam, this background is harmless. The reason is that the propagators in
Fig. 5 are both far off-shell. This is not so if one does not distinguish e$^+$ and e$^-$
charges in the final state. Then one might see an e$^+$ at small or moderate angle to
the e$^-$ beam (and the reverse). This is not obviously harmless, because the first
graph in Fig. 5 has a u-channel pole (backward scattering). However, the outcoming
lepton has small laboratory energy. An energy cut on the outgoing lepton will reduce
this problem. Fig. 6 shows the cross section as a function of E' and θ of the
scattered electron (or, as a dashed line, electron plus backscattered e$^+$ if charges
are not distinguished). With a cut on the scattered e$^-$ energy of E' $\gtrsim$ 2 GeV or
$\gtrsim$ 12 GeV the inelastic Compton background is unimportant below $\approx 20^\circ$ scattering angle.
This incorporates the whole angular range of the PLUTO "large angle" tagger.[2] The
situation is essentially the same at LEP.[2]

As a final remark, it is worth noting that F_L and F_X may be measurable. F_{L_2} is larger
in eγ scattering than in ℓN. The relevant distributions are of the form

$$[1 + (1-y)^2] F_2^\gamma - y^2 F_L^\gamma \equiv A - B$$

$$[1 + (1-y)^2] F_2^\gamma - y^2 F_L^\gamma + 2(1-y) F_X^\gamma \cos 2\phi \tag{16}$$

$$\propto 1 - a \cos 2\phi$$

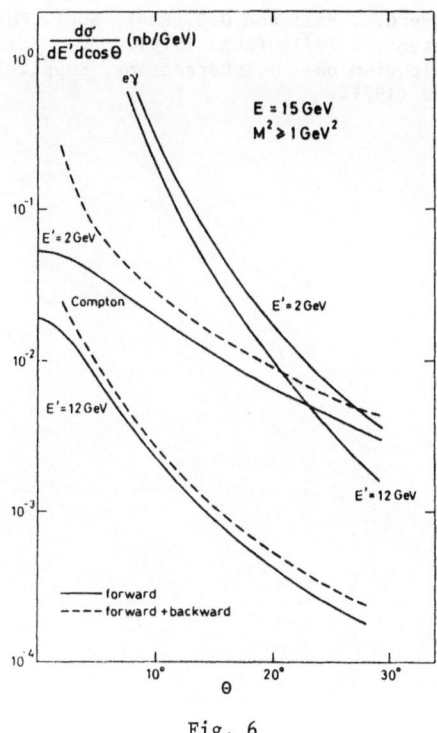

Fig. 6

Numerically, B/A and a are large enough to offer some hope that a high statistics experiment can extract $F_L^\uparrow$ and $F_3^\uparrow$.

References

1. Results of the PLUTO Collaboration, presented by C. Berger at this meeting.
2. C. Peterson, T.F. Walsh and P.M. Zerwas, NORDITA preprint 80/13 and references there.
3. I want to thank P. Kessler for emphasizing this to me.
4. K.J. Anderson et al., Chicago-Princeton report EFI-78-38.
5. T.F. Walsh and P.M. Zerwas, Phys. Lett. 44B (1973), 195;
   R.L. Kingsley, Nucl. Phys. B60 &1973) 45;
   R. Worden, Phys. Lett. 51B (1974) 57.
6. G. Köpp, T.F. Walsh and P.M. Zerwas, Nucl. Phys. B70 (1974) 461 and references there.
   According to P. Lepage and S. Brodsky (SLAC PUB-2478) the QCD asymptotic limit of $|q^2| F(q^2, k^2=o)$ can actually be calculated to be $12 f_\pi (f_\pi = 135$ MeV)
7. E. Witten, Nucl. Phys. B120 (1977) 189;
   C.H. Llewellyn Smith, Phys. Lett. 29B (1978) 83;
   W.R. Frazer and J.F. Gunion, Phys. Rev. D20 (1979) 147.
8. Ref. (2) and W.R. Frazer and G. Rossi, San Diego preprint UCSD-10-P10-211.
9. W.A. Bardeen and A.J. Buras, Phys. Rev. D20 (1979) 166.
10. K. Koller, T.F. Walsh and P.M. Zerwas, Z. Phys. C2 (1979) 197 and
    C.H. Llewellyn Smith, ref. (7).

11. A start has been made: C.T. Hill and G.G. Ross, Nucl. Phys. <u>B148</u> (1979) 373.
12. M.K. Chase, DAMTP preprint 79/14 (Aug. 1979).
13. J. Parisi, 1973 Colloquium on   Interactions, Suppl. Journal de Physique, Tome 35, Fasc. 3 C-2 (1974).

Trends In Particle Physics

C.H. Llewellyn Smith

Department of Theoretical Physics

1 Keble Rd., Oxford OX1 3NP

England.

1.   Introduction

My talk is divided into two main sections. First I shall discuss the region up to a few hundred GeV., where the gross features of the phenomena are probably well described by QCD and $SU(2)_L$ x U(1). I will then consider speculations about the region beyond 1 TeV., identifying possible alternatives and stressing experiments which may discriminate between them (or against all of them!). I shall take it for granted that gauge theories provide the correct framework.

2.   Physics Below 1 TeV.

2a   QCD

The following are facts:

1) Hadrons are made of coloured quarks and gluons (as well as accounting for the spin statistics problem, $\pi^0 \to \gamma\gamma$ and $R_{\bar{e}e}$, recall that colour is needed to remove anomalies   in $SU(2)_L$ x U(1)).

2.   Gluons have spin 1 (the evidence for this is the success of (approximate) chiral symmetry, reviewed below).

3)   Strong forces must be colour dependent (otherwise there would be eight coloured pions degenerate with the colour singlet).

The only sensible field theory which accounts for these facts is QCD, which I therefore believe must be correct (this conclusion might be wrong if quarks are composite but QCD would probably still be a true epiphenomenon).

I will now review some elementary ideas about chiral symmetry, which will be needed later, and summarise the very impressive experimental evidence. Let $\vec{A}_\mu = \bar{q}\ \vec{\tau}\gamma_\mu\gamma_5 q$ where $q = \begin{pmatrix} u \\ d \end{pmatrix}$. $\int \vec{A}_0 d^3x$ is the generator of chiral transformations and chiral symmetry would be exact if $\partial^\mu \vec{A}_\mu = 0$. For the nucleon matrix elements of $\vec{A}_\mu$ this would require

$$q^\mu\ \bar{u}\ (\gamma_\mu\gamma_5 g_A + q_\mu\gamma_5 F)\ u\ =\ \bar{u}\ \gamma_5 u\ (2M_N g_A + q^2\ F) = 0.$$

Recalling   that the pion pole contributes $\dfrac{f_\pi\ g_{np\pi}}{q^2 - M_\pi^2}$ to F, we see that there are two

ways to satisfy this relation either explicit chiral symmetry: $M_N = 0$, $2M_N g_A$ or spontaneously broken chiral symmetry: $M_\pi = 0$ (experiment: $M_\pi^2 = 0.03\ M_\rho^2$) and $\dfrac{2M_N g_A}{f_\pi g_{np\pi}} + 1 = 0$
(experiment: $0.08 \pm .02$) The second possibility is evidentally a good approximation
to the truth. In the limit of exact spontaneously broken chiral symmetry there are
many other relations which can be derived which agree quite well with experiment:

Theory	Experiment
$M_\pi a_{\pi N}^{\frac{1}{2},3/2} = 0.16,\ -0.079$	$0.17 \pm 0.005,\ -0.088 \pm 0.004.$
$\lambda_{K_{e3}}^{0} = 0.020$	$0.020 \pm 0.003.$
$\Gamma_{\pi^0} \doteq 7.87\ \text{eV}$	$7.95 \pm 0.55.$

To obtain $\partial^\mu \vec{A}_\mu = 0$ requires $m_{u,d}^0 = 0$ and either free field theory or vector/axial
vector forces. The approximate success of setting $\partial^\mu \vec{A}_\mu = 0$ in so many different
processes shows that $m_{u,d}^0$ are small and the forces are (axial) vector.

Some problems/questions about QCD are:

1) It has not been shown that QCD leads to the existence of hadrons at all, let
alone hadrons with the observed properties (spontaneously broken chiral symmetry
among them). Recently very impressive progress in understanding QCD at large dis-
tances has been made by studying lattice gauge theories. Time (and incompetence)
preclude further discussion.[1)]

2) When discussing light hadrons we can neglect all but u and d quarks and set
$m_{u,d}^0 = 0$ to a good approximation. With no bare masses, QCD has no free parameters –
a source of delight and frustration! To start calculating we introduce a unit of
energy E_0 at which by definition the coupling constant is equal to one:

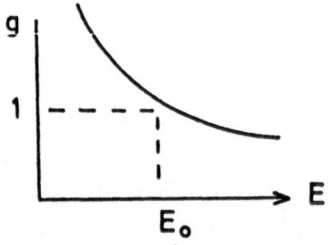

Solving, we would then (presumably!) find a massive nucleon, M_N/E_0 being
calculable(so that we can trade in E_0 for M_N as a scale). We could then calculate
$g(M_N)$! Approximations need parameters and 't Hooft has suggested that we should
use $1/N_c$ where N_c is the number of colours (finally we must set $N_c = 3$; $1/3$ may not
seem small but note, probably unfairly, that $e \simeq 1/3.3$). In the limit $N_c \to \infty$

$(g \sqrt{N_c}$ fixed), only planar Feynman diagrams without internal quark loops survive and it may be possible to make progress. In fact encouraging results can be derived in this limit. Among others,[2] there must be an infinite number of non-exotic narrow resonances with decays dominated by resonant two body final states, Zweig's rule becomes exact and the $q\bar{q}$ "sea" in the nucleon vanish.

3) As $m^0_{u,d} \to 0$, it seems naively that

$$\partial^\mu (\bar{u} \gamma_\mu\gamma_5 u + \bar{d} \gamma_\mu\gamma_5 d) = 0$$

leading to a fourth (isoscalar) massless pseudoscalar meson (with $m^0_{u,d} \neq 0$ its mass would have to be less than $\sqrt{3} M_\pi$). This is the U(1) problem. However QCD has an anomaly, which is a necessary ingredient to cure this problem - whether it is sufficient is not yet completely clear but the situation looks promising.[3]

4) In general there will be a P and T violating term $\theta\epsilon^{\mu\nu\alpha\beta} F^i_{\mu\nu} F^i_{\alpha\beta}$ (where $F^i_{\mu\nu} = \partial_\mu A^i_\nu - \partial_\nu A^i_\mu + gf_{ijk} A^j_\mu A^k_\nu$) in the QCD Lagrangian[4]. The limit on the neutron dipole moment requires $|\theta| < 10^{-8}$. Why is this term so small? Its effects could be transformed away if there was some exact chiral symmetry but this requires that either $m^0_u = 0$, which is probably excluded, or the existence of a scalar particle called the axion. In simple models the axion is light and its existence is ruled out be experiment. θ can always be adjusted to a small value, but ultimately we would like a framework in which it is naturally small. (It is naturally small in models in which CP violation is soft but such models do not account for baryon number asymmetry in the universe as easily as models with hard CP violation[4]).

5) Further work is needed exploiting perturbative QCD. Experimentally, better data on $\frac{\sigma_L}{\sigma_T}$ in lepto-production continues to be very desirable and very accurate data on R_{ee} , which probably provides the cleanest test of QCD, would be very welcome (there are hints of problems at $\sqrt{s} \sim 7$ GeV.[5]).

2b Electro-Weak Interactions

The present status is:

QED - excellent.

Charged currents - data on the Michel parameter, longitudinal electron polarization in Gamow-Teller decays etc. show that any non-standard contributions are less than or of order 10% at low energy.

Neutral currents - the data now require the couplings $\nu_\mu u$, $\nu_\mu d$, $\nu_\mu e$, and the parity violating part of eu and ed to be essentially those of the standard model. The ten parameters involved are all determined and have values close to those of the standard model (up to some ambiguities)[6].

Alternately, setting

$$L_{eff} = \frac{G}{\sqrt{2}} [J^+_\lambda J^-_\lambda + \rho(J^3_\lambda - \sin^2\theta_w J^{em}_\lambda)^2]$$

an excellent fit is obtained with $\rho = 1.004 \pm 0.019$ and $\sin^2 \theta_w = 0.235 \pm 0.016$[7]. I conclude that the $SU(2)_L \times U(1)$ model probably works well for $E < O(200 \text{ GeV})$ and expect the Z^0 and $W^{\pm}$ to exist with masses of about 88 and 78 GeV. respectively. Can we avoid this conclusion ? It is easy to construct other models which give L_{eff} above but it is not so easy to arrange that they automatically give $\rho \simeq 1$. The only simple way to do this is to take $[SU(2)]^n \times U(1) \times G$ where only one $SU(2)$ and $U(1)$ couple to the light fermions (in the standard way)[8]. The enlarged gauge structure then only influences the light fermions through mixing of the vector bosons. Suppose the neutral and charged bosons of $[SU(2)]^n$ mix in the same way, so that $M_{W_i^0} = M_{W_{\pm}^i}$, as in the standard model (this requires Higgs doublets only). If the only further mixing of the $SU(2)$ bosons is between the W_i^0 and the $U(1)$ boson B (which mixes with the other neutrals in G), then $\rho = 1$ is obtained automatically. However in such models there are two important differences from the standard model:

1) There are several Z^0's and $W^{\pm}$'s. At least one Z^0 and one $W^{\pm}$ must have a mass less than the value (88 or 78 GeV.) predicted by the standard model.

2) There is an extra term[9]

$$\frac{G}{\sqrt{2}} \quad C(J_{\lambda}^{em})^2$$

in L_{eff} which would not, of course, have been detected in neutrino experiments or parity violation experiments, but could be detected in $e^+e^- \to e^+e^-$ and $e^+e^- \to \mu^+\mu^-$. QED tests at PETRA require $C \leq 0.15$[10] (If $C = 0$ the standard model is recovered). Incidentally it would be nice to have the results of QED tests expressed in terms of g_V and g_A for a contact neutral current interaction[11].

Some problems/ questions about electro-weak interactions at modest energies are:

1) Do the $W^{\pm}$ and Z^0 really exist ? Are their masses really those predicted by the standard model ? Precise measurements of the mass would be very interesting because there are finite corrections to the lowest order values which are calculable in a renormalizable model but depend on the spectrum of heavy fermions (the mass shifts are about + 3 GeV. including three families of quarks and leptons[12]). This shift therefore tests renormalizability and/or probes the fermion mass spectrum.

2) Although the standard model is probably correct for the first and second generation of quarks and leptons, what about the third ? In particular Need the ν_{τ} exist ? Yes, the data exclude the possibility that $\tau \to \nu_{\mu}, \bar{\nu}_{\mu}, \nu_e$ or $\bar{\nu}_e$ and a new neutrino is therefore required[13].
Need the t exist ? No.

There are exotic and a non-exotic scenarios in which there is no t. Both can be established or eliminated by studying b decays. In the non-exotic case, the b decays by mixing with d and s so that the $SU(2)_L$ doublets are

$$\begin{pmatrix} u \\ \alpha d + \beta s + \gamma b \end{pmatrix}_L \quad , \quad \begin{pmatrix} c \\ \gamma d + ws + \sigma b \end{pmatrix}_L$$

Normalizing, orthogonalizing and eliminating $\Delta S = 1$ neutral currents, leaves two undetermined parameters which are constrained by Cabibbo universality and limits on the b lifetime. There are two solutions[14]:

1) $\sigma \simeq 4.2\gamma$. The b decays mainly to c and s (in ratio about 2:1 in amplitude). Decays which are completely forbidden by the GIM mechanism in the standard model include $b \to s\, e^+e^-$ (2%) and $b \to s\, \nu\bar{\nu}$ (10%), which allows $e^+e^- \to b\bar{b} \Rightarrow K^+K^- + 4\ \nu$'s (looking like a 2γ reaction)!

2) $\sigma \simeq -0.24\gamma$ and b decays mainly to u and d (in ratio about 2:1 in amplitude). Non-standard decays include $b \to d\, e^+e^-$ (1%) and $b \to d\, \nu\bar{\nu}$ (6%). $\nu d \to b\nu$ is allowed in this case but the cross-section is very small.

This non-exotic topless scheme seems contrived in the framework of $SU(2)_L \times U(1)$ but it would occur, for example, in $SU(3)_L \times U(1)$ models with left hand multiplets

$$\begin{matrix} d' - u' & & b' - c' \\ \searrow \ \nearrow & & \searrow \ \nearrow \\ s' & & l' \end{matrix}$$

where primes indicate that mixing occurs and l is a new quark of charge -1/3. In these models it is possible to eliminate $\Delta S = 1$ neutral currents naturally[15] but flavour changing currents _must_ occur in b decay. Such models are quite plaussible and attractive.

Georgi and Glashow have recently considered a whole class of exotic topless models in which the b decays only semi-leptonically[16] (Derman[17] has discussed models in which there is a t quark but $b \to qll'$ only, where q = quark and l(l') = lepton, e.g. $b \to s\, e^+\mu^-$). In these models decays such as

$$\begin{aligned} b &\to q\, l\, \tau \\ &\to \bar{q}\,\bar{q}\,\tau \ (\Rightarrow B \to \bar{p}\,\tau^+) \\ &\to q\,\tau\,\tau \\ &\to q\,\bar{l}\,\bar{l} \end{aligned}$$

occur and the b lifetime (and also the τ lifetime) are non-standard. It is quite likely that many of these models are already excluded by limits on anomalous μ production at PETRA given by the PLUTO group.

It is sometimes suggested that studying b decays is unrewarding because measuring the K M mixing angles is hard. But this assumes the standard model is correct! Any information on the decay modes and lifetime of the b would be extremely interesting. In particular, evidence for $\Delta F = 1$ neutral currents would exclude the standard model.

3)   Does the Higgs boson exist ? If so, is its mass 0(10 GeV), a prediction with some small theoretical basis[18]? Note that if the Higgs boson does not exist or has a mass of order 1 TeV or greater, W-W scattering would violate unitarity in perturbation theory at $\sqrt{s} \sim$ 1 TeV. - indicating a break down of perturbation theory and the onset of "new physics" at or before this energy[19].

4)   Is nature really intrinsically left-right asymmetric ? If we paramaterise the data using an $SU(2)_L$ x $SU(2)_R$ x $U(1)$ model, they require essentially the standard values for $SU(2)_L$ x $U(1)$ and the bosons coupled to right handed currents must have masses greater than 0(225GeV)(assuming the same structure for right and left handed fermion multiplets; otherwise they can be lighter). Better limits on right handed currents from low energy experiments would be welcome. Experiments at e p machines such as HERA would be sensitive to right handed W's of a few hundred GeV.[21]

5)   What determines the numerous arbitrary parameters in $SU(2)_L$ x $U(1)$? In particular, why is electric charge quantized ? It is well known that this requires further unification.

3.   Physics at 1 TeV. and Beyond

3a   Further Unification
       Weak and Electromagnetic

       True unification requires a simple group or a semi-simple group plus discrete symmetries so that there is just one coupling constant. In this case the electric charge (Q) and the neutral current charge ($Q^Z = I_3 - Q \sin^2\theta_w$) are generators and, since the trace of the product of two different generators vanishes,

$$\text{Tr} \quad Q \, Q^Z = 0$$

which yields

$$\sin^2 \theta_w = \frac{\Sigma \, I_3^2}{\Sigma Q^2} \quad .$$

With u, d, e and ν_e we get the famous result that $\sin^2 \theta_w$ = 3/8, which is substantially bigger than the measured value.   There are two ways to get a smaller number:

1)   Enlarge the group and/ or add new representations. For example, with particles of charge 5/3 or -4/3 but $|\vec{I}| \leq \frac{1}{2}$ we would clearly get less than 3/8. Alternately, exploiting the freedom of indentifying I_3 in a larger group almost any value can be obtained e.g. with the group $[SU(3)]^{15}$ Georgi[22] gets 3/14. No very attractive models of this sort are known to me.

2)   Argue that there is an enormous energy scale M_x so that the Clebsch Gordon

prediction of 3/8 is substantially renormalized at the energies at which data are available. This can be made to work because $\tan \theta_w = g'/g$, where g is the $SU(2)_L$ and g' the U(1) coupling constant, and g decreases with energy because SU(2) is asymptotically free, being non-Abelian, whereas g' increases. As is well known[23] 3/8 is renormalized to about 0.20 at low energy if $M_X \sim 10^{15}$ GeV (to get $\sin^2\theta_w = 0.23$ requires $M_X \sim 10^{12}$ GeV.).

Weak and Electromagnetic and Strong

At low energy $\alpha_{SU(2)_L} \equiv g^2/4\pi$ is about 1/30 which is probably substantially less than α_S. The unified prediction is

$$\frac{\alpha_{SU(2)}}{\alpha_S} = \frac{\text{Tr }\lambda^2}{\text{Tr }I_3^2}$$

which is one with the usual groups and representations. Again this can be reduced either by altering the group and/or representation content, so that there are par-ticles which contribute to I_3 but not λ (thus Georgi[22] obtains 1/4 in [SU(3)]^15), or by introducing a huge mass scale.

It seems that elegant unification probably requires an enormous mass scale. To exploit this idea it is necessary to make the bold assumption that there is no unknown structure between 100 GeV and M_X. If such structure exists, the idea of unification at M_X may still be correct - but we cannot extrapolate to M_X without knowing or assuming what happens at lower energy.

3b Grand Unification

The minimal model which unifies $SU(2)_L \times U(1) \times SU(3)_c$ is SU(5). Among its virtues are[23]:
1) It is truly unified and charge is quantized.
2) The unit of quark charge is the unit of lepton charge divided by the number of colours.
3) A V-A structure for leptons requires a V-A structure for quarks (not antiquarks).
4) $\sin^2 \theta_w$ is predicted to be 0.20. This may turn out to be a vice since it dis-agrees with experiment by two standard deviations but it is better than 3/8!
5) It predicts nucleon decay - a good feature since it has stimulated experimental interest in a fundamental topic[24] and it provides the possibility of resolving the long standing problem of the cosmic value of n_B/n_γ[25].

Among its vices:
1) The Higgs system is unnatural (I return to this below).
2) The existence of three or more "generations" of fermions is not explained.

3) One might object to the intrinsic left,right asymmetry.

There are various other models such as E(6)[26] and SO(10)[27] which share some of the vices and virtues of SU(5). SO(10) is interesting because it is the simplest of a very large class of models in which neutrinos are expected to have small masses and neutrino oscillations may occur[28]. Further experiments on ν oscillations could be very profitable.

Many of these models have interesting features and suggest important experiments but none looks like being the whole truth.

3c "Horizontal" Unification

Presumably the different "generations" of fermions would be included in one representation in an ultimate theory. Attempts to introduce discrete symmetries between generations have been made (usually with an eye to reproducing empirically successful relations for the Cabibbo angle) but they appear very ad hoc. Any continuous symmetries must be gauge symmetries[29] (otherwise there would be unwanted massless Goldstone bosons) so they would imply the existence of vector bosons of mass M_V which mediate intergeneration transitions such as $K \to e\,\mu$, $K \to \pi\,e\,\mu$, $\mu \to e\,\gamma$ etc. Existing limits require $M_V > 0(30 \text{ TeV.})$. Improvement of the limits would obviously be welcome.

Attempts to unify a horizontal group G_H with $SU(2)_L \times U(1) \times SU(3)_c$ in a "super grand unified model" (SGUM) have not proved very successful. Gell-Mann et al.[28] have considered SO(18) with fermions in the 256 dimensional representation as an example of a SGUM but with so many fermions asymptotic freedom is lost and nothing can really be calculated. Orthogonal groups, advocated by Wilczek and Zee[30], may prove more promising.

Faced with such large groups and representations it is hard not to feel that some idea or principle is missing. The continuing absence of plausible examples of SGUMs suggeststhat it may be necessary to consider the idea that quarks and leptons are composite; we return to this later.

3d Models Without Higgs Mesons

The Higgs mechanism seems unsatisfactory for several reasons:

1) It introduces a plethora of arbitrary parameters.

2) In models such as SU(5) the parameters must be adjusted to twenty places of decimals to keep some particles light compared to M_X[31].

3) Even forgetting GUM, it can be argued that absurd adjustments are needed to keep particles light in theories with elementary scalars. Because of the quadratic divergence, the bare parameters must be tuned to incredible accuracy if there is a genuine cut off at very high energy[32] (e.g. at the Planck mass). Alternatively.if,

following 't Hooft[33], we require that the qualitative properties of physics at an
energy scale E, expressed in terms of parameters defined at $E_2 \gg E_1$ should not
require adjustment of those parameters to accuracy E_1/E_2, then elementary scalars
are intolerable.

Although the idea that we should reject all theories which require "unnatural"
adjustment of parameters is interesting, it must be treated with caution. What
appears arbitrary and unnatural in a partial theory of nature may emerge quite
naturally as part of some more comprehensive theory which is not yet dreamt of.
In any case it is clear that we should seek some principle which determines the
coupling of the Higgs mesons (supersymmetry ? or the assumption that the couplings
are infrared stable fixed points[18]?) or attempt to do without them.[34]

Consider the world of $SU(3)_{C_0} \times SU(2)_L \times U(1)$ with just u and d quarks. There
is exact chiral symmetry (since $m_q = 0$ with no Higgs) but presumably it is spontaneously broken and consequently the pion is massless. Consider the vacuum polarization diagrams for the $\vec{W}$ and B fields

Because of gauge invariance, the vacuum polarization tensor $\pi_{\mu\nu}$ necessarily has the
structure $(-g_{\mu\nu}q^2 + q_\mu q_\nu)\,\pi(q^2)$. The pion pole makes a contribution $\dfrac{e^2 f_\pi^2}{q^2}$ to
π. The pole in the $J = 0$ $q_\mu q_\nu$ term removes the zero as $q^2 \to 0$ in
the $g_{\mu\nu}$ part of $\pi_{\mu\nu}$, thereby vitiating the argument that $\vec{W}$ and B remain massless.
In fact, summing vacuum polarization diagrams, the propagators contain a piece

$$\frac{-g_{\mu\nu}}{q^2 - e^2 f_\pi^2}$$

This is nice. Both $W^\pm$ and Z^0 acquire mass with (when we put in the correct factors)
$M_{W^\pm} = M_{Z^0}\cos\theta_W$, as required by the data $(\rho \simeq 1)$[35]. However, $M_W \sim O(ef_\pi) \sim O(30$
MeV) and there are no physical π mesons left – they have become the longitudinal
degrees of freedom of $W^\pm$ and Z^0.

This disaster can be avoided by introducing a new heavy doublet of quarks (U,D)
– called "techniquarks" by Susskind, whose terminology we adopt – and assuming that
they interact through a new non-Abelian force called "technicolour" which becomes
strong at an energy $\Lambda' \sim O(1\ \text{TeV.})$. It is assumed that chiral symmetry of the U,D
system is spontaneously broken giving rise to a zero mass "technipion" with

$f_{\pi'} \sim 0\left(\dfrac{\Lambda' f_\pi}{\Lambda_{QCD}}\right)$ and large masses (0(1 TeV))for U and D. It is the technipions
which now become the longitudinal degrees of freedom of $W^\pm, Z^0$ which, by construction,
acquire masses of order 80 GeV as required (unitarity already led us to anticipate
a scale of 1 TeV for whatever replaces the Higgs mechanism). A physical pion, made
of u and d, remains - but alas, it is massless since chiral symmetry is exact.
Technicolour successfully replaces the Higgs fields in giving mass to the vectors
but not in giving chiral symmetry violating masses to quarks and leptons, which is
normally done through Yukawa couplings of the Higgs fields which break chiral symmetry. Additional gauge interactions, called "extended technicolour" (ETC), seem
to be needed to explicitly break chiral symmetry (while leaving enough so that
$M_{\pi'}$ is zero). Alternatively it is possible that the light fermions are bound states
of the heavy technifermions, which acquire mass dynamically[33]. We consider this
possibility below.

In ETC the light quarks and leptons acquire mass because they are linked to
heavy technifermions by a new set of gauge bosons of mass M. The relevant diagram
is

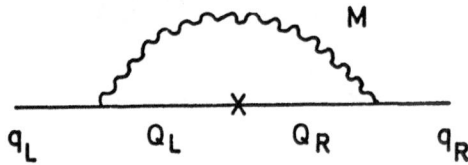

Since it is dynamically generated, the technifermion mass insertion (which flips
Q_L to Q_R) is momentum dependent, behaving as $(1\ \text{TeV})^3/Q^2$ (up to logs) for large Q^2.
This diagram therefore gives

$$m_q \sim \frac{(1\ \text{TeV})^3}{M^2}$$

requiring M $\sim$ 0(100 TeV).

No believable models have yet been constructed but all such Higgless schemes
share several general features with rather definite observable consequences. First,
left handed quarks and leptons must belong to the same irreducible representation
of the ETC gauge group; otherwise too much chiral symmetry remains and there would
be massless axion-like Goldstone bosons[36]. In the case of grand unified models
all fermions must belong to a single irreducible representation. This means
that there is some sort of intergeneration and quark-lepton unification at 0(100 TeV),
although it does not induce nucleon decay. Second, a large number of technifermions
are needed in order to give mass to the numerous light fermions. The consequences
are

1)   There are ETC interactions which link different generations giving rise to decays such as

$$K^0_L \to \mu e, \ K \to \pi \mu e, \ \mu \to e\gamma$$

The effective coupling is $G' \sim 0(10^{-5} G_F)$, so the rates may be close to the present limits.

2) Of the many possible massless technipions made from the numerous technifermions, one combination is eaten by $W^\pm$ and Z^0 and the others gain masses of order 10-40 GeV through chiral symmetry breaking ETC interactions (other techihadrons are of course much heavier). These "pseudo Goldstone bosons" (P) are quite like physical Higgs particles (H) in that their coupling to fermions (F) are proportional to M_F and they will appear point-like at energies << 100 TeV. They differ in that

a) $P^\pm$ must exist, whereas $H^\pm$ need not.

b)   They couple to $\bar{\psi} \gamma_5 \psi$ whereas H couples to $\bar{\psi}\psi$, so they decay into S wave rather than P wave $f\bar{f}$ pairs.   They can be produced (like H's) in $e^+e^- \to P^+P^-$, $(t\bar{t}) \to P^0\gamma$, $e^+e^- \to b\bar{b}P$ etc. Given the predicted masses, it would be worth searching carefully for $P^\pm$ production at PETRA and PEP.

3e   Composite Quarks and Leptons ?

The proliferation of fermion species suggests that we consider the possibility that the fermions we observe are composite. This idea faces two obvious difficulties :

1)   Why should u, d, e, μ and particularly ν_e and ν_μ be so light ? This could be ensured by having some chiral symmetry which is exact (or approximately exact e.g. it might be broken by electro-weak interactions only) which is not spontaneously broken. Manifest chiral symmetry requires the existence of physical zero mass fermions - naively these would be the fermion fields in the Lagrangian but they would have to be   bound states if the fundamental fermions were confined.
't Hooft has investigated the conditions which must be satisfied in order that an effective low energy field theory consisting only of the massless bound states would be anomaly free and hence could be unitary and renormalizable[33]. His results are discouraging, but Dimopoulos, Raby and Susskind have recently found examples which satisfy his consistency conditions[37].

2)   The magnetic moment problem.As stressed by Lipkin[38], there seem to be two difficulties in obtaining Dirac   moments for composite particles :

a)   The mass which sets the scale of the magnetic moment is not generally the mass of the composite (consider the case of an atom)

b)   The magnetic moment of a composite is not necessarily proportional to the charge (consider the magnetic moment of the neutron in the quark model), yet we want

$\mu_u/\mu_d \simeq -2$ etc. Proportionality can be obtained for a non-relativistic S wave three body system of spin $\frac{1}{2}$ fermions only if either the charges are all equal   or the wave functions is a 50% mixture of configurations in which constituents 1 and 2 couple to S = 0 and 1 (which might make problems with statistics).

However, while these problems are obviously genuine for weakly bound non-relativistic systems, they may not be in the case of very tight binding. We might suspect that for very tightly bound systems there should be an effective field theory containing only fields representing the light composites which would be well behaved for energies   up to the binding energy E_B, where the composite nature becomes apparent. If this is so, μ_{anom} (the deviation from a Dirac moment for the composite) would be bounded by $0 \, (E_B^{-1})$(up to logs) in order that unitarity be respected for $E < E_B$ in processes such as $\gamma F \rightarrow \gamma F$.

There are clearly many problems in making consituent models for quarks and leptons which deserve further study and it is probably premature to consider specific models. However, note that if quarks are 3 fermion composites we would expect the first excited state to have spin 3/2. Spin 3/2 quarks might show up in ep experiments at very high energy[39].

3f   Gravity

There are two problems in a straight forward particle theorist's approach to perturbation theory in quantum gravity:

1)   In graviton   - graviton scattering, second order diagrams

give an amplitude proportional to K^2S (where $K = \sqrt{G}$ $= (1.22 \times 10^{19} \text{GeV})^{-1}$. Unitarity is violated and hence perturbation theory fails for $\sqrt{S} \sim 0(10^{19}\text{GeV.})$.

2)   The theory is non-renormalizable because of divergences e.g. in diagrams like

It is easier to imagine curing problem 2) than problem 1). For example, consider extended supergravity which is the one known theory of gravity coupled to matter which may be finite[40]. Divergences cancel between diagrams in which particles of different spin propagate around closed loops. When we cut open these diagrams to obtain the imaginary part, cuts through loops containing different particles contribute to the lower order amplitudes for different processes. The mechanism by which the closed loop divergences cancel does not make the theory unitary. The failure of perturbative unitarity at $\sqrt{s} \sim K^{-1}$ seems almost inevitable and suggests that perhaps we should not worry if perturbative renormalizability also fails (if it holds we could at least calculate perturbatively for $S \ll K^{-1}$ which would be nice . Of course perturbative unitarity fails in QED also but only at a truly astronomical energy $\sim m_e e^{1/\alpha}$).

In any case it seems that, if we exclude particles with spin greater than two, the largest possible group allowed by extended supergravity is $O(8)$ which cannot accomodate $SU(2) \times U(1)$[40]. Supergravity may therefore be a dead end (unless we think $SU(2) \times U(1)$ etc. is an epiphenomenon and apply $O(8)$ at a deeper level).

Apart from the hint from unitarity, there are two other arguments which suggest that gravity may be relevant at energies $\ll K^{-1}$:

1)   Einstein's equation reads

$$R_{\mu\nu} - \tfrac{1}{2}g_{\mu\nu}R + \lambda g_{\mu\nu} = 8\pi G T_{\mu\nu}.$$

λ is the cosmological constant, which Einstein added because without it a homogeneous isotropic universe cannot be static, a step which he later described as "the biggest blunder of my life". λ is thought to be < 0 $(10^{-56} cm^{-2})$[41]. In $SU(2)_L \times U(1)$ with the standard Higgs mechanism $< 0|T_{\mu\nu}|0 > = -g_{\mu\nu} < 0|L|0> =$ const $g_{\mu\nu}$, giving rise to an effective cosmological constant about 50 orders of magnitude too big[42] (assuming $M_\phi \gtrsim O(1 \text{ GeV})$). This can be cured by introducing a cancelling cosmological constant (tuned to 50 decimal places). If we take this idea seriously, this constant would have been uncancelled at very early times, before ϕ developed a vacuum expectation value, and the evolution of the universe would have been greatly effected[43]. A more likely message is that our understanding of gravity is incomplete. In any case it seems that gravity "knows" about physics on the $SU(2)_L \times U(1)$ scale of 100 GeV.

2)   The source of the unitarity/ divergence problem is that R has dimensions. However, Einstein's theory is invariant under general coordinate transformations which include scale transformations. de Alfaro , Fubini and Furlan have pointed out that it must therefore be possible to write the Lagrangian in such a way that K does not enter[44]. The usual formulation is only obtained when an expansion is made about flat space-time

$$g_{\mu\nu}(x) = K^{-2}\delta_{\mu\nu} + K^{-1}\phi_{\mu\nu}$$

This expansion is certainly very good at low energies but we should not necessarily be surprised if it fails at large energy (just as we would only expect to be able to expand in f_π^{-1} at low energy in chiral dynamics). In models in which gravity is coupled to matter, de Alfaro et al. find classical instanton and meron like solutions which differ radically from the expansion about $\delta_{\mu\nu}$ (they might be obtained by an infinite summation of terms in that expansion). In these solutions $g_{\mu\nu}$ and matter fields are of the same order of magnitude. This leads them to suggest that gravity may even play a role in confinement !

If these suggestions and speculations have any truth, we continue to neglect gravity in particle physics at our peril. It certainly appears likely that gravity may hold the clue to the next big step forward.

4.   Summary/ Conclusions/ Problems/ Questions

QCD - is probably correct but eventually we must show that QCD predicts hadrons.
SU(2)$_L$ x U(1) - is probably correct but the W and Z must be found. SU(2)$_L$ x U(1)
could be just part of a larger electro-weak group (e.g. SU(3)$_L$ x U(1)).
In particular the t quark need not exist, in which case b decays must be non-standard.

Real Unification - probably requires the introduction of an enormous mass scale M_X.
To explore this idea quantitativelywe must assume the absence of new gauge structure below M_X (the failure of this assumption would not imply that the idea is wrong). SU(5) is an attractive example of a grand unified model but it is presumably not the end even if it is correct. The "generation" problem is not satisfactorily solved by any known model.

The Higgs mechanism - has several unattractive features. A further principle is needed to specify the Higgs interactions (supersymmetry ?). Alternatively, perhaps there are no Higgs mesons, in which case exotic decays (K $\rightarrow$ μe) and detectable pseudo-Goldstone bosons must exist.

Composite quarks and leptons - might be the answer to the fermion proliferation problem but there are many unsolved difficulties with this idea.

Gravity - remains an enigma. There are hints that it may become important at relatively low energy. In any case it is probably foolish to ignore it while seeking a grand synthesis.

There are obviously many open theoretical problems and questions. However, we should remember that the leading questions of ten years ago are solved (Are quarks "real" in any sense at all ? What is the origin of the strong force ? Is it possible to make a renormalizable field theory of weak interactions?). We are only just beginning to come to grips with a new generation of more profound problems (they amount to: What are the fundamental degrees of freedom ? What is the origin of mass ? How can we incorporate gravity ?).

As well as the obvious experimental questions (searches for further quarks and leptons, W's, Z's and Higgs bosons) there are many slightly less obvious. In particular, we would welcome

- better low energy limits on right handed currents
- better neutral current data
- any information on b decay
- better data/limits on CP violation
- further study of rare decays such as $K_L^0 \rightarrow \mu e$, $\mu \rightarrow e\gamma$ and $K \rightarrow \pi e \nu$.
- study of possible neutrino oscillations
- searches for nucleon decay on $n - \bar{n}$ oscillations.

Clearly there is much to occupy us while waiting do do two photon physics at LEP.

Acknowledgements

I am grateful to Graham Ross for numerous discussions and to him and Ian Aitchison for comments on the manuscript.

Footnotes and References

1)   I am referring to work by M. Creutz and K. Wilson which is apparently not yet published. Some of the results are quoted by C. Callan et al. Phys. Rev. Lett. 44, 435, 1980.

2)   For a lucid review see E. Witten Harvard preprint 79/A007;to be published in Nucl. Phys. B and Harvard preprint HUTP 79/A078, lectures at 1979 Cargèse institute.

3)   For a review see R.J.Crewther Riv. Nuovo Cimento 2, 63, 1979 and CERN TH 2791, 1979. Because of the anomaly the gauge invariant isoscalar axial current is not conserved but there is another (gauge dependent) isoscalar axial current which is conserved, which must therefore be coupled to zero mass particles. However, since the current is not observable, being gauge dependent, there might be two mass-less particles with opposite metric, allowing them to cancel exactly in observable amplitudes. This happens in the two dimensional Schwinger model (J. Kogut and L. Susskind Phys. Rev. D 11, 3594, 1974). 't Hooft (Phys. Rev. Lett. 37, 8, 1976 and Phys. Rev. D 14, 3432, 1976, (E) D 18, 2199, 1978) claims that it happens in QCD also. Crewther (loc. cit.) argues that in the approximation considered, in which gauge fields become pure gauges at ∞, isovector chiral symmetry would not be spontaneously broken and claims that this invalidates 't Hooft's conclusion. Further-more, he stresses the difficulty of satisfying all the Ward identities if the U(1) boson decouples. However, it seems that they are satisfied in the $1/N_c$ expansion (see P. Di Vecchia Phys. Lett. 85B, 357, 1979 and references therein).

4)   For a review see R.D. Peccei, Munich preprint MPI-PAE/ P Th 59/78. For a re-cent brief summary of attempts to explain the small value of θ see M. Gaillard, Fermilab-Conf. 79/87-Thy. It might seem that this term has no effect because it can be written as a total divergence, whose contribution to the action can be expressed as a surface integral (of a gauge dependent current). However this is not true because of the rich vacuum structure in QCD. Even if the $F_{\mu\nu}^i$ vanish at infinity, which is presumably not the case in a confining theory,there are configurations of non-zero winding number, in which the A_μ^i become pure gauges at infinity but cannot be simultaneously transformed away in all directions. These configurations con-tribute to the surface term. There are two contributions to θ: 1) It enters as a coupling constant (even if it is zero initially this term will be needed to can-cel divergences induced by the weak interactions) 2) It enters the effective Lagrangian as a reflection of the phase of the relative contributions of configura-tions of different winding number to Green functions (expressed as path integrals), which is an (arbitrary ?) parameter. These two contributions must combine to give a net θ which is small.

5)   R.M. Barnett, M. Dine and L. Mclerran.   SLAC- PUB - 2475, 1980.

6)   R.Q. Hung and J.J. Sakurai Phys. Lett. 88B, 91, 1979.

7)   P. Langacker et al. BNL-26498, 1979. See also I. Liede and M. Roos. Helsinki

preprint HU-TFT-79-27.

8) See e.g. G.G. Ross and T. Weiler. Journal of Physics G5, 733, 1979. H. Georgi and S. Weinberg Phys. Rev. D. 17, 275, 1978. E.H. de Groot, G.J. Gounaris and D. Schildkneckt Phys. Lett. 85B, 399, 1979, and Bielefeld preprints B1-TP 79137, and B1-TP 79-39.

9) There is a simple way to see this. For light fermions we only need the W^0-B propagator matrix in which the new neutrals change $M_B^2 - q^2$ to $M_B^2 (q^2)$. Expanding $M_B^2 (q^2) = (M_B^0)^2 + \lambda q^2 + \lambda'(q^2)^2 + \ldots$ we can rescale the B field to obtain $\lambda = 1$. Dropping terms of $O(q^2)^2)$ and higher we obtain exactly the same propagator matrix as originally in $SU(2)_L \times U(1)$. However, although elsewhere we can neglect it, we must include the effects of the $\lambda'(q^2)^2$ term in calculating the photon eigenvalue of the inverse propagator matrix which becomes $q^2(1 - \lambda' \cos^2\theta_w q^2) + O((q^2)^3)$. This changes $\frac{1}{q^2} (J_\lambda^{em})^2$ to $\frac{1}{q^2} (J_\lambda^{em})^2 + \lambda' \cos^2\theta_w (J_\lambda^{em})^2 + O(q^2)$ in the effective Lagrangian at low q^2.

10) de Groot et al., loc. cit.

11) The relevant formulae are given, for example, by C.H. Llewellyn Smith and D.V. Nanopoulos Nucl.Phys. B78, 205, 1974, with important corrections in Nucl Phys. B83, 544, 1974.

12) F. Antonelli et.al. Phys. Lett. 91B, 90, 1980. M. Veltman Phys. Lett. 91B, 95, 1980.

13) D. Horn and G.G. Ross Phys. Lett. 67B, 460, 1977. G. Altarelli et al. Phys. Lett. 67B, 463, 1977.

14) V. Barger and S. Pakvasa Phys. Lett. 81B, 195, 1979.

15) H. Georgi and A. Pais Phys. Rev. D 19, 2746, 1979. There can of course be a third triplet and hence another quark with Q = 2/3 in this model.

16) H. Georgi and S.L. Glashow. Harvard preprint HUTP79/A073.

17) E. Derman Phys. Rev D19, 317, 1979.

18) This value is obtained if symmetry breaking is due entirely to radiative corrections. For a discussion of the phenomenology of a 10 GeV Higgs boson and references see J. Ellis et.al.Phys. Lett. 83B, 339, 1979 (see also footnote 31). The mass of the Higgs meson is also predicted if the couplings lie in the domain of attraction of infrared stable fixed points (B.J. Pendleton and G.G. Ross, Oxford preprint in preparation).

19) C.E. Vayonakis Nuovo Cimento Lett 17, 383, 1977. B.W. Lee, C. Quigg and H.B. Thacker Phys. Rev. Lett. 38, 883, 1977 and Phys. Rev. D.16, 1519, 1977.

20) For a recent analysis and references see G. Ecker Vienna preprint UWTh Ph 79-30, to be published in Proc. 1979 Visegrad symposium.

21) ECFA 80/42 DESY-HERA 80/01.

22) H. Georgi Harvard preprint HUTP 79/ A036.

23) For recent reviews and references see S.L. Glashow, lectures at the 1979 Cargèse institute (Harvard preprint HUTP79/A059) and J. Ellis, CERN-2723 (to be pub-

lished in Proc. 1979 EPS conference).

24) Indirect searches for nucleon decay may produce very interesting limits. See J.C. Evans and R.I. Steinberg. Science, 2 Sept. 1977, p.989. K.W. Allen (private communication) has proposed an interesting experiment with tellurium. If a neutron in Te_{52}^{130} decays it yields Te_{52}^{129} whereas if a proton decays it yields Sb_{51}^{129} which β decays in 10^4 secs. to give Te_{52}^{129} also. Te_{52}^{129} then β decays in 10^4 secs. to I_{53}^{129} which then β decays in 1.7×10^7 years to Xe_{54}^{129}, which is stable. Any primordial I_{53}^{129} will have decayed by now so the detection of I_{53}^{129} in Te ore could be attributed to the accumulated effect of nucleon decay over 10^7 years (provided it exceeds the amount produced by cosmic ray induced inverse β decay etc.) A nucleon lifetime of 10^{30} years would give 10^3 atoms of I^{129}/Kg. I_{129} provides the only negative ion of mass 129 and can be detected using a Van de Graff as a mass spectrometer.

25) For an introductory account see M.S. Turner and D.N. Schramm Physics Today, p. 42, Sept. 1979.

26) For recent discussions and references see R. Barbieri and D.V. Nanopoulos, CERN TH 2810, 1980 and H. Ruegg and T. Shücker Nucl. Phys. B. 161, 388, 1979.

27) For recent discussions see H. Georgi and D. Nanopoulos Nucl. Phys. B. 155, 52, 1979.

28) M. Gell-Mann, P. Ramond and R. Slansky (unpublished) have suggested a mechanism which naturally gives ν_R a large Majorana mass leaving ν_L nearly massless. Witten has shown (Phys. Lett. 91B, 81, 1980) that in the minimal 0(10) model the mass of ν_L would be $10^{0 \pm 2}$ eV. Note that existing data are compatible with substantial mixing angles and neutrino masses in the eV. range, giving oscillations to which reactor and accelerator experiments would be sensitive (for a recent discussion see A. De Rujula et. al. CERN preprint TH 2788-1979). The low solar neutrino flux may be evidence for oscillations. Barger et.al. (Wisconsin preprint COO-881-135, 1980, to be published in Physics Letters) claim reactor data show evidence for oscillations.

29) See F. Wilczek and A. Zee Phys. Rev. Lett. 42, 421, 1979 and P. Ramond Caltech preprint CALT-68-709-1979 for examples.

30) F. Wilczek and A. Zee Princeton preprint "Spinors and Families", 1979.

31) This can be done (A. Buras, J. Ellis and M. Gaillard Nucl. Phys. B. 135, 66, 1978). Weinberg (Phys. Rev. Lett. 82B, 387, 1979) has shown that if we view the adjustment in terms of the effective potential and tune the ϕ^2 term to be zero (which could possibly be required by some symmetry in a future theory), then radiative corrections give rise to the required heirarchy of masses of order e^{1/g^2}. The mass of the light Higgs is of 0(10 GeV) in this case. J. Ellis et. al. (Nucl. Phys. B. 164, 253, 1980) argue that M_X itself might be understood in terms of the Planck mass in this case.

32) This argument is discussed by L. Susskind (ref. 34) who attributes it to

K. Wilson.

33) G. 't Hooft. Lecture at the 1979 Cargèse institute.

34) This relatively old idea has been vigorously investigated by S. Weinberg (Phys. Rev. D 13, 974, 1975 and D 19, 1277, 1979). L. Susskind and collaborators (Phys. Rev. D 20, 2619, 1979, Nucl. Phys. B 155, 237, 1979 and Phys. Rev. D 20, 3404, 1979) and E.Eichten and K. Lane (Harvard preprint HUTP-79/A062).

35) M. Weinstein (Phys. Rev. D 8, 2511, 1973) was apparently the first to point out that this "weak $\Delta I = \frac{1}{2}$ rule" depends on the representation content of the unphysical Goldstone bosons and not on whether they are elementary or composite. Care must be taken to ensure that this relation is maintained when isospin breaking sufficient to obtain the correct value of m_u/m_d is introduced. See P. Sikivie et. al. Stanford preprint 1TP-661, 1980 and A. Carter and H. Pagels, Rockefeller report COO-2232B-187, 1979.

36) E. Eichten and K. Lane (loc. cit.) whose discussion of the general implications of these models we follow. See also M. Beg et.al., Rockefeller report COO-2732B-189, 1979 and S. Dimopoulos, invited talk at the 1979 EPS conference for a discussion of the properties of the pseudo Goldstone bosons.

37) S. Dimopoulos, S. Raby and L. Susskind Stanford preprint 1TP-662, 1980.

38) H. Lipkin Riv. del Nuovo Cimento 1, 134, 1969 and FermiLab-Conf 79/60-THY. See also M. Gluck Phys. Lett. 87B, 247, 1979.

39) Possibly we could argue that $E_B > 0$ (μ_{anom}^{-1}), giving enormous binding energies for leptons but leaving open the possibility that quarks will appear composite at 10's of GeV.

40) For a review see P. van Niewenhuisen Physics Reports (in press).

41) C.W. Misner, K.S. Thorne and J.A. Wheeler "Gravitation", Freeman, 1973.

42) A. Linde JETP Lett. 19, 183, 1974. M. Veltman Phys. Rev. Lett. 34, 777, 1975.

43) I understand that A. Guth has investegated the consequences.

44) V. de Alfaro, S. Fubini and G. Furlan CERN TH 2799, 1979.

SUMMARY AND CONCLUDING REMARKS

G. Altarelli

Istituto di Fisica dell'Università - ROMA (Italy)

Istituto Nazionale di Fisica Nucleare - Sezione di Roma

It is certainly not simple to summarize in a short talk the great amount of impressive work presented at this meeting. I must start by saying that, as any other apparatus, I also have a limited acceptance and solid angle plus many systematic biases, so that if you shall find that my talk was missing a lot, this is due to my limitations of which I am well aware.

The topics discussed can be divided into three broad classes: a) Experimental and theoretical techniques for the preparation and interpretation of γ-γ measurements b) Experimental results and their theoretical implications c) Prospects for the near future of γ-γ physics.

The main conclusion of the workshop on experimental techniques, led by Barbiellini, as emphasized for example by Courau was the following. With the improved quality of present and future central region detectors and the increase of accelerator energy, it becomes easier to separate the three components of single γ, γ-γ and background events. For example the reaction energy of beam-gas events increases as $2E_1 M$, where M is the molecule mass, i.e. linearly with the beam energy, while the energy of interesting events increases as $2E_1 E_2$, i.e. quadratically. As a consequence, if not at DCI, but already at PETRA energies and better beyond, tagging becomes less and less important to dig out γ-γ physics. While tagging appeared in the past years of fundamental importance, now it looks that in future experiments tagging will only be necessary for selecting highly virtual photons for deep inelastic physics on photon targets.

On the theoretical side we all know that the Weizsäcker-Williams result remains the physical basis for the possibility of extracting two photon physics from $e^+ e^-$ collisions. A lot of valuable contributions on this subject have been brought up at this meeting by Kessler, Carimalo and J.Parisi. The region of validity of the method has been further clarified in different physical conditions, and the relative merits and /or shortcomings of existing variations of the approach have been quantitatively compared. On the other hand it has been convincingly shown at this meeting by the really impressive talks of Vermaseren and Smith that modern computer techniques (algebraic programming and Montecarlo multidimensional integration) make by now possible to treat a given process exactly with all the particular cuts in acceptance properly taken into account. It is clear that both methods are important and they most usefully complement each other. The first is remarkable for simplicity and transparency , the second is exact and complete and can be adapted to the details of a given experiment. The same remarks apply to the problem of radiative corrections. While general studies, like the one presented by Cochard, are important for an estimate of the possible orders of magnitude for the effects, it remains true that for each experiment a detailed numerical evaluation is necessary (and in fact possible). This is because it often happens that radiative corrections are concentrated in small regions of phase space and therefore a realistic evaluation and subtraction must carefully take into account the actual acceptance of the experiment.

At this meeting experimental results have been presented of a quality and quantity that clearly show we are out of the pioneering epoch of γ-γ physics and entering a period of valuable returns for all the intellectual and material investments in the field. A first collection of results is on the radiative widths of pseudoscalar and tensor mesons, mainly up to now from the Mark II experiment at SPEAR (presented by Jenni) and from the Pluto collaboration at PETRA (illustrated here by Berger). More results are about to come from the Tasso group (as discussed by Hilger). The theoretical basis for these measurements was established in 1960 by F.Low in a classic paper.

The most recent results are summarized as follows:

$$\Gamma_{\gamma\gamma} (\eta') = 5.8 \pm 1.1 (\pm 20\% \text{ syst.}) \text{ KeV} \qquad \text{(Mark II)}$$
$$\Gamma_{\gamma\gamma} (f_0) = 2.3 \pm 0.5 \text{ KeV} \qquad \text{(Pluto)}$$
$$\Gamma_{\gamma\gamma} (A_2) < 2.5 \text{ KeV} \qquad \text{(Mark II)}$$
$$\Gamma_{\gamma\gamma} (f') \cdot B (f' \rightarrow K^+ K^-) < 0.6 \text{ KeV} \qquad \text{(Mark II)}$$

The η' width completes the set of pseudoscalar widths into 2γ. The $\pi^0 \rightarrow 2\gamma$ width can be predicted starting from three principles which are all first class in modern elementary particle physics: the axial anomaly, PCAC and colored quarks. The predicted value (which would be nine times smaller with no color) is $\Gamma_{\gamma\gamma}(\pi^\circ) \sim$ 7.6 eV to be compared with the experimental width $\Gamma_{\gamma\gamma}(\pi^\circ)_{EXP} \simeq 7.95\pm0.55$ eV$_{theory}$ and this agreement is an important achievement of our present theoretical framework. The η and η' widths can be related to $\Gamma_{\gamma\gamma}(\pi^\circ)$ if one assumes nonet symmetry, a pseudoscalar mixing angle $\theta_p = -11^\circ$ and fractionally charged quarks. Only the last of these assumptions is first class, while the other two are more questionable. In QCD nonet symmetry for wave functions is as good as SU(3), but SU(3) breaking effects are particularly important for pseudoscalar mesons as is clear from the values of their masses. The value for θ_p is obtained empirically from the assumed validity of the quadratic Gell-Mann-Okubo mass formula. From these assumptions, being the amplitudes for 2γ decays proportional to the squared charges of the quarks one derives the ratios of rates (excluding momentarily phase space corrections):

$$\Gamma_3 : \Gamma_8 : \Gamma_1 = 3 : 1 : 8$$

where $\Gamma_{3,8,1}$ refer to the SU(3) components tranforming as the third and eight members of an octet and as a singlet respectively (that is $\eta_3 = \frac{\bar{u}u - \bar{d}d}{\sqrt{2}}$, $\eta_8 = \frac{\bar{u}u + \bar{d}d - 2\bar{s}s}{\sqrt{6}}$, $\eta_1 = \frac{\bar{u}u + \bar{d}d + \bar{s}s}{\sqrt{3}}$). From this, by using $\theta_p = -11^\circ$, p^3 phase space corrections and $\Gamma_{\gamma\gamma}(\pi^\circ) = 7.95$ eV, one obtains $\Gamma_{\gamma\gamma}(\eta)=414$ eV and $\Gamma_{\gamma\gamma}(\eta') = 6.3$ KeV. These values are to be compared with the experimental values $\Gamma_{\gamma\gamma}(\eta)=324\pm46$ eV and $\Gamma_{\gamma\gamma}(\eta')=5.9\pm1.6\pm1.2$ KeV, which leads to a quite reasonable agreement. The central values of the three experiments are in fact reproduced by $\theta_p = -7.7^\circ$ and an 8% deviation from nonet symmetry.

An important observation is that the measured value of $\Gamma_{\gamma\gamma}(\eta')$ is a blow against integrally charged quarks à la Han-Nambu. In fact in this model the electromagnetic current has a color singlet and a color octet component:

$$J_\mu = J_\mu^1 + J_\mu^8 \tag{1}$$

J_μ^1, the color singlet part, is the same as for fractionally charged quarks, while J_μ^8 is a color octet, flavor singlet, additional component. For a color singlet hadron the matrix element: $\langle h | J_\mu^8 \, J^{\mu 8} | 0 \rangle$ is not necessarily zero, provided h is a flavor singlet. In fact it turns out that:

$$\langle \eta_1 | J_\mu^8 \, J^{\mu 8} | 0 \rangle = \langle \eta_1 | J_\mu^1 \, J^{\mu 1} | 0 \rangle \tag{2}$$

It follows that the amplitude for η_1 decay is twice the value for fractionally charged quarks. Then the η' rate is about four times larger, and would be exactly four
times larger for $\theta_p = 0$, i.e. for $\eta' = \eta_1$. The by now few remaining supporters of
integrally charged quarks could blame nonet symmetry and the assumed value of θ_p
for this failure of their preferred model, but it is in any case true that the most
direct and plausible interpretation of the data strongly favors fractionally charged quarks. Note that the previous argument strongly relies on the fact that the axial
anomaly is a short distance effect that can be viewed as a subtraction at infinite
q^2, so that no pushing up of the color threshold can change the result. Otherwise
one could argue, as usual in similar cases, that the matrix element on the l.h. side
of eq.2 can only be different from zero above threshold for production of intermediate colored states.

Going now to tensor mesons a similar argument can be used for relating $\Gamma_{\gamma\gamma}(A_2)$,
$\Gamma_{\gamma\gamma}(f_o)$ and $\Gamma_{\gamma\gamma}(f'_o)$. In this case no absolute prediction for any of the three
rates can be given from an anomaly theorem. However nonet symmetry is expected to
be a better approximation because of the smaller relative impact of SU(3) breaking
for heavier mesons and likewise the ideal mixing ansatz for tensor mesons is well
grounded (for example it makes little difference whether a linear or quadratic mass
formula is assumed and further support arises from the decay rates). Ideal mixing
means that $A_2 \approx \dfrac{\bar{u}u - \bar{d}d}{2}$, $f_o = \dfrac{\bar{u}u + \bar{d}d}{2}$, $f'_o = \bar{s}s$. One then predicts :
(with no phase space corrections):

$$\Gamma_{A_2} : \Gamma_{f_o} : \Gamma_{f'_o} = 9 : 25 : 2$$

This prediction remains for coming experiments to test, in that up to now only the
f_o width has been measured (note that Γ_{f_o} is predicted to be the largest width).
Many models have been proposed in the past for the absolute width of the f_o. As discussed in detail by Greco the a priori sufficiently reliable predictions based on
approximate saturation of super convergence relations tend to lead to a larger $\Gamma_{\gamma\gamma}(f_o)$
than measured (by a factor of order two). It is important in this respect to wait
for the result on $\Gamma_{\gamma\gamma}(f_o)$ from the Tasso collaboration which should come out
pretty soon.

As for the prospects of measuring the radiative widths of heavier mesons, like
η_c or η_B they look rather hopeless because increasing the energy even up to the
LEP range does not make the situation better.

The data on the $\gamma + \gamma \rightarrow$ hadron total cross section are by now sufficient to provide a precise determination of this quantity in the range of c.o.m. energy between 1 and 7 GeV. The values at large energy agree reasonably well with the Regge pole expectation obtained through factorization. At low energies the measured value is higher then the extrapolated Regge fit, indicating an additional contribution falling down as 1/s. This can be attributed to non leading Regge contributions (beyond the Pomeron and the tensor trajectories) and to the known existence of fixed singularities in this channel. Greco and Srivastava presented a model for this contribution in terms of the parton box diagram. Although the observed magnitude of the 1/s contribution is reproduced the model is too simplistic to be true.

The data on (quasi) real $\gamma - \gamma$ cross sections into $\mu\mu$ and ee, as presented by PLUTO, TASSO, DCI are in agreement with QED. This tells us that the experiments are good but what else? Incidentally it was impressive to me that the same cross sections were measured at ISR as illustrated here by Vannucci, showing that $\gamma - \gamma$ physics is also doable with proton beams!

Finally the data presented by Pluto at large $P_\perp^2$ and Q^2 are remarkable as the first examples of events in the hard parton region. The plot of $Q^4\sigma$ versus E^2, that should be linear in the scaling limit at high Q^2, although still very limited in statistics and not conclusive, yet it shows the right order of magnitude for the expected slope and establishes the possibility of completing this test in the near future.

The most promising prospect for $\gamma\gamma$ physics is the possibility of measuring the γ structure functions at large Q^2. The importance of this measurement as discussed by Walsh is that the γ structure functions are asymptotically computable. The argument was given by Witten in 1977 in a very important paper. For a general target the structure functions are determined once the quark densities are known:

$$\frac{F_2}{x} \simeq 2F_1 \simeq \sum_i e_i^2 \, q_i \,(x,t) + O(\alpha_s(t)) \tag{3}$$

where $t = \ln Q^2/\mu^2$. The Q^2 dependence of the quark densities is ruled by evolution equations of the form:

$$\frac{dq}{dt} \simeq \alpha_s(t) \left[q \otimes P_{qq} + G \otimes P_{qG} \right] + \alpha_{EM} \, \gamma \otimes P_{q\gamma} \tag{4}$$

and similarly for the gluon, where $q \otimes P_{qq}$ etc. are well known convolution integrals with the kernels P_{qq}, P_{qG} and $P_{q\gamma}$ all known. The last term, proportional to $\alpha_{EM} \sim 1/137$ is negligible for hadronic targets, as is also the case for γ, the density of photons in the hadron. But for G photon target q and G (the densities of quarks and gluons in the photon) are themselves of order α_{EM} and $\gamma \sim \delta(1-x) +$ + O(α_{EM}). Thus in this case the last term is as important as the other ones. Going to moments:

$$Q^N(t) = \int_o^1 dx\, x^{N-1} q(x,t) \tag{5}$$

and restricting the discussion to non singlet densities for simplicity, eq.(4) becomes:

$$\frac{dQ^N}{dt} \simeq \alpha_S(t)Q^N(t)\, A_{qq}^N + \alpha_{EM}\, A_{q\gamma}^N \tag{6}$$

where A_{qq}^N and $A_{q\gamma}^N$ are obtained as moments of P_{qq} and $P_{q\gamma}$. The general solution of this inhomogeneous equation can be written as a sum of the general solution of the homogeneous equation plus a particular solution of the full equation:

$$Q^N(t) \simeq Q_o^N \left(\frac{\alpha_S}{\alpha_S(t)}\right)^{A_{qq}^N/b} + \frac{\alpha_{EM}\, A_{q\gamma}^N/b}{1-A_{qq}^N/b}\, \frac{1}{\alpha_S(t)} \tag{7}$$

where $\alpha_S(t) = 1/b\ln Q^2/\Lambda^2$. The first term either dies away for $N>1$ or stays constant for N=1, while the last term increases logarithmically, so that it eventually dominates. The first term contains non perturbative effects and in particular all vector meson dominance effects, which become negligible at sufficiently large Q^2. Non leading corrections of order $\alpha_S(t)$ to eq.(7) (and its generalization to the singlet case) have also been computed recently. The verification of this subtle QCD effect for photon structure functions which are computable but yet not given by the free field result is, in my opinion, the main target to aim to in $\gamma\gamma$ physics. We have learnt that within two years F_2 will be known at average Q^2 of 5 GeV^2. At this value of Q^2, log $Q^2/\Lambda^2 \sim 3\div5$ so that the leading term although not very large should however be visible. Within ten years physics at LEP, as discussed by Field, will measure F_2 at average Q^2 of order 20 GeV^2 and jet-ology from $\gamma\gamma$ processes will also be studied with all its phenomenological richness (as described in the talk by Gunion and in the workshop led by Kajantie).

To conclude, being the last speaker, I am glad to extend our appreciation and gratefulness to the Organizing Commitee for having made possible for all of us to spend these pleasant and fruitful days here at Amiens.

Topics in the theory of photon-photon processes

Summary of the theoretical parallel sessions

K. Kajantie

Department of Theoretical Physics, University of Helsinki, Helsinki, Finland

1. Introduction

Various theoretical topics related to photon-photon processes were discussed in the theoretical parallel sessions. A selection of the abstracts of the talks presented in the parallel talks will appear at the end of this summary. The aim of the brief summary is to present in concise form the issues discussed in the parallel sessions and, in particular, to connect them with questions raised in the plenary session talks.

2. Equivalent photon formulas

The question of determining the density $N(\omega, \theta)$ of photons in an electron of energy E as a function of the photon energy ω and angle θ relative to the electron and applying the resulting formula in different experimental situations has been studied for many years in photon-photon physics. Since one is dealing with effects which are subleading in log (E/m), the question is quite non-trivial and requires careful consideration. The present state of the art was reviewed by Kessler [1] and commented on by Field [2]. The situation is now well under control and the differences between formulas are understood.

Since the equivalent photon formulas anyway are approximate (although often very accurate), it is also important to develop techniques which allow one to go beyond them. The most promising new technique is Monte Carlo integration, and the plenary sessions showed that it has recently become an effective and widely used tool.

3. Resonance production

In the plenary sessions several experimental groups from DESY and SLAC reported results on resonance M production in the channel $\gamma\gamma \rightarrow M$. This has motivated the theorists to resume the discussion of the decay widths $\Gamma(M \rightarrow \gamma\gamma)$, already started nearly ten years ago. New measurements and tests of theories are in sight. New theoretical results were reported by Dunbar [3] if M is f or A_2^0 and by Goldberg [4] if M is a glueball. The tensor meson f has already been experimentally observed;

the detection of a glueball in this channel would be highly interesting, but the signal is probably not very distinctive.

4. Radiative corrections

Recently it has been clearly realized that quite a lot of effort has to be devoted to careful computation of electromagnetic radiative corrections. They may not be too small, but depend in any case sensitively on the experimental arrangement. This was shown in the talks given by Cochard in the plenary session, and by Defrise [5] and Ong [6] in the parallel session.

5. Photon-photon collisions from proton-proton processes

The idea of calculating the amount of photons in an electron can in a natural way be extended to a calculation of photons in a proton. However, since the proton is not a pointlike object, the calculation is much more involved and requires several phenomenological assumptions. The resulting formula can then be used, e. g., to compute the rate of producing muon pairs in proton-proton collisions via the two-photon mechanism. Interest in these calculations has been particularly increased due to the recent experimental observation of these muon pairs at the CERN ISR, also reported in a plenary session in the course of the workshop.

Theoretical calculations appropriate to photon-photon collisions in proton-proton processes were reported in the theoretical parallel session by Kessler [7] and F. Schrempp [8]. In the course of the discussion the experimentalists were excited by the idea of tagging at the very high energy Isabelle storage (400 GeV + 400 GeV) on the process $p\ p \longrightarrow p\ p \gamma\gamma \longrightarrow p\ p\ q\ \bar{q}$. This might make it possible to observe quark jets in a very clean process without any disturbing quark debris from the initial protons.

List of abstracts

(1) C. Carimalo, P. Kessler and J. Parisi: On various equivalent-photon formulas.

(2) J.N. Field: Luminosity functions for two-photon processes in $e^+ e^-$ collisions.

(3) I.H. Dunbar: Tensor meson production in $\gamma\gamma$ collisions.

(4) H. Goldberg: Pseudoscalar glueball production in $\psi \longrightarrow \gamma X$, $\Upsilon \longrightarrow \gamma X$, $e^+ e^- \longrightarrow e^+ e^- X$ and p p $\longrightarrow X$.

(5) M. Defrise: Radiative corrections for two-photon processes with limited acceptance of the central detectors.

(6) S. Ong: Radiative corrections to photon-photon collisions induced by $e^+ e^-$ processes.

(7) C. Carimalo, P. Kessler and J. Parisi: Equivalent-photon spectrum for the proton.

(8) B. Schrempp and F. Schrempp: Muon pair production in p p collisions.

ABSTRACTS OF SHORT CONTRIBUTIONS

PRESENTED IN PARALLEL SESSIONS

(THEORY)

On various equivalent-photon formulas

C. Carimalo, P. Kessler and J. Parisi

Laboratoire de Physique Corpusculaire, Collège de France, Paris

A comparison is made between various equivalent-photon formulas. It is explained, and shown numerically, why the Dalitz-Yennie (or Brodsky-Kinoshita-Terazawa) version of the equivalent-photon approximation becomes unappropriate for photon-photon collision processes when the scattered electrons are tagged at finite angles.

Luminosity functions for two-photon processes in e^+ e^- collisions

J.H. Field

DESY, Hamburg

An analysis is given of the QED factors relating the cross section for e e $\longrightarrow$ e e X to the virtual 2-photon collision $\gamma^* \gamma^* \longrightarrow$ X. Only transverse photons are considered, but no kinematical approximations are made. The cases where none, one or both of the scattered electrons are detected at angles $\gg m_e/E$ (E = beam energy) are separately considered. A full discussion is given of the kinematical restrictions necessary to arrive at factorizable Equivalent-Photon Approximation formulae, and quantitative comparisons are given. Also discussed are the rapidity distribution of the produced system X and restrictions on the effective two-photon luminosity due to angular cuts on produced particles.

Tensor meson production in $\gamma\gamma$ collisions

I.H. Dunbar

School of Mathematical and Physical Sciences, University of Sussex, Brighton, Sussex

The prediction that the production and decay of the neutral tensor mesons f and A_2^o is an important source of prompt leptons and photons in hadronic collisions is reviewed. The vector dominance model on which this prediction is based can be tested in $\gamma\gamma$ collisions, and the helicity structure of the decays can be elucidated. Present data on $\gamma\gamma \longrightarrow$ hadrons are consistent with an important A_2^o contribution.

Pseudoscalar Glueball Production in $\psi \rightarrow \gamma$ X, $\Upsilon \rightarrow \gamma$ X, $e^+ e^- \rightarrow e^+ e^-$ X, and $p\,p \rightarrow$ X [+)]

H. Goldberg

Department of Physics, Northeastern University, Boston, Massachusetts

The chiral SU(3) x SU(3) Ward identities (incorporating the U(1) axial anomaly), in conjunction with a final state interaction model, are used in order to estimate the contribution of a heavy (≈ 2 GeV) 0^- "glueball" to the radiative decay of the $\psi(3.1)$. It is shown that this contribution could entirely saturate the branching ratio $(\psi \rightarrow \gamma + \text{hadrons})/(\psi \rightarrow \text{hadrons}) \approx 8\%$ predicted by QCD, as well as produce the observed γ ray spectrum. As a result, a predominance of $K\bar{K}\pi$, $\eta(\eta')\pi\pi$ and/or 4π final states are expected for the radiative decay. If this saturation indeed takes place, then it is predicted that $\sigma(e^+e^- \rightarrow e^+e^-G) \approx (1/2)\,\sigma(e^+e^- \rightarrow e^+e^-\eta') \approx 0.3$ nb at a beam energy of 3 GeV. Finally, there is a discussion of the role of G production in the radiative decay $\Upsilon \rightarrow \gamma + \text{hadrons}$, and in the rise of σ_{pp}^{total} at ISR energies.

+) Work supported in part by the National Science Foundation

Radiative corrections for two-photon processes with limited acceptance of the central

detectors

M. Defrise

Theoretische Natuurkunde, Vrije Universiteit Brussel, Brussels

The radiative corrections for the cross section of $e^+ e^- \longrightarrow e^+ e^- A^+ A^-$, differentiated with respect to the energy E_X of the $A^+ A^-$ system, to its longitudinal momentum K_X, and to the transverse momentum k_{+T} of A^+, are calculated in the framework of the Williams-Weizsäcker approximation, for the case where the outgoing electrons are not tagged. The integration over the $A^+ A^-$ system's transverse momentum is performed up to a cut-off K_{Tmax}. In the region of phase space where $W \simeq (E_X^2 - K_X^2)^{1/2}$ (calling W the invariant mass of the $A^+ A^-$ system) and $k_{+T} \gg K_{Tmax}$, the elastic corrections are cancelled within 4% by the hard-photon corrections. That region does not yield the dominant contribution to the total cross section, but corresponds to the production of particles of high transverse momentum.

Double equivalent-photon approximation including radiative corrections for γγ experiments without electron tagging

S. Ong

Laboratoire de Physique des Particules, Université de Picardie, Amiens

and

Laboratoire de Physique Corpusculaire, Collège de France, Paris

It is shown that, at lowest order in perturbation theory and in first approximation, radiative corrections in γγ collision experiments without electron tagging can be estimated by applying a double equivalent-photon approximation of the same type as (but obviously much more complicated than) that used for the computation of cross sections of γγ processes without radiative corrections. In such an approximation, radiative corrections - for a given beam energy E_o and a given invariant mass M produced - become independent of the specific process γγ⟶X considered.

The invariant-mass spectrum, thus corrected for radiation, will be written quite generally in the form:

$$d\sigma^{corr}/dM = (1 + \delta) \; d\sigma^o/dM$$

where $d\sigma^o/dM$ is the uncorrected invariant-mass spectrum. Values obtained for δ are shown for various beam energies and invariant masses considered.

Equivalent-photon spectrum of the proton

C. Carimalo, P. Kessler and J. Parisi

Laboratoire de Physique Corpusculaire, Collège de France, Paris

An equivalent-photon spectrum is given for the proton, taking account of the inelastic contribution (for the latter, the quark-parton model is used). That spectrum is then used in order to compute the contribution of the $\gamma\gamma$ mechanism to the process p p (or p $\bar{\text{p}}$) $\longrightarrow \ell^+ \ell^-$ X. It is shown that this contribution should be of the same order as the Drell-Yan effect at s $\simeq 10^6$ GeV^2 and M$^2 \simeq$ 10 - 1000 GeV^2 (M being the invariant mass of the lepton pair produced).

The $\gamma\gamma$ contribution to $p\,p \longrightarrow \mu^+ \mu^-$ X at large transverse momenta

B. Schrempp [+) ++)] and F. Schrempp [++)]

University of Durham

Over a large range of energies $\sqrt{s}$, the $\gamma\gamma$ contribution to $p\,p \longrightarrow \mu^+\mu^-$ X is calculated as a function of the dilepton transverse momentum $Q_\perp$, as well as of the dilepton mass $\sqrt{Q^2}$ and of Feynman x_F. An important issue in this paper is the study of the $Q_\perp$ distributions of the $\gamma\gamma$ mechanism, in particular at large values of $Q_\perp$, where QCD predictions are reliable. In that region the effective-photon approximation is not applicable and thus the calculations had to be performed essentially without using any kinematical approximations. The main results are as follows. At small $Q_\perp$, the $Q_\perp$ distribution is extremely steep. It flattens out drastically, however, at large $Q_\perp$. Up to ISR energies the $\gamma\gamma$ contribution turns out to be small (of the order of 1%) over regions of phase space where data are available so far. For large $Q_\perp$ a detailed comparison of the $\gamma\gamma$ mechanism with a prototype QCD calculation is presented. With increasing energy the $\gamma\gamma$ contribution becomes more and more important over an increasing region of $\tau = Q^2/s$ and $x_\perp = 2\,Q_\perp/\sqrt{s}$. It reaches $\approx 10\%$ of the QCD contribution typically for $\tau > 0.1$, $x_\perp \gtrsim 0.05$ $(1 - \tau)$ and $\sqrt{s} \gtrsim 60$ GeV, and even dominates the reference QCD cross section in the extreme corner of large τ and $x_\perp$. With minor modifications our calculation may provide information on the possibly interesting case of two-jet production $(\gamma^* \gamma^* \longrightarrow q\,\bar{q})$ at large transverse momenta in $p\,p$ or $p\,\bar{p}$ collisions for high s.

+) Heisenberg Foundation fellow

++) Address after May 1, 1980: II. Institut für theoretische Physik, Universität Hamburg (Germany)

EXPERIMENTAL WORKING GROUPS ON γγ PHYSICS

Convenor: Guido Barbiellini
Laboratori Nazionali INFN
Frascati, Italy

The participants in the experimental working groups at the 1980 Photon-Photon Conference held at Amiens considered the following subjects of study:

1) Development of γγ physics at the existing e^+e^- facilities (DCI, DORIS, SPEAR, PETRA, PEP). Coordinators: P. Jenni, C. Berger

2) Dedicated facility for γγ physics. Coordinator: P. Waloscek

3) γγ physics at LEP. Coordinator: J. Field

4) γγ physics at the $p\bar{p}$ and pp colliders. Coordinators: W. Hofmann, F. Vannucci

5) γγ physics at the e-p machines. Coordinator: G. Coignet

Some contributions of the study groups will appear in the proceedings as short abstracts written by the coordinators. The results of Working Group II will not be included since the conclusion drawn during the meeting was that all possible improvements on γγ facilities are generally useful for e^+e^- machines and in the near future γγ physics will develop in parallel with e^+e^- annihilation experiments.

As an introduction to these presentations I will outline the main features foreseen for the short and long-term development of photon-photon physics.

The γγ physics of resonances at the existing machines has overcome the main background problems and after achieving the present interesting results on the η'(958) and the f_o(1250) other important contributions are expected from the machines in the 3-10 GeV range (i.e., DCI, DORIS, SPEAR). Future improved detectors at these machines will allow a better signal-to-noise separation. The existing e^+e^- colliders, operating at the highest available energy will also facilitate a study of resonance production but most of the emphasis will be on jet production with energy of the two jets between 7 and 20 GeV.

γγ physics at LEP was presented in the plenary session and the working group coordinated by J. Field essentially reviewed this material.

Working Groups IV and V showed that with the high energy available in hadron-hadron or lepton-hadron colliders, γγ physics has shown up in these machines and some experimental effort will soon be completely devoted to this field.

The experimental results presented at the plenary sessions have shown that the quality improvement of the central detectors for annihilation physics has been fully applied to γγ physics.

Jet γγ physics will require detectors covering more forward-backward angular regions. The present instrumentation for electron tagging is adequate but with increased machine energy smaller angular regions have to be covered if tagging is required. G.P. Murtas offered some comments on small angle tagging, including the possibility of using electron interactions in crystal to measure the electron angle and to reject hadrons falling in the angular region useful for tagging.

TWO—PHOTON RESONANCE PRODUCTION
AT EXISTING e^+e^- COLLIDING BEAM MACHINES

Peter Jenni[*]

Stanford Linear Accelerator Center

Stanford University, Stanford,

California 94305

Contribution to the specialized working

group on machines and detectors at the

International Workshop on $\gamma\gamma$ collisions

Amiens, France, April 8-12, 1980

[*] now at : CERN, Geneva, Switzerland

About one year ago the first hadron resonance production in two-photon interaction, $e^+e^- \to e^+e^- \eta'$ (958), was observed by the MARK II Collaboration [1] at SPEAR. Even though much of the current interest has shifted towards the study of the photon structure functions and the investigation of multi-jet production, both topics extensively covered at this workshop, the hunt for resonances remains a fascinating aspect of two-photon physics. Only the η'(958) and f(1270) mesons have been observed[1-3] so far in two-photon interactions. Systematic studies of the radiative widths of the "old" neutral mesons like the scalars, the pseudoscalars and the tensors have not yet been done [4,5]. Another type of mesons which is producible in two-photon collisions is the "glueball", a quarkless state expected to exist in QCD [6,7]. During the Parallel Sessions of the Specialized Working Group on Machines and Detectors we have summarized the experimental situation on resonance production and we have tried to guess from which kind of experiments and machines one may expect within the next few years, further results in this field.

Resonances have been seen only without tagging more or less as a byproduct of e^+e^- annihilation experiments [1-3]. Only the decay products of the resonance X, from the reaction $e^+e^- \to e^+e^- X$, have been observed in the central detector with no additional final state particles detected. This method provides acceptances bigger by factors typically larger than 5 than the ones possible in measurements with single or even double tagging at finite electron scattering angles ($\geq$ 20 mrad in most experiments). Tagging at zero degree electron scattering angle would not lose much of this advantage [8], however early dedicated two-photon experiments [9] using this technique have not been successful in observing meson resonances and were very difficult to realize.

The background from e^+e^- annihilation events with part of the final state particles undetected can be strongly eliminated by simple kinematical cuts [1]. In fact, in the case of the f(1270) $\to \pi^+\pi^-$ signal the most severe background has been found to be e- and μ-pair production by the two-photon process itself [1-3], this background would not be suppressed by the tagging technique. None of the three experiments [1-3] has suffered from beam-gas background in their resonance signals.

It seems therefore very promising for a resonance search to concentrate on the central detector alone without tagging the outgoing electrons. Besides the obvious requirement of a large solid angle coverage the main features needed are charged particle identification and photon detection down to low momenta. The transverse momentum of a resonance from a two-photon interaction is in general very small which is then also the case for its decay products, especially for multiprong decay modes. A good efficiency for low transverse momentum particles is already necessary at the stage of the event trigger in a two-photon experiment without tagging. Two central detectors currently in preparation, the MARK III experiment for SPEAR and ARGUS for the DORIS machine, are expected to have significant improvements in these features over the experiments discussed in Ref. 1-3.

When comparing the described type of experiment at different existing e^+e^- colliding beam machines one has to consider the following points. On one hand the two-photon resonance production cross sections (for masses of about 1 GeV/c^2) increase by a factor of 3 to 4 from beam energies around 3 GeV (SPEAR, DORIS) to around 15 GeV (PETRA, PEP). However, the increase in cross section occurs mainly in a rapidity (y) range usually not covered by the central detector [1,10] (in the extreem forward and backward regions). For example, the rapidity interval covered by the MARK II detector[1] is only about $-0.5 \leq y \leq 0.5$. The resulting reduction in acceptance tends to cancel the increase in cross section for higher energies. This means that the determining factors are the integrated luminosity available to an experiment and the background conditions at the trigger level in order to be able to accumulate data with a low transverse momentum threshold on the tracks. Both these requirements seem to favour for the moment the lower energy machines, especially SPEAR, as one can deduce from the data presented at this workshop (MARK II [1] at SPEAR : $\int \mathcal{L} dt =$ 18 pb^{-1}, $p_T \geq 100$ MeV/c, PLUTO [2] and TASSO [3] at PETRA : $\int \mathcal{L} dt \cong 3$ pb^{-1}, $p_T \geq 300$ MeV/c).

In conclusion, it seems to be a plausible guess to expect in the near future new results on two-photon resonance production mostly from the new generation of experiments without tagging at the medium energy machines. Fortunately however, these conclusions are weak, and different approaches over the energy range from DCI to the PETRA/PEP machines may hopefully lead to significant results as well.

It is a pleasure to acknowledge the many fruitful discussions with my colleagues from the SLAC-LBL Berkeley MARK II Collaboration, especially with Valery Telnov, who had a decisive impact on the feasibility to study with MARK II two-photon resonance production without tagging.

REFERENCES AND FOOTNOTES

1 - G.S. Abrams et al., Phys. Rev. Lett. $\underline{43}$, 477 (1979)
   P. Jenni, SLAC-PUB-2421 (1979), invited talk at the International Conference on two-photon interactions, Lake Tahoe, California, Aug. 30- Sept. 1, 1979.
   P. Jenni, MARK II Collaboration, invited talk at this Workshop.

2 - C. Berger, PLUTO Collaboration, invited talk at this Workshop.

3 - R. Hilger, TASSO Collaboration, invited talk at this Workshop.

4 - F.J. Gilman, SLAC-PUB-2461 (1980), invited talk at the International Conference on two-photon interactions, Lake Tahoe, California, Aug. 30- Sept. 1, 1979.

5 - M. Greco, invited talk at this Workshop.

6 - J.D. Bjorken, Lectures at the SLAC Summer Institute on Particle Physics, July 1979.

7 - H. Goldberg, contribution to this Workshop.

8 - A. Courau, invited talk at this Workshop.

9 - G. Barbiellini, invited talk at this Workshop.

10 - D. Burke, MARK II, invited talk at this Workshop.

2γ process in high energy ep collisions
G.Coignet (LAPP)

ABSTRACT.

At the squared center of mass energy $s \simeq 10^5$ GeV^2 which could be reached with HERA (30 GeV e × 800 GeV p) or LEP-SPS (80 GeV e × 300 GeV p), the 2γ process is not negligible. Taking into account the luminosity (10^{31}-10^{32} cm^{-2}s^{-1}) expected for these machines, the events rate would be comparable to the one observed at PETRA. However, the untagged 2γ contribution seems difficult to isolate from the competitive processes, except for very special kinematical regions : for example, when the produced hadrons are going on the electron side.

The detection of the tagged events looks more promising : Due to the motion of the e-p center of mass, the tagging of the electron would be easier (larger angle emission) than in e$^+$e$^-$ machines. The tagging of the proton would imply its detection very near ($\simeq$ 1 mrad) the beam line. One way to solve this problem would be to accelerate deuterons instead of protons. The simultaneous detection of the position and the energy of both the final proton, using the machine magnets as a spectrometer, and the spectator neutron emitted at zero degree would insure the nucleon tagging. The events rate reduction, due to the half s value available when using deuterons, would be largely compensated by the background suppression.

DIRECTIONAL EFFECTS ON HIGH ENERGY ELECTRONS
AND POSITRONS SHOWERING INTO A SILICON CRYSTAL

R. Del Fabbro[*] and G.P. Murtas

INFN - Laboratori Nazionali di Frascati

ABSTRACT

Coherent bremsstrahlung of electrons and pair production in mono-
crystals have already been studied theoretically and experimentally
in many laboratories.

Consequently, polarized and quasimonochromatic γ ray beams were
obtained in many electrosynchrotrons using diamond or silicon thin crysta

Actually we think that it is interesting and feasible to look at the
shower produced in a thick crystal by high energy electron in order to
obtain a very compact electron and positron detector with high directiona
properties and probably with very low sensitivity for other particles.

In our mind, we have the problem of detecting electrons and/or
positrons at angles of ten milliradians with respect to an intense beam,
as the LEP case, to tag virtual photons in the photon-photon experimental
physics.

We have already obtained promising preliminary experimental results
using a thick silicon crystal.

We recall briefly the properties of radiation of electron and
positron impinging on a crystal at a small angle with respect to the
crystal axis. Successively we describe our experimental apparatus and
the results obtained. We then discuss the work line we intend to
follow in the future.

[*]   INFN - Sezione di Pisa

H. M. Pilkuhn

Relativistic Particle Physics

1979. 85 figures. XII, 427 pages
(Texts and Monographs in Physics)
ISBN 3-540-09348-6

Contents: One-Particle Problems. – Two-Particle Problems. – Radiation and Quantum Electrodynamics. – The Particle Zoo. – Weak Interactions. – Analyticity and Strong Interactions. – Particular Hadronic Processes. – Particular Electromagnetic Processes in Collisions with Atoms and Nuclei. – Appendices.

Beginning with Maxwell's equation and Lorentz invariance, the book comes quickly to the quantum theory of relativistic particles. The first two chapters are on relativistic one- and two-body equations and provide a basis for understanding atomic physics. The following chapters cover quantum electrodynamics, the particle zoo, the theory of β-decay and other weak interactions and its applications to nuclear physics. The general formalism of quantum field theory is kept to the necessary minimum, leaving ample space for practical calculations. Important recent developments such as the Salam-Weinberg model and the discovery of charmed particles are included. The book will be useful both for the student who wants to learn the facts and methods of this quickly broadening field and for the research worker in atomic, nuclear, medium energy, high energy, and cosmic ray physics.

Springer-Verlag
Berlin
Heidelberg
New York

Selected Issues from

Lecture Notes in Mathematics

Lecture Notes in Physics